ROLE OF LIPID EXCIPIENTS IN MODIFYING ORAL AND PARENTERAL DRUG DELIVERY

THE WILEY BICENTENNIAL–KNOWLEDGE FOR GENERATIONS

Each generation has its unique needs and aspirations. When Charles Wiley first opened his small printing shop in lower Manhattan in 1807, it was a generation of boundless potential searching for an identity. And we were there, helping to define a new American literary tradition. Over half a century later, in the midst of the Second Industrial Revolution, it was a generation focused on building the future. Once again, we were there, supplying the critical scientific, technical, and engineering knowledge that helped frame the world. Throughout the 20th Century, and into the new millennium, nations began to reach out beyond their own borders and a new international community was born. Wiley was there, expanding its operations around the world to enable a global exchange of ideas, opinions, and know-how.

For 200 years, Wiley has been an integral part of each generation's journey, enabling the flow of information and understanding necessary to meet their needs and fulfill their aspirations. Today, bold new technologies are changing the way we live and learn. Wiley will be there, providing you the must-have knowledge you need to imagine new worlds, new possibilities, and new opportunities.

Generations come and go, but you can always count on Wiley to provide you the knowledge you need, when and where you need it!

WILLIAM J. PESCE
PRESIDENT AND CHIEF EXECUTIVE OFFICER

PETER BOOTH WILEY
CHAIRMAN OF THE BOARD

ROLE OF LIPID EXCIPIENTS IN MODIFYING ORAL AND PARENTERAL DRUG DELIVERY

BASIC PRINCIPLES AND BIOLOGICAL EXAMPLES

Edited by

KISHOR M. WASAN

University of British Columbia
Vancouver, British Columbia

WILEY-INTERSCIENCE

A JOHN WILEY & SONS, INC., PUBLICATION

Published by John Wiley & Sons, Inc., Hoboken, New Jersey
Published simultaneously in Canada

Library of Congress Cataloging-in-Publication Data:

Role of lipid excipients in modifying oral and parenteral drug delivery : basic principles and biological examples / [edited by] Kishor M. Wasan.
p. ; cm.
Includes bibliographical references and index.
ISBN-13: 978-0-471-73952-4 (cloth)
ISBN-10: 0-471-73952-9 (cloth)
1. Lipids—Therapeutic use. 2. Excipients. 3. Drugs—Dosage forms. I. Wasan, Kishor M.
[DNLM: 1. Excipients—therapeutic use. 2. Administration, Oral. 3. Drug Compounding—methods. 4. Drug Delivery Systems. 5. In-fusions, Parenteral. 6. Pharmaceutical Solutions.
QV 800 R745 2007]
RS201.E87R65 2007
615′.7—dc22

2006019129

CONTENTS

PREFACE vii

CONTRIBUTORS ix

CHAPTER 1 *INTERACTION OF DRUG TRANSPORTERS WITH EXCIPIENTS* 1

K. Sandy Pang, Lichuan Liu, and Huadong Sun

CHAPTER 2 *FORMULATION ISSUES AROUND LIPID-BASED ORAL AND PARENTERAL DELIVERY SYSTEMS* 32

Seong Hoon Jeong, Jae Hyung Park, and Kinam Park

CHAPTER 3 *LIPID-BASED PARENTERAL DRUG DELIVERY SYSTEMS: BIOLOGICAL IMPLICATIONS* 48

Vladimir P. Torchilin

CHAPTER 4 *PRINCIPLES IN THE DEVELOPMENT OF INTRAVENOUS LIPID EMULSIONS* 88

Joanna Rossi and Jean-Christophe Leroux

CHAPTER 5 *PROTEIN ADSORPTION PATTERNS ON PARENTERAL LIPID FORMULATIONS: KEY FACTOR DETERMINING THE IN VIVO FATE* 124

Rainer H. Müller and Torsten M. Göppert

CHAPTER 6 *NANOPARTICLE TARGETING FOR DRUG DELIVERY ACROSS THE BLOOD–BRAIN BARRIER* 160

James Egbert, Werner Geldenhuys, Fancy Thomas, Paul R. Lockman, Russell J. Mumper, and David D. Allen

CHAPTER 7 *LIPID-COATED PERFLUOROCARBON STRUCTURES AS PARENTERAL THERAPEUTIC AGENTS* 170

Evan C. Unger, Terry O. Matsunaga, and Reena Zutshi

INDEX 197

PREFACE

The primary roles of traditional excipients were to bind and provide bulk to the dosage form, to facilitate or control drug release from the excipient matrix, and to facilitate product manufacturing on high-speed, automated, production equipment. However, lipid excipients, unlike their traditional counterparts, have the ability to solubilize hydrophobic drugs within the dosage form matrix. This often results in the case of oral drug delivery, improved drug absorption, which is primarily mediated by a reduction in the barriers of poor aqueous solubility, and slow drug dissolution rate in the gastrointestinal (GI) fluids. Some of these excipients also have desirable self-emulsifying properties, readily forming fine dispersions of lipid-solubilized drug in the aqueous contents of the GI tract and creating optimal conditions for absorption.

The pivotal activities involved in the development of any oral dosage form (a conventional solid or lipid-based formulation) include: (1) physiochemical and biopharmaceutical understanding of the drug substance, which would guide initial excipient selection and subsequent design of a prototype dosage form; (2) product stability and dissolution testing, which demonstrates physical and chemical stability of the drug substance during the shelf-life of the product; (3) formulation scale-up to production size batches; (4) development of a discriminating dissolution test method, to provide assurance of product quality and batch-to-batch consistency; and (5) justification of the formulation rationale to regulatory agencies. However, although the pivotal activities associated with the development of a lipid-based formulation are similar to those for a conventional oral solid, the manner in which pharmaceutical scientists achieve these goals will be different. The ability of pharmaceutical scientists in re-defining these pivotal development activities for lipid-based formulations, both oral and parenteral, may in fact determine the future success of this technology.

A number of important questions will need to be more fully addressed in order to provide pharmaceutical scientist formulators with consistent guidelines for the development of novel oral and parenteral lipid-based formulations, e.g. what role will dissolution testing play in the development and evaluation of liquid and semi-solid lipid-based dosage forms? Should dissolution testing be performed at all? If so, what parameters are important and how will the data be interpreted? How is stressed stability testing performed on semi-solid dosage forms, which melt at elevated temperatures? Will the physical state of drugs in matrices change upon aging and how might this impact drug delivery? What types of chemical incompatibilities are peculiar to lipid excipients? How will these excipients affect the integrity of gelatin capsules?

The availability of a wide variety of pharmaceutical grade lipid excipients has coincided with a recent advance in encapsulation technology, which now allows hard gelatin encapsulation of both liquid and semi-solid formulations. This advance, along with the fact that almost half of all new chemical entities fit the category of 'poorly water soluble' has created a window of opportunity for the rapid introduction of oral lipid-based drug formulations into the marketplace.

As we begin to unravel the intricacies of the GI processing of lipid excipients, further improvements in the performance of lipid-based delivery systems can be expected, e.g. an increasing body of evidence has shown that certain lipids are capable of inhibiting both pre-systemic drug metabolism and P-glyoprotein-mediated drug efflux by the gut wall. And it is well known that lipids are capable of enhancing lymphatic transport of hydrophobic drugs, thereby reducing drug clearance resulting from hepatic first-pass metabolism. This book addresses not only formulation issues, but also these physiological and biopharmaceutical aspects of oral lipid-based drug delivery.

As a new and evolving discipline, lipid-based drug delivery has attracted considerable attention from academia to industry. Over the past few years, academic and industrial interests in this area have been evident from the increase in national and international symposiums and workshops on many different aspects of lipid-based drug delivery systems. Many universities are now offering comprehensive courses in lipid-based drug delivery systems. Although these courses are highly effective, frequent requests for a standard textbook on oral lipid-based drug delivery systems prompted me to this current project in editing a book on lipid-based drug delivery. At the present, there is no comprehensive textbook available and various graduate level courses on this interesting topic were taught by professors with materials gathered from diverse sources. To satisfy this urgent need, we plan to assemble a group of experienced investigators/educators who are on the frontline of pharmaceutical sciences to develop such a textbook. This textbook is intended for both pharmaceutical scientists and trainees in the field of drug delivery, formulation development, and drug discovery, and presents fundamental principles and biological examples in the use of lipid excipients to develop both oral and parenteral drug delivery systems.

Kishor M. Wasan

CONTRIBUTORS

David D. Allen, RPh, PhD, FASHP
Dean and Professor, Northeastern Ohio Universities, College of Pharmacy, Rootstown, Ohio, USA

James Egbert
Northeastern Ohio Universities, College of Pharmacy, Rootstown, Ohio, USA

Werner Geldenhuys
Northeastern Ohio Universities, College of Pharmacy, Rootstown, Ohio, USA

Torsten M. Göppert
Department of Pharmaceutical Technology, Biotechnology and Quality Management The Free University of Berlin, Germany

Seong Hoon Jeong
Departments of Pharmaceutics and Biomedical Engineering, Purdue University, West Lafayette, Indiana, USA

Jean-Christophe Leroux
Canada Research Chair in Drug Delivery, Faculty of Pharmacy, University of Montreal, Canada

Paul R. Lockman
Northeastern Ohio Universities, College of Pharmacy, Rootstown, Ohio, USA

Terry O. Matsunaga, PharmD, PhD
ImaRx Therapeutics, Inc., Tucson, Arizona, USA

Rainer H. Müller
Professor, Department of Pharmaceutical Technology, Biotechnology and Quality Management, The Free University of Berlin, Germany

Russell J. Mumper
Northeastern Ohio Universities, College of Pharmacy, Rootstown, Ohio, USA

K. Sandy Pang
Leslie Dan Faculty of Pharmacy, University of Toronto, Ontario, Canada

Jae Hyung Park
Departments of Pharmaceutics and Biomedical Engineering, Purdue University, West Lafayette, Indiana, USA

Kinam Park, PhD
Purdue University, School of Pharmacy
West Lafayette, Indiana, USA

Joanna Rossi
Faculty of Pharmacy, University of Montreal, Canada

Fancy Thomas
Northeastern Ohio Universities, College of Pharmacy, Rootstown, Ohio, USA

Vladimir P. Torchilin
Professor, Department of Pharmaceutical Sciences
Northeastern University, Boston, Massachusetts, USA

Evan C. Unger, MD
ImaRx Therapeutics, Inc., Tucson, Arizona, USA

Reena Zutshi, PhD
ImaRx Therapeutics, Inc., Tucson, Arizona, USA

CHAPTER 1

INTERACTION OF DRUG TRANSPORTERS WITH EXCIPIENTS

K. Sandy Pang, Lichuan Liu, and Huadong Sun

1.1 INTRODUCTION

1.2 INTESTINAL ABSORPTIVE TRANSPORTERS

1.3 INTESTINAL EFFLUX TRANSPORTERS

1.4 MODULATION OF DRUG TRANSPORTERS BY EXCIPIENTS AND SURFACTANTS

1.5 CONCLUSION

1.1 INTRODUCTION

The intestine is the most important site for drug absorption and regulates the extent of orally administered drug that reaches the circulation. Oral drug absorption or bioavailability relates to the net amount of dose absorbed, and occurs mainly via the small intestine where the surface area is much greater than that in the stomach. The first major obstacle is crossing the intestinal epithelium and survival from intestinal and liver metabolism. There are essentially two routes: the paracellular and transcellular pathways in which compounds permeate across the intestinal membrane. For small hydrophilic, ionized compounds, absorption may occur via the paracellular pathway. The transcellular transport processes include passive diffusion, membrane permeation via transporters that are primary, secondary, and tertiary in terms of ATP requirements, and include co-transporters (symport) or exchangers (antiport) (Figure 1.1). Transcellular absorption of drugs occurs from the lumen to blood, and necessitates uptake across the apical membrane, and then the drug exits across the basolateral membrane.

Lipophilic drugs readily diffuse across the apical membrane from the lumen, and their subsequent passage across the basolateral membrane into blood is also by diffusion. Within the small intestine, phase I and II enzymes are present to

Role of Lipid Excipients in Modifying Oral and Parenteral Drug Delivery, Edited by Kishor M. Wasan

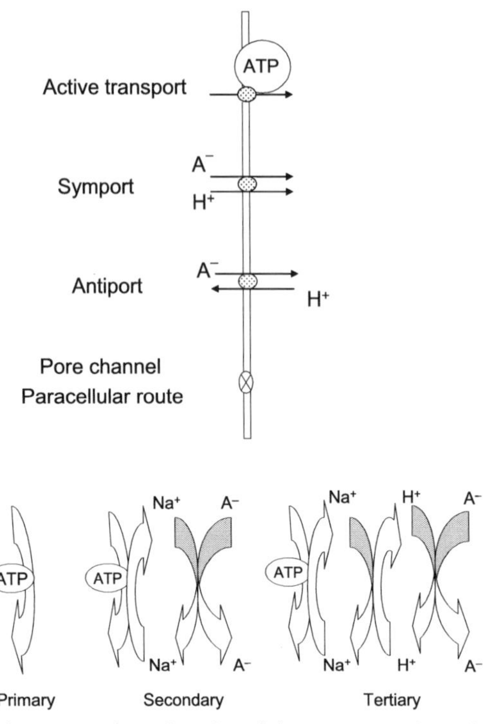

Figure 1.1 Schematic presentation of modes of drug transport (upper) and energy-dependent transport.

effect drug removal in a competitive fashion, and these also compete with efflux transporters at the apical and basolateral membranes [1, 2]. Although the activities of these drug-metabolizing enzymes of the small intestine are usually lower than those of the liver, the intestine is a portal tissue that regulates the level of substrate reaching the liver. A thorough understanding of the interactions between transport and metabolism in the small intestine is achieved only through the understanding that transporters and enzymes are pathways competing for the substrate [1–4].

At the apical membrane of the intestine, solute carrier transporters (SLCs) mediate the absorption of chemical entities from the lumen, and then the drugs exit at the basolateral membrane to enter the circulation [5–11]. The membrane transporters include the SLC transporter family and the ATP-binding cassette (ABC) transporter family (see www.gene.ucl.ac.uk/nomenclature). Currently, the SLC transporter family includes about 50 families and more than 360 transporter genes. A transporter is assigned to a specific SLC family if at least 20–25% of its amino acid sequence identifies with other members of that family. The SLC transporters are involved in uptake, whereas the ABC transporters are efflux proteins located at the apical membrane of the small intestine to redirect the absorbed drug to re-enter the lumen (Figure 1.2). Both transporters and enzymes are under regulation of orphan nuclear receptors (see the literature [12–25]). These include the aryl hydro-

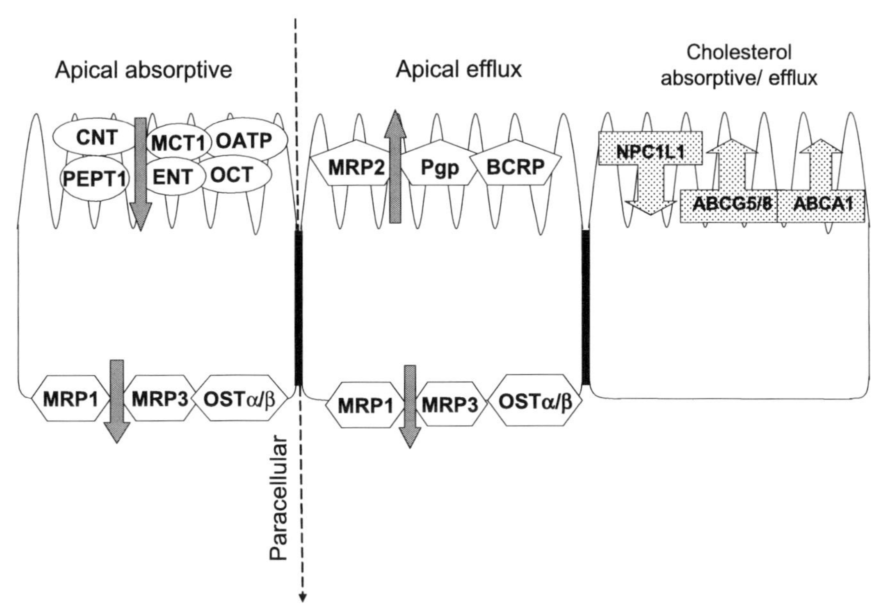

Figure 1.2 Intestinal apical absorptive (left) and efflux (middle) transporters, and basolateral transporters of solutes and absorptive and efflux transporters of cholesterol (left panel).

carbon receptor (AHR), pregnane X receptor (PXR), constitutive androstane receptor (CAR), farnesoid X-receptor (FXR), the nuclear factor-E2 p45-related factor 2 (Nrf2), hepatocyte nuclear factor 1α and 4α HNF-1α and HNF-4α, liver receptor homolog 1 (LRH-1], liver X-receptor (LXR), small heterodimer partner-1 (SHP-1), the glucocorticoid receptor (GR) and the vitamin D receptor (VDR).

The regulation of both transporters and enzymes by nuclear orphan receptors and the associated complexities, such as coordinate regulation and cross-talk, are under intense investigation [18, 22]. However, the topic of regulation of transporters and enzymes by nuclear receptors is beyond the scope of this chapter, the focus of which reviews some of the transporters involved in drug disposition. In addition the modulation of the transporters in the intestine by excipients is described.

1.2 INTESTINAL ABSORPTIVE TRANSPORTERS

The intestine is known to absorb drugs as a result of the increased surface area caused by the presence of villi and microvilli. The existence of various transporters on the apical membrane has been reviewed [8–11]. Nutrients such as amino acids are broken down from protein by peptide digestion, and are absorbed via by a set of amino acid transporters that differ with regard to sodium dependence, substrate specificity, driving force, and genetic classification. Vectorial transport is ensured by other basolateral amino amid transporters. Various transporters exist for nutrients:

amino acid [26, 27], glucose (SGLT and GLUT, Na^+-dependent and -independent glucose transporters) [28], ascorbic acid (SLC23A1), thiamine (SLC19A2), folic acid (SLC19A1, SLC19A3), and fatty acid (FATP4 or SLC27A4, and FABpm). Various fatty acids transporters are known to facilitate fat transport [29]: the fatty acid translocase (FAT/CD36) [30, 31], fatty acid-transporting protein (FATP41) [32], and fatty acid-binding protein (FABPpm) [33].

Transporters exist specifically for vitamin absorption. The vitamin C transporter (SVCT1) prefers L-ascorbinic acid over the D-isoascorbinic acid [34]. The reduced folate transporters (RFC1 or SLC19A1), delivering folate for purine and pyrimidine nucleotide synthesis, transport folate via a pH-dependent, DIDS (4,4′-disothiocyanostilbene-2-2-′disulfonic acid)-sensitive, electroneutral system across the human colonic basolateral membrane [35]. Thiamine (SLC19A2 and SLC19A3) transporters are temperature and energy dependent, pH sensitive, and Na^+ independent, and transport is inhibited by the thiamine structural analogs, amprolium and oxythiamine, but not by unrelated organic cations, choline tetraethylammonium, and *N*-methylnicotinamide [36]. Thiamine uptake was also found on the brush border side in brush border membrane vesicles (BBMV) of human ileum of a Na^+-independent, pH-dependent, amiloride-sensitive, electroneutral, carrier-mediated mechanism in native, human small intestinal BBMV [37]. At the basolateral membrane, there is also a proton gradient-dependent, specialized carrier-mediated exchange mechanism for thiamine transport across the human jejunum basolateral membranes [38]. A host of important transporters is known to transport important organic anions and cations. These are summarized in Table 1.1.

1.2.1 ASBT (SLC10A2)

The apical Na^+-dependent bile acid transporter or ASBT is highly concentrated at the apical membrane of the distal ileum for bile acid absorption [39, 40]. ASBT demonstrates a stringent substrate specificity that is reserved mostly for natural as well as synthetic bile acids. The uptake of bile acids is a Na^+-dependent but chloride-independent process. Analysis of the substrate specificity in transfected COS7 (transformed African Green Monkey kidney fibroblast) cells or CHO (Chinese hamster ovary) cells showed that both conjugated and unconjugated bile acids are efficiently transported [41]. The transporter reclaims almost completely (29.4 g of the 30 g excreted bile acids every 24 h for a 70 kg man) all of the bile acids excreted by the biliary tract [42]. The presence of the transporter is highly important for health and disease because malabsorption or polymorphism (the SLC10A2 missense mutations and 5′-flanking sequence) results in familial hypertriglyceridemia (FHTG); the disease is associated with impaired intestinal absorption of bile acids and characterized by elevated plasma levels of the very-low-density lipoprotein triglyceride [43]. A congenital form of idiopathic intestinal bile acid malabsorption that is associated with dysfunctional mutations is known to cause steatorrhea [44, 45]. ILBP – the ileal lipid-binding protein or ileal bile acid-binding protein – may facilitate vectorial transport of bile acids [46, 47]. The basolateral half-transporters, OSTα and OSTβ [48] or the multidrug resistance-associated protein 3 (MRP3) [49] then mediate the transport of the bile acids at basolateral membrane to re-enter the circulation.

TABLE 1.1 Summary of substrates for absorptive transporters expressed in apical membrane of human intestine

Transporters	Substrates	Reference
PEPT1 (SLC15)	β-Lactam antibiotics (cefaclor, cephalexin, cephradine, etc.)[a]	[54, 55, 56]
	ACE inhibitors: dipeptides	[57]
	Valaciclovir[a]	[59, 60]
OATP 2B1 (OATP-B, SLC21A9)	Estrone 3-sulfate[a]	[68, 69]
	Benzylpenicillin[a]	[66]
	Fexofenadine[a]	[68]
	Prostaglandin E_2[a]	[66]
	Glibenclamide[a]	[70]
	Dehydroepiandrosterone sulfate (DHEAS)[a]	[70]
	Pravastatin[a]	[68]
	Fluvastatin[a]	[71]
MCT1 (SLC16A1)	L-Lactate propionate, and butyrate[a]	[79, 82, 84]
	short-chain fatty acids: acetate[a]	[81]
	Benzoic acid[a]	[80]
CNT1 and CNT2 (SLC28A1 and SLC28A2)	Zidovudine[a]	[93]
	5′-Deoxy-5-fluorouridine[a]	[99]
	2′,3′-Dideoxyinosine (DDI)[a]	[93, 94]
	Floxuridine[a]	[100]
	Gemcitabine[a]	[101]
ENT2 (SLC29A2)	Zidovudine, zalcitabine, and DDI[a]	[10]
OCT1 (SLC22A1)	Aciclovir, ganciclovir[a]	[109]
OCTN2 (SLC22A5)	L-Carnitine[a]	[110]
	Verapamil, quinidine[a]	[112]
	Cephaloridine[a]	[112]
	Cephaloridine, cefoselis, cefepime, and cefluprenam[a]	[113]

[a]Uptake/drug accumulation in transfected cells or their vesicles.

1.2.2 PEPT1 (SLC15A1)

The oligopeptide transporter 1 (PEPT1) operates as an electrogenic proton/peptide symporter. The mode of transport is enantio selective and involves a variable proton-to-substrate stoichiometry for uptake of neutral and mono- or polyvalently charged peptides [50]. PEPT1, but not PEPT2, is found in the duodenum [51]. Substrates of PEPT1 include the di- and tripeptides, peptidomimetics of fewer than four amino acids, including the β-lactam antibiotics, the aminocephalosporins and aminopenicillins, amino acid-conjugated, nucleoside-based, antiviral agents, and some angiotensin-converting enzyme inhibitors [52–56]. Strategies have been targeted towards the L-valyl ester prodrugs, e.g. acyclovir and ganciclovir to valaciclovir, which increased the bioavailability three to five times [54, 57–60]. For the same reason, the 5′-amino acid ester prodrug of zidovudine (AZT), intended to increase absorption, was designed for PEPT1 [61]. The same applies to the midodrine antihypertensive prodrug and its active metabolite, 1-[2′,5′-dimethoxyphenyl)-2-aminoethanol [62].

A review of the literature shows a wide range of affinity constants between 2 μmol/L and 30 mmol/L. Affinity constants for substrates or inhibitors of PEPT1 lower than 0.5 mmol/L were viewed as high affinity, those between 0.5 and 5 mmol/L, as medium affinity, and those above 5 mmol/L as low affinity [63]. The bioavailability of peptidomimetic drugs is reduced only in a low-frequency PEPT1-F28Y variant among the 38 single nucleotide polymorphisms; this variant displayed significantly reduced cephalexin uptake, and is explained by the increased Michaelis constant, K_m [64]. The mRNA and protein for this transporter were found in villous, but not crypt, cells from the normal rabbit intestine, and were unaltered with inflammation, although a decrease in the affinity of the co-transporter for the dipeptide was found [65].

1.2.3 OATP (SLCOA9)

The OATP 2B1 (or OATP-B), together with OATP-D and OATP-E, were found in the intestine [66, 67] (see Table 1.1). OATP-B, immunohistochemically localized as present at the apical membrane of intestinal epithelial cells in humans, exhibits transport towards estrone-3-sulfate [66]. OATP-B-mediated uptake is independent of sodium, chloride, bicarbonate, or glutathione, whereas the proton ionophore carbonylcyanide *p*-trifluoromethoxyphenylhydrazone exhibits a pH-dependent inhibitory effect, suggesting that a proton gradient is a driving force for OATP-B. Moreover, transport of taurocholic acid and pravastatin by OATP-B has been observed only at acidic pH [68]. Uptake of estrone-3-sulfate and pravastatin by OATP-B at pH 5.5 is higher than that at pH 7.4 [69], and glibenclamide is also a substrate [70, 71]. The transport of fexofenadine is, however, inhibited by citrus juices [70]. OATP-A mRNA was further found to be present in the intestine [72], and OATP-A transports fexofenadine [73] and saquinavir [74]. This transport is also reduced by grapefruit juice [75]. Other inhibitory interactions with herbal products – extracts of bilberry, echinacea, green tea, banana, grape seed, ginkgo, and soybean on estrone-3-sulfate transport – were found [76]. Other OATP transporters were found present in the intestine, as mRNA levels of OATP2A1 (or the prostaglandin transporter, PGT), OATP3A1 (OATP-D) and 4A1 (OATP-E) were detected [77]. The roles of these transporters on solute transport have not been clearly identified.

1.2.4 MCT1 (SLC16A)

A monocarboxylate transporter isoform type 1 (MCT1), comprising the proton symport, *trans* acceleration, and sensitivity to α-cyanocinnamates, was first cloned in hamster [78]. Inhibition by pyruvate, butyrate, propionate, and acetate, but not by Cl^- and SO_4^{2-}, was identified for MCT1 on the luminal membrane of human colon [79]. The transporter mediates the transport of L-lactate and short-chain fatty acids: acetate, propionate, and butyrate [80–82]. MCT1 requires the membrane-spanning glycoprotein, CD147, at the apical plasma membrane for the correct plasma membrane expression and function [83, 84]. Uptake was stimulated in the presence of an outward-directed anion gradient at an extravesicular pH of 5.5 [79]. At least five isoforms of MCT were found in Caco-2 cells [82]. Immunoblotting studies detected a protein band of about 39kDa for MCT1, predominantly localized in the apical membranes. The relative abundance of MCT1 mRNA and protein was found to increase along the length of the human intestine. Immunohistochemical studies confirmed that human MCT1 antibody labeling was confined to the apical membranes, whereas MCT5 antibody staining was restricted to the basolateral membranes of the colonocytes [85].

1.2.5 Phosphate Transporters (SLC34A4)

Apically located Na^+-dependent symports, normally found in the kidney [86], exist for the transport of phosphate (Na^+/P_i) in the intestine. By expression cloning using oocytes of *Xenopus laevis*, the Na^+-dependent phosphate transporter has been identified in the small intestine [87]. Evidence has been obtained that the cloned Na^+/P_i symport is localized in the apical membrane of proximal tubular or small intestinal epithelial cells [87–89]. The involvement of the phosphate transporter in the uptake of phosphonoformic acid and forscarnet, demonstrated in the rat intestine [90], was Na^+ dependent [91].

1.2.6 ENT and CNT (SLC29A and SLC28A)

Nucleosides are hydrophilic molecules and require specialized transport proteins for permeation across cellular membranes [92–96]. There are two types of nucleoside transport processes: equilibrative bidirectional processes driven by chemical gradients (hENT – es for equilibrative/sensitive and ei for equilibrative/insensitive to nitrobenzylthioinosine inhibition), and an inwardly directed concentrative process driven by the Na^+ electrochemical gradient (hCNT). The uptake transporters are of broad specificity, play an important role in adenosine-mediated regulation of many physiological processes, including neurotransmission and platelet aggregation, and are a target for coronary vasodilator drugs. The human jejunal brush border membrane expresses both the concentrative Na^+-dependent nucleoside transporters, CNT1, and CNT2 [97]. CNT1 is purine selective, CNT2 is pyrimidine selective and CNT3–CNT5 exhibit variable selectively for both purine and pyrimidine nucleosides [93–95, 98].

CNT1–CNT2 transport a wide range of physiological purine and pyrimidine nucleosides, and antineoplastic and antiviral nucleoside drugs. CNT1 transports 5′-deoxy-5-fluorouridine (5′-DFUR), an intermediate metabolite of capecitabine [99]. CNT1 is potentially involved in both the intestinal absorption of purine nucleosides (including adenosine), uridine and purine nucleoside drugs [93, 94] and in the mediation of small, but significant, fluxes of the antiviral purine nucleoside analog 2′,3′-dideoxyinosine (DDI). The hCNT2 transports inosine. Among several pyrimidine nucleosides, hCNT1 and hCNT2 favorably interacted with the uridine analog floxuridine [100] and gemcitabine [101]. The expressed Na^+-dependent nucleoside transporters mediated the transport of their respective nucleoside substrates with a high affinity and a low capacity, whereas the ENT transporters mediated the transport of nucleosides with a low affinity and a high capacity [102]. Zidovudine, zalcitabine and DDI are substrates of the hENT2 [103]. The mRNA expression of hCNT2 in human duodenum was 15-fold greater than that of hCNT1 or hENT2. Na^+-dependent nucleoside transporters are present on the brush-border membranes of the enterocytes along the entire length of the fetal and adult small intestines. The activity of these transporters was higher in the proximal than in the distal small intestine [104].

1.2.7 OCT and OCTN2 (SLC22)

The OCT transporters belong to a superfamily of transporters that mediate electrogenic transport of small organic cations with different molecular structures, sometimes independently of Na^+ and H^+ gradients [105, 106]. The current knowledge of the distribution and functional properties of cloned cation transport systems and of cation transport is that they exist in the intestine. Immunocytochemical analyses of human jejunum revealed that hOCT3 is localized to the brush-border membrane, whereas hOCT1 immunolabeling was mainly observed at the lateral membranes of the enterocytes [107]. Intestinal apical uptake of the neurotoxin, MPP(+) or 1-methyl-4-phenylpyridinium, a substrate of OCT1, is regulated by phosphorylation/dephosphorylation mechanisms, being most probably active in the dephosphorylated state [108]. Some antiviral drugs are taken up by OCT1 [109]. Na^+-dependent uptake of L-carnitine in Caco-2 cells is mediated by the recently identified organic cation/carnitine transporter, OCTN2, at the brush-border membrane [110, 111]. It also transports the organic cation tetraethylammonium (TEA) in an Na^+-independent manner. Both Na^+-dependent and -independent pathways are involved in OCTN2 uptake of cephaloridine, cefoselis, cefepime, and cefluprenam [112, 113]. OCTN3, a Na^+-independent L-carnitine transporter was further found in basolateral membrane of enterocytes of chickens and mice [114].

1.2.8 Niemann–Pick C1-like 1 Protein (NPC1L1)

Very recently, the Niemann–Pick C1-like 1 protein has been identified as a key player in cholesterol absorption by the small intestine and may represent a target of the cholesterol absorption inhibitor, ezetimibe [115]. The identification of the NPC1L1 protein as a sterol transporter in the gut has focused attention on sterol transport processes in the small intestine and liver. NPC1L1 expression is enriched in the small intestine and is in the brush-border membrane of enterocytes, localized in jejunal enterocytes, which are critical for intestinal cholesterol absorption [116]. Patients with diabetes had more NPC1L1 mRNA than the control individuals [117]. The identification of this new cholesterol-transporting protein has fueled new activities towards cholesterol-lowering endeavors.

1.3 INTESTINAL EFFLUX TRANSPORTERS

The diversity of apical absorptive transporters in the intestine is counteracted by the presence of ABC drug efflux transporters that delimit drug absorption and reduce bioavailability (see Figure 1.2). The ABC transporters are primary transporter gene products that are based on a highly conserved ATP-binding cassette [118], and represent the largest family of transmembrane proteins classified, based on the sequence and organization of their ATP-binding domain(s), also known as nucleotide-binding folds. These ABC efflux transporters are gateways to the gut because they transport exogenous and endogenous substances, and they reduce the body load of potentially harmful compounds [119]. A list of the putative human ABC transporters is given on the website www.nutrigene.4t.com. The subfamily of ABC transporters includes ABCA to ABCG [120] (Table 1.2).

TABLE 1.2 Summary of substrates for transporters expressed in apical membrane of human intestine

Transporters	Substrates	Reference
Pgp (MDR1, ABCB1)	*Anti-cancer drugs*	
	Daunorubicin[d]	[127]
	Doxorubicin[c,d]	[128]
	Actinomycin D[d]	[129]
	Etoposide[c,d]	[128]
	Vinblastin[d]	[128]
	Vincristine[d]	[130]
	Taxol (paclitaxel)[c]	[131]
	Topotecan[c]	[131]
	Mitoxantrone[d]	[132]
	Colchicine[a]	[133]
	HIV protease inhibitors	
	Indinavir[d]	[134]
	Nelfinavir[d]	[134]
	Saquinavir[d]	[134]
	Ritonavir[c]	[135]
	Amprenavir[c]	[136]
	Immunosuppressant	
	Cyclosporin[d]	[137]
	Tacrolimus[c]	[131,138]
	FK506[d]	[137]
	Antibiotics	
	Levofloxacin[d]	[139]
	Sparfloxacin[d]	[140]
	Erythromycin[d]	[141]
	Hormones	
	Cortisol, aldosterone[d]	[142]
	Dexamethasone[d]	[142]
	Estrone[d]	[143]
	Estradiol[d]	[143]
	Ethinyl estradiol[d]	[143]
	Methylprednisolone[c]	[144]

(Continued)

TABLE 1.2 Continued

Transporters	Substrates	Reference
	Cardiovascular drugs	
	Digoxin[c,d]	[128]
	Quinidine[c,d]	[128]
	Celiprolol[c]	[145]
	Losartan[c,d]	[246]
	Diltiazem[d]	[147]
	H_2-receptor antagonists	
	Ranitidine[c]	[148]
	Famotidine[c]	[148]
	Cimetidine[c,d]	[149]
	Others	
	L-Dopa[a]	[150]
	Emetine[a,d]	[129,151]
	Fexofenadine[d]	[73]
	Metkephamid[c]	[152]
	Domperidone[c]	[153]
	Ivermectin, loperamide[d]	[154]
	Terfenadine[d]	[155]
	Vecuronium[d]	[156]
	Flesinoxan[d]	[157]
	Morphine[c,f]	[158, 159]
MRP2 (ABCC2)	Grepafloxacin[c]	[189]
	Methotrexate[a]	[190]
	Benzylpenicillin[a]	[190]
	Pravastatin, cervastatin[e]	[187, 188]
	Saquinavir, ritonavir, and indinavir[d]	[183, 191]
	Enalapril and enalaprilat[e]	[192]
	Dehydroepiandrosterone sulfate (DHEAS)[e]	[193]
	Etoposide[d]	[194]
	Sulfinpyrazone[d]	[186]
MRP3 (ABCC3)	Methotrexate[a]	[197]
	Leukovorin[a]	[197]
	Etoposide[a]	[200]
BCRP (ABCG2)	Mitoxantrone[b]	[223]
	Topotecan[b]	[224]
	SN-38[b]	[224]
	Irinotecan[b]	[224]
	Prazosin[b]	[223]
	Estrone-3-sulfate[e]	[188]
	Cerivastatin, pravastatin[e]	[188]
	Imatinib[b]	[225]
	Flavopiridol[a]	[226]
	Methotrexate[a,b]	[227]

[a]Uptake/drug accumulation in transfected cells or their vesicles.
[b]Uptake/drug accumulation in cells overexpressing target transporters.
[c]Transport in Caco-2 cells.
[d]Transport in transfected cells (LLC-PK_1, MDCKII).
[e]Transport in double-transfected cells.
[f]Pharmacokinetic studies.

ABC pumps are mostly unidirectional. Most of the known functions of ABC transporters involve the shuttling of hydrophobic compounds either within the cell as part of a metabolic process or outside the cell for transport to other organs or secretion from the body [118]. The subfamilies of ABC transporters for drug transport carry out specific transport: ABCCA is limited for lipids; ABCB, in which Pgp (P-glycoprotein, ABCB1) belongs, is for lipophilic species; and ABCC or MRP2 (multidrug resistance-associated protein 2, ABCC2) and ABCG or breast cancer resistance protein (BCRP, ABCG2) are transporter families for more polar entities. Much of the work has been facilitated by use of the Caco-2 cell system, knockout animals, and polarized cells transfected with the cDNA (copy DNA) of the transporter genes. Clinically relevant examples have been reported to be associated with multidrug resistance of cancer cells and with in cystic fibrosis.

1.3.1 MDR1 (ABCB1)

The 170-kDa P-glycoprotein, Pgp, also known as the multidrug resistance protein 1, is a ubiquitous transporter that is present not only in the intestine, but also in the brain, kidney, liver, and testes [121]. Human Pgp confers multidrug resistance to cancer cells by ATP-dependent extrusion of a great many structurally dissimilar hydrophobic compounds. The list contains chemotherapeutic agents, analgesics, antibiotics, cardiac glycosides, neuroleptics, antiemetics, calcium channel blockers, corticoids, immunosuppressants, HIV protease inhibitors, tricyclic antidepressants, insecticides, and toxicants, and is constantly increasing [121–123]. The functional unit of Pgp consists of two nucleotide-binding domains (NBDs) and two *trans*-membrane domains, which are involved in the transport of drug substrates. Considerable progress has been made in recent years in characterizing these functionally and spatially distinct domains of Pgp.

The topology has been augmented by the resolution of structures of several non-mammalian ABC proteins, revealing the specific conserved amino acids in ATP hydrolysis, the dimensions of the drug-binding pocket, and a current understanding of the mechanisms of coupling between energy derived from ATP binding and/or hydrolysis and efflux of drug substrates [124]. There is presence of at least two non-identical substrate interaction sites in Pgp [125, 126]. This transporter is an energy-dependent multidrug efflux pump involved in the mechanism of multidrug resistance in cancer. The list of substrate of Pgp is impressive (see Table 1.2). So far, several MDR1 SNPs have been identified, and mutations at positions 2677 and 3435 were associated with alteration of Pgp expression and/or function. However, the impact of MDR1 polymorphisms on pharmacokinetics and pharmacodynamics of Pgp substrates is unimpressive [160].

Much of the low bioavailability in oral drug absorption has been blamed on the presence of MDR1, which effluxes absorbed drugs back to the lumen. As a result of the important implication of Pgp in drug absorption, Caco-2 cells have been developed to study intestinal epithelial transport and used as in vitro screens for ready identification of Pgp substrates as well as drug–drug interactions [161, 162]. The in vitro data correlate well in predicting drug absorption and interactions in vivo, as shown for the model Pgp substrate, digoxin (increase in digoxin C_{max} in humans) and its interactants, talinolol, omeprazole, verapamil, quinidine, and cyclosporin [163]. The strategy in drug development is towards the avoidance of drugs that are good Pgp substrates because these would exhibit poor bioavailability and exert the potential of drug–drug interactions. In parallel fashion, much effort has been devoted to the identification of new chemical entities as Pgp inhibitors to enhance drug absorption. Both direct inhibition and allosteric inhibition were found among drug candidates [164].

The first strategy to inhibit Pgp function relied on the identification of non-chemotherapeutic agents as competitors. These are exemplified by valspodar (PSC833), quinine-like compounds, tariquidar (XR-9576), and mitotane (NSC-38721) [123]. Other approaches have included the use of hammerhead ribozymes against the *MDR-1* gene and MDR-1-targeted antisense oligonucleotides [165]. Newer agents employ molecular targeting, such as immunoliposomes using antibody-directed binding and internalization, exemplified by encapsulation of liposomal anthracyclines in stable and non-reactive carriers in some clinical settings involving resistant tumors. These agents selectively deliver the drug to the tumor cells via efficient internalization for intracellular drug release, and can potentially enhance both efficacy and safety [166]. Although modulation of drug resistance has yet to be proven clinically effective, much has been learned about drug resistance [165].

Coincidentally, some flavonoids – flavonols (quercetin and kaempferol) and isoflavones (genistein and daidzein) – markedly increase drug accumulation and the sensitivity of the multidrug-resistant human cervical carcinoma KB-V1 cells (high Pgp expression) to vinblastine and paclitaxel [167, 168]. In several extensive reviews, various herbal products affect Pgp [169, 170]. Some herbal constituents (e.g. hyperforin and kava) were shown to activate pregnane X receptor, an orphan nuclear receptor acting as a key regulator of MDR1 and many other genes, exemplified by the induction effects of St John's wort on intestinal expression of Pgp in vitro and in vivo. Certain natural flavonols (e.g. kaempferol, quercetin, and galangin) are potent stimulators of the Pgp-mediated efflux of the carcinogen, 7,12-dimethylbenz(a)-anthracene. Curcumin, ginsenosides, piperine, some catechins from green tea, and silymarin from milk thistle were found to be inhibitors of Pgp, whereas some catechins from green tea increased Pgp-mediated drug transport by heterotropic allosteric mechanisms.

Many of these herbal constituents, in particular flavonoids, are reported to modulate Pgp by directly interacting with the ATP-binding site, the steroid-binding site, or the substrate-binding site. The stimulatory/inhibitory role of the flavonoid may be concentration dependent as a result of the presence of dual binding sites, so that allosteric induction or competitive inhibition may occur. The inhibition of Pgp by herbal constituents may provide a novel approach for reversing multidrug resistance in tumor cells. Modulation of Pgp activity and expression by the herb constituents may result in altered absorption and bioavailability of drugs that are Pgp substrates [169, 170]. In addition to the inhibition of Pgp activity, bergamottin from grapefruit juice was reported to modulate cytochrome P450 activities [171–175].

1.3.2 MRP2 (ABCC2) and MRP3 (ABCC3)

Apart from Pgp, another extensively studied ABC member is the MRP, or the multidrug resistance-associated protein, family [121, 176, 177]. MRP2 transcripts are abundant in the human jejunum [178]. A parallelism is drawn for substrate specificity between rat and human MRP2, because of the ease of study of mutant rats, the TR^- and Eisai hyperbilirubinemic rats, which lack the functional Mrp2 and cause conjugated hyperbilirubinemia [179] as a result of reduced excretion of conjugated bilirubin [180]. The loss of MRP2 function in humans is associated with the Dubin–Johnson syndrome [181, 182]. Besides the organic molecules being conjugated to negatively charged ligands, these proteins also transport cytotoxic drugs for which no negatively charged conjugates are known to exist. Lipophilic substrates such as HIV protease inhibitors, saquinavir, ritonavir and indinavir [183], and paclitaxel [184] are substrates. The direct examination of substrates of MRP2 is rendered more difficult when substrates do not enter the basolateral side readily. Membrane vesicles, prepared from recombinant, human, MRP2 stably transfected cells, showed ATP-dependent transport of estradiol 17β-glucuronide, leukotriene C4, etoposide, vincristine, cisplatin, doxorubicin, and epirubicin [185]; glutathione export was found in association with the transporter [186]. The examination has recently been facilitated by the development of double transfectants with a basolateral transporter, bringing the substrate into the cell for excretion [187, 188]. A list of the MRP2 substrates may be found in Table 1.2.

Human MRP3, an ATP-binding cassette transporter cloned from Caco-2 cells, was found in the human small intestine and colon [195]. MRP3 is 56% identical to MRP1 and 45% identical to MRP2 [196]. MRP3 is expressed in the basolateral membrane of enterocytes, and has the ability to transport bile salts and amphiphilic anions [197]. MRP3 is able to confer resistance to anticancer agents, such as etoposide, and to transport lipophilic anions, such as bile acids and glucuronides. MRP3 mRNA was found to be induced threefold in human colon cells by chenodeoxycholic acid [198]. These capabilities, along with the induction of the MRP3 protein on hepatocyte sinusoidal membranes in cholestasis and the expression of MRP3 in enterocytes, have led to the hypothesis that MRP3 may function in the body to protect cholestatic hepatocytes from endobiotics, and to facilitate bile-acid reclamation from the gut [199]. Etoposide is a substrate of MRP3 [200] (see Table 1.2).

1.3.3 BCRP (ABCG2)

The breast cancer resistance protein is a recently described ABC transporter originally identified by its ability to excrete mitoxantrone and confer drug resistance that is independent of MRP1 and Pgp. Unlike MRP1 and Pgp, ABCG2 is a half-transporter that must homodimerize to acquire transport activity. ABCG2 is found in a variety of stem cells and may protect them from exogenous and endogenous toxins. ABCG2 expression is upregulated under low oxygen conditions, consistent with its high expression in tissues exposed to low oxygen environments. ABCG2 interacts with heme and other porphyrins, and protects cells and/or tissues from protoporphyrin accumulation under hypoxic conditions [201]. There appears to be a role for BCRP in folate homeostasis [202]. The protein is expressed at high levels in the human intestine [203], has a broad substrate specificity, and actively extrudes a wide variety of drugs, carcinogens, and dietary toxins from cells [204]. BCRP functions as an efflux transporter for mitoxantrone, methotrexate [205, 206], estrone sulfate [207], estradiol 17β-glucuronide, irinotecan [208], topotecan [203], nitrofurantoin [209], the fluoroquinolone antibiotics (ciprofloxacin, ofloxacin, and norfloxacin) [210], rosuvastatin [211], cimetidine [212], and anthelmintic agents ethylcarbamate benzimidazoles (albendazole, fenbendazole, and their respective sulfoxide derivatives) [213]. In animal studies, Bcrp1 is found to extrude sulfate and glucuronide conjugates [214].

Common natural allelic variants of BCRP have been identified that did not influence interindividual variation in expression of BCRP mRNA in human intestine, but this remains to be tested for effect on BCRP function [215]. In a preliminary fashion, the heterozygous CA allele observed in two patients was associated with a 1.34-fold increased oral bioavailability of topotecan compared with the bioavailability in 10 patients with the wild-type allele (42.0% versus 31.4%). It is suggested that the high frequency of the A allele in certain ethnic groups may have therapeutic implications for individuals treated with topotecan or other ABCG2 substrates [203]. It was confirmed that the ABCG2 421 $C > A$ genotype significantly affected the pharmacokinetics of diflomotecan [216]. High-throughput-based assays for ABCG2 inhibitors exist [217]. Isothiocyanates inhibit BCRP in BCRP-overexpressing and BCRP-negative human breast cancer (MCF-7) and large-cell lung carcinoma (NCI-H460) cells [218]. The flavonoids again inhibit BCRP in mice [219, 220]. HIV protease inhibitors, ritonavir, saquinavir, and nelfinavir, are effective inhibitors of BCRP [221]. Novobiocin is also found to be an effective BCRP-reversal agent [222].

1.3.4 ABCG5/G8 and ABCA1

Two members of the ABCG subfamily, namely ABCG5 and ABCG8, have been implicated in the efflux of dietary sterols from intestinal epithelial cells back to the gut lumen, and from the liver to the bile duct [228]. Furthermore, the ABCG5 and ABCG8 half-transporters, functioning as heterodimers, represent apical sterol export pumps that promote active efflux of cholesterol and plant sterols from enterocytes back into the intestinal lumen for excretion [24, 25]. Interestingly, mutation of ABCG5 and/or ABCG8 genes in humans causes sitosterolemia, a rare genetic disease characterized by massive absorption of plant sterols and premature arteriosclerosis. The identification of defective structures in the ABCG5 or ABCG8 transporters in patients with the rare disease of sitosterolemia elucidated their role as sterol efflux pumps, regulating at least in part the intestinal sterol absorption and the hepatic sterol output. ABCG5 and ABCG8 themselves are regulated by cholesterol via liver X receptors (LXRs). However, at the level of the enterocyte, the cellular distribution patterns of ABCG5 and ABCG8 differed, such that ABCG5 was more diffuse, but ABCG8 was principally apical [229].

Mice fed diets containing high cholesterol markedly increased the expression of ABCG5/8 mRNA in liver and intestine, and these are increased on LXR activation [230, 231], although the magnitude of this increase was generally less if the mice were given SCH 58053, an analog of ezetimibe [231]. ABCG5 and ABCG8 mRNA was also induced by cholesterol in rat ileum, but not mouse ileum [232]. The best hypothetical drug to lower cholesterol should decrease serum lipoprotein cholesterol levels, increase biliary cholesterol secretion, and fecal elimination, favoring at the same time gallbladder emptying to prevent gallstone formation [233]. Expression of ABCG5 and ABCG8 mRNA was lower in patients with diabetes [117]. At least during the early phase after transplantation, hepatic ABCG5/ABCG8 expression is not directly related to biliary cholesterol secretion in humans [234]. This finding suggests the existence of alternative pathways for the hepatobiliary transport of cholesterol that are not controlled by ABCG5/ABCG8.

Evidence has accumulated over the past few years to suggest that a subgroup of 12 structurally related 'full-size' transporters, referred to as ABC A subfamily transporters, mediates the transport of a variety of physiologic lipid compounds [235–237]. ABCA1, previously implicated in the control of cholesterol absorption, was found to be dramatically upregulated in jejunal enterocytes on exposure to LXR agonists [230]. Cholesterol feeding increased the relative mRNA level for ATP-cassette transporter A1 (ABCA1), but the increase was lessened if the mice were given SCH 58053, an analog of ezetimibe [233]. Another new chemical, aramchol, is a fatty acid–bile acid conjugate designed to treat a range of lipid disorders and increased cholesterol efflux in a dose-dependent fashion [238]. ABCA1-disrupted sv129/C57BL/6 hybrid mice showed a significant reduction in intestinal cholesterol absorption. ABCA1 plays a pivotal role in reversing cholesterol transport by mediating the cellular efflux of phospholipid and cholesterol, and mediates efflux of 25-hydroxycholesterol [239]. ABCA1 is found to play a major role in cholesterol homeostasis and high-density lipoprotein (HDL) metabolism, binding to apolipoprotein A-I and cellular cholesterol and phospholipids, mainly phosphatidylcholine, and loading on to apoA-I to form pre-β HDL [240]. Heterozygosity for an *ABCA1* mutation (K776N) conferred a two- to threefold risk of ischemic heart disease [241].

1.4 MODULATION OF DRUG TRANSPORTERS BY EXCIPIENTS AND SURFACTANTS

Akin to the presence of juices and flavonoids that result in changes in transporters and enzyme activities and cause changes in bioavailability, it has been found that lipid excipients and surfactants enlisted in lipid-based formulations are also able to modulate the activity of drug efflux transporters that delimit drug absorption. The mechanism of modulation of transporter activity by lipid excipients has not been fully elucidated. Lipids are a collection of fats and fat-like substances and encompass fatty acids, glycolipids, lipoproteins, phospholipids, and steroids. Being ubiquitously distributed as a basic constituent of cell structure, lipids serve as a source of fuel and exert important biological functions. Lipid excipients typically include long- or medium-chain fatty acids, mono-, di- and triglyceride lipids, supplemented with individual or mixed surfactants, and various hydrophilic solvents in various liquid emulsion and microemulsion formulations and semisolid/solid dispersions [242]. Lipid-based formulations have been used to increase the bioavailability of hydrophobic entities because the latter constitute a predominant proportion of chemicals that are already available on the market or are new chemical entities considered for development. The mechanism of increased bioavailability traditionally lies in improved solubility via formation of micelles or facilitated lymphatic drug absorption [243, 244].

The mechanism of modulation of transporter activity by surfactants is not fully understood in vivo. Most of the studies about the effect of excipients/surfactants on transporters were carried out in vitro with cell lines, such as Caco-2 or MDCK cells, or tissues, such as rat everted gut sac or rat jejunal tissue chamber. The impact may or may not be extrapolated readily to the in vivo situation. However, the finding of interaction of transporters with excipients, including the inhibition of transporters by Pluronic block copolymers, Tween 80, Cremophor EL and tocopheryl polyethylene glycol succinate (TPGS), makes it possible to understand the pharmacokinetic mechanism and optimize the formulation design. In most instances, Pgp is inhibited (Table 1.3). The best studied is the Pluronic block copolymers (Pluronic P85) that are potent sensitizers of multidrug-resistant (MDR1 or Pgp) cancer cells. The inhibition of Pgp is a consequence of two reactions. One is intracellular ATP depletion [262], and the second is the inhibition of ATPase activity of drug efflux proteins such as Pgp, MRP1, and MRP2 that excrete vinblastine, and leukotriene C4 [263]. The V_{max} decreased whereas the apparent Michaelis constants increased in the presence of various concentrations of P85. The changes may be the result of conformational changes of the transporter as a result of membrane fluidization and/or nonspecific steric hindrance of the drug-binding sites by P85 chains embedded in cellular membranes. Hence, the Pluronic block copolymer in the formulation may sensitize the cancer cells to cause an increase in cytotoxicity compared with unformulated dosage form, as exemplified in doxorubicin which results in apoptosis in resistance cancer cells [264]. Furthermore, Pluronic block copolymers abolish drug sequestration in acidic vesicles as well as inhibiting the glutathione/glutathione *S*-transferase detoxification system. All these mechanisms induced by Pluronic block copolymers are potential reasons for chemosensitization of these cells, leading to formulations containing drugs (doxorubicin) and Pluronic mixture (L61 and F127), SP1049C, in phase I clinical trials [265].

Studies conducted in the Caco-2 cell culture system, polyoxyethylene sorbitan fatty acid esters (e.g. Tween 80) and polyethoxylated castor oil (e.g. Cremophor EL) showed inhibition of Pgp-mediated drug efflux in cell culture systems at concentrations below their critical micellar concentrations, leading to the suggestion that the excipients may exert a relatively specific, inhibitory effect on the microenvironment of the Pgp transporter. The expected outcome is enhanced drug absorption. The alteration of cell membrane fluidity may be necessary but not sufficient for modulation of transporter activity of nonionic surfactants [246]. D-α-Tocopheryl poly(ethylene glycol) 1000 succinate (TPGS 1000 or TPGS), a widely used form of vitamin E as a solubilizer, an emulsifier, and a vehicle for lipid-based drug delivery

TABLE 1.3 Summary of effects of various excipients/surfactants inhibiting various transporters

Transporter	Excipient/Surfactant	Experimental system	Reference
Pgp ↓	Tween 80	I[a]	[245, 246]
		II	[247, 248]
		III	[249]
		IV	[250]
	Polyethoxylated castor oil (Cremophor EL)	I	[245, 246, 251]
		II	[248, 252]
		III	[249]
	TPGS	I	[246, 253, 254]
		II	[252]
		X	[254]
		XI	[254]
	PEG-300	I	[245, 255]
		VIII	[245]
	Pluronic block copolymer	I	[256]
	P85	VII	[257]
		VI	
	Tween 20	II	[247]
		V	[258]
	Phospholipids	I;II	[259]
	PEG-400	VI	[257]
	Peceol	I	[260]
	Pluronic block copolymers L81, P85, and F68	I	[251]
	Myrj 52	II	[247]
	Brij 30	II	[247]
	Labrasol	II	[252]
	Imwitor 742	II	[252]
	Acconon E	II	[252]
	Softigen 767	II	[252]
	Miglyol	II	[252]
	Solutol HS 15	II	[252]
	Sucrose monolaurate	II	[252]
	Triton X-100	V	[258]
	Nonidet P-40	V	[258]
MRPs ↓	PEG-300	I	[255]
	Pluronic block copolymer (L81, P85, and F108)	IX	[261]
	Tween 80	I;II	[247]
	Tween 20	I;II	[247]
	Myrj 52	I;II	[247]
	Brij 30	I;II	[247]
PEPT1 ↓	Tween 80	I	[246]
MCT1 ↓	Polyethoxylated castor oil (Cremophor EL)	I	[246]

[a]System I: Caco-2 cell; system II: rat everted gut sac; system III: isolated rat intestinal membrane; system IV: in vivo rat; system V: Chinese hamster ovary AA8 cell and its Pgp-overexpressing EmtR1 subline; system VI: rat jejunal tissue chamber; system VII: bovine brain microvessel endothelial cell (BBMEC); system VIII: MDCK cell; system IX: human pancreatic adenocarcinoma cell (Panc-1); system X: NIH 3T3 cell; system XI: HCT-8 cell.

formulations, increased the sensitivity of Pgp-expressing cells to several cytotoxic Pgp substrates in vitro, and effectively blocked polarized transport of Pgp model substrates such as rhodamine 123 (RHO) and paclitaxel [254], and HIV protease inhibitors [266]. Callnot et al. [253] reported that the polyethyleneglycol (PEG) chain length (200, 238, 400, 600, 1000, 2000, 3400, 3500 4000, and 6000) in TPGS analogs influenced RHO transport in Caco-2 monolayers. Large increases in mucosal-to-serosal but decreases in serosal-to-mucosal transport in Caco-2 cell line and everted sacs of jejunum and ileum were observed with epirubicin (a high-affinity Pgp substrate) in encapsulated liposomes, mixtures of empty liposomes, or free epirubicin. However, free phospholipids failed to change the transport of epirubicin [259]. The observation suggests that phospholipids in the form of liposomes are able to interact with cell membrane and modulate Pgp activity.

Lipid excipients and surfactants, such as PEG-300, Tween 80, and Tween 20 are suggested to have an effect on the MRPs [247] (see Table 1.3). The Pluronic block system also affects the MRPs [261]. The effect of P85 on Pgp ATPase activity is considerably less on the MRP1 and MRP2 ATPases compared with that for the Pgp ATPases [263]. In addition, the intestinal absorptive transporters, such as peptide transporter (PEPT1) and monocarboxylic acid transporter (MCT1), could be affected by excipients/surfactants, although the mechanism is mostly unknown. In an in vitro Caco-2 cell study, Tween 80 inhibited the PEPT1, as measured by glycylsarcosine permeability [246]. Similarly, MCT1 was inhibited by Cremophor EL, as measured by benzoic acid permeability [246]. These results suggest that surfactants can inhibit multiple transporters although the changes in membrane fluidity may not be a generalized mechanism to reduce transporter activity. A recent study reported the enhanced uptake of doxorubicin in multidrug-resistant breast cancer cells by polymer–lipid hybrid nanoparticle (PLN) system, presumably as a result of inhibition of BCRP [267].

1.5 CONCLUSION

This mode of drug transport has been described in reference to a list of transporters responsible for solute absorption and efflux in the small intestine of humans. Substrate specificity differs among the solute absorptive transporters, but there is some overlap with the ABC transporters, especially in relation to drug resistance. Excipients, in particular the Pluronic copolymers, and other surfactants inhibit the functions of transporters, leading to increased drug absorption or retention, and increased bioavailabilty.

ACKNOWLEDGMENT

Supported by CIHR MOP64350 and CIHR net grant.

REFERENCES

1. Pang, K.S. Modeling of intestinal drug absorption: Roles of transporters and metabolic enzymes (for the Gillette Review Series). *Drug Metab Dispos* 2003;**31**:1507–1519.
2. Doherty, M., Pang, K.S. The role of the intestine in the absorption and metabolism of drugs. *Drug Chem Toxicol* 1997;**20**:329–344.
3. Tam, D., Tirona, R.G., Pang, K.S. Segmental intestinal transporters and metabolic enzymes on intestinal drug absorption. *Drug Metab Dispos* 2003;**31**:373–383.
4. Lin, J.H., Chiba, M., Baillie, T.A. Is the role of the small intestine in first-pass metabolism overemphasized? *Pharmacol Rev* 1999;**51**:125–158.
5. Trauner, M., Boyer, J.L. Bile salt transporters: molecular characterization, function, and regulation. *Physiol Rev* 2003;**83**:633–671.
6. Eloranta, J.J., Meier, P.J., Kullak-Ublick, G.A. Coordinate transcriptional regulation of transport and metabolism. *Methods Enzymol* 2005;**400**:511–530.
7. Huang, Y., Anderle, P., Bussey, K.J., Barbacioru, C., Shankavaram, U., Dai, Z., et al. Membrane transporters and channels: role of the transportome in cancer chemosensitivity and chemoresistance. *Cancer Res* 2004;**64**:4294–4301.
8. Tsuji, A., Tamai, I. Carrier-mediated intestinal transport of drugs. *Pharm Res* 1996;**13**:962–977.
9. Mizuno, N., Niwa, T., Yotsumoto, Y., Sugiyama, Y. Impact of drug transporter studies on drug discovery and development. *Pharmacol Rev* 2003;**55**:425–461.
10. Steffansen, B., Nielsen, C.U., Broden, B., Eriksson, A.H., Andersen, R., Frokjaer, S. Intestinal solute carriers: an overview of trends and strategies for improving oral drug absorption. *Eur J Pharm Sci* 2004;**21**:3–16.
11. Anderle, P., Huang, Y., Wolfgang, S. Intestinal membrane transport of drugs and nutrients: genomics of membrane transporters using expression microarrays. *Eur J Pharm Sci* 2004;**21**:17–24.
12. Gerk, P.M., Vore, M. Regulation of expression of the multidrug resistance-associated protein 2 (MRP2) and its role in drug disposition. *J Pharmacol Exp Ther* 2002;**302**:407–415.
13. Goodwin, B., Redinbo, M.R., Kliewer, S.A. Regulation of cyp3a gene transcription by the pregnane x receptor. *Annu Rev Pharmacol Toxicol* 2002;**42**:1–23.
14. Willson, T.M., Kliewer, S.A. PXR, CAR and drug metabolism. *Nat Rev Drug Discov* 2002;**1**:259–266.
15. Kliewer, S.A., Goodwin, B., Willson, T.M. The nuclear pregnane X receptor: a key regulator of xenobiotic metabolism. *Endocr Rev* 2002;**23**:687–702.
16. Wang, H., LeCluyse, E.L. Role of orphan nuclear receptors in the regulation of drug-metabolising enzymes. *Clin Pharmacokinet* 2003;**42**:1331–1357.
17. Staudinger, J.L., Madan, A., Carol, K.M., Parkinson, A. Regulation of drug transporter gene expression by nuclear receptors. *Drug Metab Dispos* 2003;**31**:523–527.
18. Bock, K.W., Kohle, C. Coordinate regulation of drug metabolism by xenobiotic nuclear receptors: UGTs acting together with CYPs and glucuronide transporters. *Drug Metab Rev* 2004;**36**:595–615.
19. Haimeur, A., Conseil, G., Deeley, R.G., Cole, S.P. The MRP-related and BCRP/ABCG2 multidrug resistance proteins: biology, substrate specificity and regulation. *Curr Drug Metab* 2004;**5**:21–53.
20. Catania, V.A., Sanchez Pozzi, E.J., Luquita, M.G., Ruiz, M.L., Villanueva, S.S., et al. Co-regulation of expression of phase II metabolizing enzymes and multidrug resistance-associated protein 2. *Ann Hepatol* 2004;**3**:11–17.
21. Runge-Morris, M., Kocarek, T.A. Regulation of sulfotransferases by xenobiotic receptors. *Curr Drug Metab* 2005;**6**:299–307.
22. Eloranta, J.J., Kullak-Ublick, G.A. Coordinate transcriptional regulation of bile acid homeostasis and drug metabolism. *Arch Biochem Biophys* 2005;**433**:397–412.
23. Tirona, R.G., Kim, R.B. Nuclear receptors and drug disposition gene regulation. *J Pharm Sci* 2005;**94**:1169–1186.
24. Lammert, F., Wang, D.Q. New insights into the genetic regulation of intestinal cholesterol absorption. *Gastroenterology* 2005;**129**:718–734.
25. Plosch, T., Kosters, A., Groen, A.K., Kuipers, F. The ABC of hepatic and intestinal cholesterol transport. *Handbook Exp Pharmacol* 2005;**170**:465–482.

26. Terada, T., Shimada, Y., Pan, X., Kishimoto, K., Sakurai, T., Doi, R., et al. Expression profiles of various transporters for oligopeptides, amino acids and organic ions along the human digestive tract. *Biochem Pharmacol* 2005;**70**:1756–1763.
27. Anderson, C.M., Grenade, D.S., Boll, M., Foltz, M., Wake, K.A., Kennedy, D.J., et al. H^+/amino acid transporter 1 (PAT1) is the imino acid carrier: An intestinal nutrient/drug transporter in human and rat. *Gastroenterology* 2004;**127**:1410–1422.
28. Bell, G.I., Kayano, T., Buse, J.B., Burant, C.F., Takeda, J., Lin, D., et al. Molecular biology of mammalian glucose transporters. *Diabetes Care* 1990;**13**:198–208.
29. Stahl, A. A current review of fatty acid transport proteins (SLC27). *Plügers Arch* 2005;**447**:772–777.
30. Hirch, D., Stahl, A., Lodish, H.F. A family of fatty acid transporters conserved from mycobacteria to man. *Proc Natl Acad Sci USA* 1998;**95**:8625–8629.
31. Abumed, N.A., el-Maghrabi, M.R., Amri, E.Z., Lozep, E., Grimaldi, P.A. Cloning of a rat adipocyte membrane protein implicated in binding or transport of long chain fatty acids that is induced during preadipocyte differentiation. Homology with human CD36. *J Biol Chem* 1993;**268**:17665–17668.
32. Brinckmann, J.F.F., Abumrad, N.A., Ibrahimi, A., Vusse, G.V.D., Glatz, J.F. New insights in long chain fatty acid uptake by heart muscle: crucial role for fatty acid translocase/CD36. *Biochem J* 2002;**367**:561–570.
33. Besnard, P., Niot, I., Bernard, A., Carlier, H. Cellular and molecular aspects of fat metabolism in the small intestine. *Proc Nutr Soc* 1996;**55**:19–37.
34. Takanaga, H., Mackenzie, B., Hediger, M.A. Sodium-dependent ascorbic acid transporter family SLC23. *Pflügers Arch* 2004;**447**:677–682.
35. Dudeja, P.K., Kode, A., Alnounou, M., Tyagi, S., Torania, S., Subramanian, V.S., Said, H.M. Mechanism of folate transport across the human colonic basolateral membrane. *Am J Physiol Gastrointest Liver Physiol* 2001;**281**:G54–G60.
36. Said, H.M., Ortiz, A., Kumar, C.K., Chatterjee, N., Dudeja, P.K., Rubin, S. Transport of thiamine in human intestine: mechanism and regulation in intestinal epithelial cell model Caco-2. *Am J Physiol* 1999;**277**(4 Pt 1):C645–C651.
37. Dudeja, P.K., Tyagi, S., Kavilaveettil, R.J., Gill, R., Said, H.M. Mechanism of thiamine uptake by human jejunal brush-border membrane vesicles. *Am J Physiol Cell Physiol* 2001;**281**:C786-C792.
38. Dudeja, P.K., Tyagi, S., Gill, R., Said, H.M. Evidence for a carrier-mediated mechanism for thiamine transport to human jejunal basolateral membrane vesicles. *Dig Dis Sci* 2003;**48**:109–115.
39. Wong, M.H., Rao, P.N., Pettenati, M.J., Dawson, P.A. Localization of the ileal sodium-bile acid cotransporter gene (SCL10A2) to human chromosome 13q33. *Genomics* 1996;**33**:536–540.
40. Shneider, B.L., Setchell, K.D., Crossman, M.W. Fetal and neonatal expression of the apical sodium-dependent bile acid transporter in the rat ileum and kidney. *Pediatr Res* 1997;**42**:189–194.
41. Craddock, A.L., Love, M.W., Daniel, R.W., Kirby, L.C., Walters, H.C., Wong, M.H., Dawson, P.A. Expression and transport properties of the human ileal and renal sodium-dependent bile acid transporter. *Am J Physiol* 1998;**274**(1 Pt 1):G157–G169.
42. Small, D.M., Dowling, R.H., Redinger, R.N. The enterohepatic circulation of bile salts. *Arch Intern Med* 1972;**130**:552–573.
43. Love, M.W., Craddock, A.L., Angelin, B., Brunzell, J.D., Duane, W.C., Dawson, P.A. Analysis of the ileal bile acid transporter gene, SLC10A2, in subjects with familial hypertriglyceridemia. *Arterioscler Thromb Vasc Biol* 2001;**21**:2039–2045.
44. Oelkers, P., Kirby, L.C., Heubi, J.E., Dawson, P.A. Primary bile acid malabsorption caused by mutations in the ileal sodium-dependent bile acid transporter gene (SLC10A2). *J Clin Invest* 1997;**99**:1880–1887.
45. Montagnani, M., Love, M.W., Rossel, P., Dawson, P.A., Qvist, P. Absence of dysfunctional ileal sodium-bile acid cotransporter gene mutations in patients with adult-onset idiopathic bile acid malabsorption. *Scand J Gastroenterol* 2001;**236**:1077–1080.
46. Lewis, M.C., Brieaddy, L.E., Root, C. Effects of 2164U90 on ileal bile acid absorption and serum cholesterol in rats and mice. *J Lipid Res* 1995;**36**:1098–1105.
47. Xu, G., Shneider, B.L., Shefer, S., Nguyen, L.B., Batta, A.K., Tint, G.S., et al. Ileal bile acid transport regulates bile acid pool, synthesis, and plasma cholesterol levels differently in cholesterol-fed rats and rabbits. *J Lipid Res* 2000;**41**:298–304.

48. Dawson, P.A., Hubbert, M., Haywood, J., Craddock, A.L., Zerangue, N., Christian, W.V., Ballatori, N. The heteromeric organic solute transporter alpha-beta, Ostalpha-Ostbeta, is an ileal basolateral bile acid transporter. *J Biol Chem* 2005;**280**:6960–6968.
49. Shoji, T., Suzuki, H., Kusuhara, H., Watanabe, Y., Sakamoto, S., Sugiyama, Y. ATP-dependent transport of organic anions into isolated basolateral membrane vesicles from rat intestine. *Am J Physiol Gastrointest Liver Physiol* 2004;**287**:G749–G756.
50. Daniel, H. Molecular and integrative physiology of intestinal peptide transport. *Annu Rev Physiol* 2004;**66**:361–384.
51. Fei, Y.J., Kanai, Y., Nussberger, Y., Ganapathy, V., Leibach, F.H., Romero, M.F., et al. Expression cloning of a mammalian proton coupled oligopeptide transporter. *Nature* 1994;**368**:563–566.
52. Boll, M., Markovich, D., Weber, W.M., Korte, H., Daniel, H., Murer, H. Expression cloning of a cDNA from rabbit small intestine related to proton-coupled transport of peptides, beta-lactam antibiotics and ACE-inhibitors. *Pflügers Arch* 1994;**429**:146–149.
53. Horie, R., Tomita, Y., Katsura, T., Yasuhara, M., Inui, K., Takano, M. (Transport of bestatin in rat renal brush-border membrane vesicles. *Biochem Pharmacol* 1993;**45**:1763–1768.
54. Ganapathy, M.E., Brandsch, M., Prasad, P.D., Ganapathy, V., Leibach, F.H. Differential recognition of beta-lactam antibiotics by intestinal and renal peptide transporters PEPT1 and PEPT2. *J Biol Chem* 1995;**270**:25672–25677.
55. Ganapathy, M.E., Prasad, P.D., Mackenzie, B., Ganapathy, V., Leibach, F.H. Interaction of anionic cephalosporins with the intestinal and renal peptide transporters PEPT 1 and PEPT 2. *Biochim Biophys Acta* 1997;**1324**:251–262.
56. Han, H.K., Rhie, J.K., Oh, D.M., Saito, G., Hsu, C.P., Stewart, B.H., Amidon, G.L. CHO/hPEPT1 cells overexpressing the human peptide transporter (hPEPT1) as an alternative in vitro model for peptidomimetic drugs. *J Pharm Sci* 1999;**88**:347–350.
57. Buyse, M., Berlioz, F., Guilmeau, S., Tsocas, A., Voisin, T., Peranzi, G., et al. PepT1-mediated epithelial transport of dipeptides and cephalexin is enhanced by luminal leptin in the small intestine. *J Clin Invest* 2001;**108**:1483–1494.
58. Beauchamp, L.M., Orr, G.F., Miranada, P., Burnette, T., Krenistsky, T.A. Amino ester prodrugs of aciclovir. *Antiviral Chem Chemother* 1992;**3**:157–164.
59. Ganapthy, M.E., Huang, W., Wng, H., Ganapthay, V. Valaciclovir: a substrate for the intestinal and renal peptide transporter PepT1 and PepT2. *Biochem Biophys Res Commun* 1998;**246**:470–475.
60. Steingrimsdottir, H., Gruber, A., Palm, C., Grimfors, G., Kahn, M., Eksborg, S. Bioavailability of aciclovir after oral administration of aciclovir and its prodrug valaciclovir to patients with leucopenia after chemotherapy. *Antimicrob Agents Chemother* 2000;**44**:207–209.
61. Han, H., de Vrueh, R.L., Rhie, J.K., Covitz, K.M., Smith, P.L., Lee, C.P., et al. 5′-Amino acid esters of antiviral nucleosides, acyclovir and AZT are absorbed by the intestinal PEPT1 peptide transporter. *Pharm Res* 1998;**15**:1154–1159.
62. Tsuda, M., Terada, T., Irie, M., Katsura, T., Niida, A., Tomita, K., et al. Transport Characteristics of a novel PEPT1 substrate, antihypotensive drug midodrine, and its amino acid derivatives. *J Pharmacol Exp Ther* 2006;Apr 5 [Epub ahead of print].
63. Brandsch, M., Knutter, I., Leibach, F.H. The intestinal H^+/peptide symporter PEPT1: structure-affinity relationships. *Eur J Pharm Sci* 2004;**21**:53–60.
64. Anderle, P., Nielsen, C.U., Pinsonneault, J., Krog, P.L., Brodin, B., Sadee, W. Genetic variants of the human dipeptide transporter PEPT1. *J Pharmacol Exp Ther* 2006;**316**:636–646.
65. Sundaram, U., Wisel, S., Coon, S. Mechanism of inhibition of proton: dipeptide co-transport during chronic enteritis in the mammalian small intestine. *Biochim Biophys Acta* 2005;**1714**:134–140.
66. Tamai, I., Nezu, J., Uchino, H., Sai, Y., Oku, A., Shimane, M., Tsuji, A. Molecular identification and characterization of novel members of the human organic anion transporter (OATP) family. *Biochem Biophys Res Commun* 2000;**273**:251–260.
67. Hagenbuch, B., Meier, P.J. The superfamily of organic anion transporting polypeptides. *Biochim Biophys Acta* 2003;**1609**:1–18.
68. Nozawa, T., Imai, K., Nezu, J., Tsuji, A., Tamai, I. Functional characterization of pH-sensitive organic anion transporting polypeptide OATP-B in human. *J Pharmacol Exp Ther* 2004;**308**:438–445.

69. Kobayashi, D., Nozawa, T., Imai, K., Nezu, J., Tsuji, A., Tamai, I. Involvement of human organic anion transporting polypeptide OATP-B (SLC21A9) in pH-dependent transport across intestinal apical membrane. *J Pharmacol Exp Ther* 2003;**306**:703–708.
70. Satoh, H., Yamashita, F., Tsujimoto, M., Murakami, H., Koyabu, N., Ohtani, H., Sawada, Y. Citrus juices inhibit the function of human organic anion-transporting polypeptide OATP-B. *Drug Metab Dispos* 2005;**33**:518–523.
71. Kopplow, K., Letschert, K., Konig, J., Walter, B., Keppler, D. Human hepatobiliary transport of organic anions analyzed by quadruple-transfected cells. *Mol Pharmacol* 2005;**68**:1031–1038.
72. Kullak Ublick, G.A., Beuers, U., Fahney, C., Hagenbuch, B., Meier, P.J., Paumgartner, G. Identification and functional characterization of the promoter region of the human organic anion transporting polypeptide gene. *Hepatology* 1997;**26**:991–997.
73. Cvetkovic, M., Leake, B., Fromm, M.F., Wilkinson, G.R., Kim, R.B. OATP and P-glycoprotein transporters mediate the cellular uptake and excretion of fexofenadine. *Drug Metab Dispos* 1999;**27**:866–871.
74. Su, Y., Zhang, X., Sinko, P.J. Human organic anion-transporting polypeptide OATP-A (SLC21A3) acts in concert with P-glycoprotein and multidrug resistance protein 2 in the vectorial transport of Saquinavir in Hep G2 cells. *Mol Pharm* 2004;**1**:49–56.
75. Dresser, G.K., Kim, R.B., Bailey, D.G. Effect of grapefruit juice volume on the reduction of fexofenadine bioavailability: possible role of organic anion transporting polypeptides. *Clin Pharmacol Ther* 2005;**77**:170–177.
76. Fuchikami, H., Satoh, H., Tsujimoto, M., Ohdo, S., Ohtani, H., Sawada, Y. Effects of herbal extracts on the function of human organic anion-transporting polypeptide OATP-B. *Drug Metab Dispos* 2006;**34**:577–582.
77. Lu, R., Kanai, N., Bao, Y., Schuster, V.L. Cloning, in vitro expression, and tissue distribution of a human prostaglandin transporter cDNA (hPGT). *J Clin Invest* 1996;**98**:1142–1149.
78. Garcia, C.K., Goldstein, J.L., Pathak, R.K., Anderson, R.G., Brown, M.S. Other monocarboxylates: implications for the Cori cycle. *Cell* 1994;**76**:865–873.
79. Ritzhaupt, A., Wood, I.S., Ellis, A., Hosie, K.B., Shirazi-Beechey, S.P. Identification and characterization of a monocarboxylate transporter (MCT1) in pig and human colon: its potential to transport L-lactate as well as butyrate. *J Physiol* 1998;**513**(Pt 3):719–732.
80. Tamai, I., Sai, Y., Ono, A., Kido, Y., Yabuuchi, H., Takanaga, H., et al. Immunohistochemical and functional characterization of pH-dependent intestinal absorption of weak organic acids by the monocarboxylic acid transporter MCT1. *J Pharm Pharmacol* 1999;**51**:1113–1121.
81. Stein, J., Zores, M., Schroder, O. Short-chain fatty acid (SCFA) uptake into Caco-2 cells by a pH-dependent and carrier mediated transport mechanism. *Eur J Nutr* 2000;**39**:121–125.
82. Hadjiagapiou, C., Schmidt, L., Dudeja, P.K., Layden, T.J., Ramaswamy, K. Mechanism(s) of butyrate transport in Caco-2 cells: role of monocarboxylate transporter 1. *Am J Physiol Gastrointest Liver Physiol* 2000;**279**:G775–G780.
83. Halestrap, A.P., Price, N.T. The proton-linked monocarboxylate transporter (MCT) family: structure, function and regulation. *Biochem J* 1999;**343**(Pt 2):281–299.
84. Buyse, M., Sitaraman, S.V., Liu, X., Bado, A., Merlin, D. Luminal leptin enhances CD147/MCT-1-mediated uptake of butyrate in the human intestinal cell line Caco2-BBE. *J Biol Chem* 2002;**277**:28182–28190.
85. Gill, R.K., Saksena, S., Alrefai, W.A., Sarwar, Z., Goldstein, J.L., Carroll, R.E., et al. Expression and membrane localization of MCT isoforms along the length of the human intestine. *Am J Physiol Cell Physiol* 2005;**289**:C846–C852.
86. Murer, H., Markovich, D., Biber, J. Renal and small intestinal sodium-dependent symporters of phosphate and sulphate. *J Exp Biol* 1994;**196**:167–181.
87. Xu, H., Bai, L., Collins, J.F., Ghishan, F.K. Molecular cloning, functional characterization, tissue distribution, and chromosomal localization of a human, small intestinal sodium-phosphate (Na^+-Pi) transporter (SLC34A2). *Genomics* 1999;**62**:281–284.
88. Feild, J.A., Zhang, L., Brun, K.A., Brooks, D.P., Edwards, R.M. Cloning and functional characterization of a sodium-dependent phosphate transporter expressed in human lung and small intestine. *Biochem Biophys Res Commun* 1999;**258**:578–582.

89. Markovich, D., Regeer, R.R., Kunzelmann, K., Dawson, P.A. Functional characterization and genomic organization of the human Na(+)-sulfate cotransporter hNaS2 gene (SLC13A4). *Biochem Biophys Res Commun* 2005;**326**:729–734.
90. Swaan, P.W., Tukker, J.J. 1995 Carrier-mediated transport mechanism of foscarnet (trisodium phosphonoformate hexahydrate) in rat intestinal tissue. *J Pharmacol Exp Ther* 1995;**272**:242–247.
91. Loghman-Adham, M., Motock, G.T., Levi, M. Enhanced bioavailability of phosphonoformic acid by dietary phosphorus restriction. *Biochem Pharmacol* 1994;**48**:1455–1458.
92. Griffiths, M., Beaumont, N., Yao, S.Y., Sundaram, M., Boumah, C.E., Davies, A., et al. Cloning of a human nucleoside transporter implicated in the cellular uptake of adenosine and chemotherapeutic drugs. *Nat Med* 1997;**3**:89–93.
93. Ritzel, M.W., Yao, S.Y., Huang, M.Y., Elliott, J.F., Cass, C.E., Young, J.D. Molecular cloning and functional expression of cDNAs encoding a human Na+-nucleoside cotransporter (hCNT1). *Am J Physiol* 1997;**272**(2 Pt 1):C707–C714.
94. Ritzel, M.W., Yao, S.Y., Ng, A.M., Mackey, J.R., Cass, C.E., Young, J.D. Molecular cloning, functional expression and chromosomal localization of a cDNA encoding a human Na^+/nucleoside cotransporter (hCNT2) selective for purine nucleosides and uridine. *Mol Membr Biol* 1998; **15**:203–211.
95. Ritzel, M.W., Ng, A.M., Yao, S.Y., Graham, K., Loewen, S.K., Smith, K.M., et al. Recent molecular advances in studies of the concentrative Na^+-dependent nucleoside transporter (CNT) family: identification and characterization of novel human and mouse proteins (hCNT3 and mCNT3) broadly selective for purine and pyrimidine nucleosides (system cib). *Mol Membr Biol* 2001;**18**:65–72.
96. Carpenter, P., Chen, X.Z., Karpinski, E., Hyde, R.J., Baldwin, S.A., Cass, C.E., Young, J.D. Molecular identification and characterization of novel human and mouse concentrative Na+-nucleoside cotransporter proteins (hCNT3 and mCNT3) broadly selective for purine and pyrimidine nucleosides (system cib). *J Biol Chem* 2001;**276**:2914–2927.
97. Patil, S.D., Unadkat, J.D. Sodium-dependent nucleoside transport in the human intestinal brush-border membrane. *Am J Physiol* 1997;**272**:1314–1320.
98. Wang, J., Su, S.F., Dresser, M.J., Schaner, M.E., Washington, C.B., Giacomini, K.M. $Na^{(+)}$-dependent purine nucleoside transporter from human kidney: cloning and functional characterization. *Am J Physiol* 1997;**273**(6 Pt 2):F1058–F1065.
99. Mata, J.F., Garcia-Manteiga, J.M., Lostao, M.P., Fernandez-Veledo, S., Guillen-Gomez, E., Larrayoz, I.M., et al. Role of the human concentrative nucleoside transporter (hCNT1) in the cytotoxic action of 5′-deoxy-5-fluorouridine, an active intermediate metabolite of capecitabine, a novel oral anticancer drug. *Mol Pharmacol* 2001;**59**:1542–1548.
100. Shin, H.C., Landowski, C.P., Sun, D., Vig, B.S., Kim, I., Mittal, S., et al. Functional expression and characterization of a sodium-dependent nucleoside transporter hCNT2 cloned from human duodenum. *Biochem Biophys Res Commun* 2003;**307**:696–703.
101. Graham, K.A., Leithoff, J., Coe, I.R., Mowles, D., Mackey, J.R., Young, J.D., Cass, C.E. Differential transport of cytosine-containing nucleosides by recombinant human concentrative nucleoside transporter protein hCNT1. *Nucleosides Nucleotides Nucleic Acids* 2000;**19**:415–434.
102. Chandrasena, G., Giltay, R., Patil, S.D., Bakken, A., Unadkat, J.D. Functional expression of human intestinal Na^+-dependent and Na^+-independent nucleoside transporters in Xenopus laevis oocytes. *Biochem Pharmacol* 1997;**53**:1909–1918.
103. Yao, S.Y., Ng, A.M., Sundaram, M., Cass, C.E., Baldwin, S.A., Young, J.D. Transport of antiviral 3′-deoxy-nucleoside drugs by recombinant human and rat equilibrative, nitrobenzylthioinosine (NBMPR)-insensitive (ENT2) nucleoside transporter proteins produced in *Xenopus* oocytes. *Mol Membr Biol* 2001;**18**:161–167.
104. Ngo, L.Y., Patil, S.D., Unadkat, J.D. Ontogenic and longitudinal activity of $Na^{(+)}$-nucleoside transporters in the human intestine. *Am J Physiol Gastrointest Liver Physiol* 2001;**280**:G475–G481.
105. Gründemann, D., Gorboulev, V., Gambaryan, S., Veyhl, M., Koepsell, H. Drug excretion mediated by a new prototype of polyspecific transporter. *Nature* 1994;**372**:549–552.
106. Koepsell, H. Organic cation transporters in intestine, kidney, liver, and brain. *Annu Rev Physiol* 1998;**60**:243–266.

107. Muller, J., Lips, K.S., Metzner, L., Neubert, R.H., Koepsell, H., Brandsch, M. Drug specificity and intestinal membrane localization of human organic cation transporters (OCT). *Biochem Pharmacol* 2005;**70**:1851–1860.
108. Martel, F., Keating, E., Calhau, C., Azevedo, I. Uptake of (3)H-1-methyl-4-phenylpyridinium ($^{(3)}$H-MPP(+)) by human intestinal Caco-2 cells is regulated by phosphorylation/dephosphorylation mechanisms. *Biochem Pharmacol* 2002;**63**:1565–1515.
109. Takeda, M., Khamdang, S., Narikawa, S., Kimura, H., Kobayashi, Y., Yamamoto, T., et al. Human organic anion transporters and human organic cation transporters mediate renal antiviral transport. *J Pharmacol Exp Ther* 2002;**300**:918–924.
110. Tamai, I., Ohashi, R., Nezu, J., Yabuuchi, H., Oku, A., Shimane, M., Sai, Y., Tsuji, A. Molecular and functional identification of sodium ion-dependent, high affinity human carnitine transporter OCTN2. *J Biol Chem* 1998;**273**:20378–20382.
111. Elimrani, I., Lahjouji, K., Seidman, E., Roy, M.J., Mitchell, G.A., Qureshi, I. Expression and localization of organic cation/carnitine transporter OCTN2 in Caco-2 cells. *Am J Physiol Gastrointest Liver Physiol* 2003;**284**:G863–G871.
112. Ohashi, R., Tamai, I., Nezu Ji, J., Nikaido, H., Hashimoto, N., Oku, A., et al. Molecular and physiological evidence for multifunctionality of carnitine/organic cation transporter OCTN2. *Mol Pharmacol* 2001;**59**:358–366.
113. Ganapathy, M.E., Huang, W., Rajan, D.P., Carter, A.L., Sugawara, M., Iseki, K., et al. Beta-lactam antibiotics as substrates for OCTN2, an organic cation/carnitine transporter. *J Biol Chem* 2000;**275**:1699–1707.
114. Duran, J.M., Peral, M.J., Calonge, M.L., Ilundain, A.A. OCTN3: A Na^+-independent L-carnitine transporter in enterocytes basolateral membrane. *J Cell Physiol* 2005;**202**:929–935.
115. Altmann, S.W., Davis, H.R., Jr, Zhu, L.J., Yao, X., Hoos, L.M., Tetzloff, G., et al. Niemann–Pick C1 Like 1 protein is critical for intestinal cholesterol absorption. *Science* 2004;**303**:1201–1204.
116. Davis, H.R., Jr, Zhu, L.J., Hoos, L.M., Tetzloff, G., Maguire, M., Liu, J., et al. Niemann-Pick C1 Like 1 (NPC1L1) is the intestinal phytosterol and cholesterol transporter and a key modulator of whole-body cholesterol homeostasis. *J Biol Chem* 2004;**279**:33586–33592.
117. Lally, S., Tan, C.Y., Owens, D., Tomkin, G.H. Messenger RNA levels of genes involved in dysregulation of postprandial lipoproteins in type 2 diabetes: the role of Niemann-Pick C1-like 1, ATP-binding cassette, transporters G5 and G8, and of microsomal triglyceride transfer protein. *Diabetologia* 2006;**49**:1008–1016.
118. Higgins, C.F. ABC transporters: from microorganism to man. *Annu Rev Cell Biol* 1992;**8**:67–113.
119. Dietrich, C.G., Geier, A., Oude Elferink, R.P. ABC of oral bioavailability: transporters as gatekeepers in the gut. *Gut* 2003;**52**:1788–1795.
120. Dean, M., Hamon, Y., Chimini, G. The human ATP-binding cassette (ABC) transporter superfamily. *J Lipid Res* 2001;**42**:1007–1017.
121. Schinkel, A.H., Jonker, J.W. Mammalian drug efflux transporters of the ATP binding cassette (ABC) family: an overview. *Adv Drug Deliv Rev* 2003;**55**:3–29.
122. Ambudkar, S.V., Dey, S., Hrycyna, C.A., Ramachandra, M., Pastan, I., Gottesman, M.M. Biochemical, cellular, and pharmacological aspects of the multidrug transporter. *Annu Rev Pharmacol Toxicol* 1999;**39**:361–398.
123. Hoffmann, U., Kroemer, H.K. The ABC transporters MDR1 and MRP2: multiple functions in disposition of xenobiotics and drug resistance. *Drug Metab Rev* 2004;**36**:669–701.
124. Ambudkar, S.V., Kim, I.W., Sauna, Z.E. The power of the pump: mechanisms of action of P-glycoprotein (ABCB1). *Eur J Pharm Sci* 2006;**27**:392–400.
125. Dey, S., Ramachandra, M., Pastan, I., Gottesman, M.M., Ambudkar, S.V. Evidence for two nonidentical drug-interaction sites in the human P-glycoprotein. *Proc Natl Acad Sci USA* 1997;**94**:10594–10599.
126. Gottesman, M.M., Ling, V. The molecular basis of multidrug resistance in cancer: the early years of P-glycoprotein research. *FEBS Lett* 2006;**580**:998–1009.
127. Takara, K., Tanigawara, Y., Komada, F., Nishiguchi, K., Sakaeda, T., Okumura, K. Cellular pharmacokinetic aspects of reversal effect of itraconazole on P-glycoprotein-mediated resistance of anticancer drugs. *Biol Pharm Bull* 1999;**22**:1355–1359.

128. Troutman, M.D., Thakker, D.R. Novel experimental parameters to quantify the modulation of absorptive and secretory transport of compounds by P-glycoprotein in cell culture models of intestinal epithelium. *Pharm Res* 2003;**20**:1210–1224.
129. Polli, J.W., Wring, S.A., Humphreys, J.E., Huang, L., Morgan, J.B., Webster, L.O., Serabjit-Singh, C.S. Rational use of in vitro P-glycoprotein assays in drug discovery. *J Pharmacol Exp Ther* 2001;**299**:620–628.
130. Lecureur, V., Sun, D., Hargrove, P., Schuetz, E.G., Kim, R.B., Lan, L.B., Schuetz, J.D. Cloning and expression of murine sister of P-glycoprotein reveals a more discriminating transporter than MDR1/P-glycoprotein. *Mol Pharmacol* 2000;**57**:24–35.
131. Collett, A., Tanianis-Hughes, J., Hallifax, D., Warhurst, G. Predicting P-glycoprotein effects on oral absorption: correlation of transport in Caco-2 with drug pharmacokinetics in wild-type and mdr1a$^{(-/-)}$ mice in vivo. *Pharm Res* 2004;**21**:819–826.
132. Taipalensuu, J., Tavelin, S., Lazorova, L., Svensson, A.C., Artursson, P. Exploring the quantitative relationship between the level of MDR1 transcript, protein and function using digoxin as a marker of MDR1-dependent drug efflux activity. *Eur J Pharm Sci* 2004;**21**:69–75.
133. Ueda, K., Cardarelli, C., Gottesman, M.M., Pastan, I. Expression of a full-length cDNA for the human 'MDR1' gene confers resistance to colchicine, doxorubicin, and vinblastine. *Proc Natl Acad Sci USA* 1987;**84**:3004–3008.
134. Kim, R.B., Fromm, M.F., Wandel, C., Leake, B., Wood, A.J., Roden, D.M., Wilkinson, G.R. The drug transporter P-glycoprotein limits oral absorption and brain entry of HIV-1 protease inhibitors. *J Clin Invest* 1998;**101**:289–294.
135. Alsenz, J., Steffen, H., Alex, R. Active apical secretory efflux of the HIV protease inhibitors saquinavir and ritonavir in Caco-2 cell monolayers. *Pharm Res* 1998;**15**:423–428.
136. Polli, J.W., Jarrett, J.L., Studenberg, S.D., Humphreys, J.E., Dennis, S.W., Brouwer, K.R., Woolley, J.L. Role of P-glycoprotein on the CNS disposition of amprenavir (141W94), an HIV protease inhibitor. *Pharm Res* 1999;**16**:1206–1212.
137. Saeki, T., Ueda, K., Tanigawara, Y. Human P-glycoprotein transports cyclosporin A and FK506. *J Biol Chem* 1993;**268**:6077–6080.
138. Goto, M., Masuda, S., Saito, H., Inui, K. Decreased expression of P-glycoprotein during differentiation in the human intestinal cell line Caco-2. *Biochem Pharmacol* 2003;**66**:163–70.
139. Ito, T., Yano, I., Tanaka, K., Inui, K.I. Transport of quinolone antibacterial drugs by human P-glycoprotein expressed in a kidney epithelial cell line, LLC-PK1. *J Pharmacol Exp Ther* 1997;**282**: 955–960.
140. Naruhashi, K., Tamai, I., Inoue, N., Muraoka, H., Sai, Y., Suzuki, N., Tsuji, A. Active intestinal secretion of new quinolone antimicrobials and the partial contribution of P-glycoprotein. *J Pharm Pharmacol* 2001;**53**:699–709.
141. Sun, H., Huang, Y., Frassetto, L., Benet, L.Z. Effects of uremic toxins on hepatic uptake and metabolism of erythromycin. *Drug Metab Dispos* 2004;**32**:1239–1246.
142. Ueda, K., Okamura, N., Hirai, M., Tanigawara, Y., Saeki, T., Kioka, N., et al. Human P-glycoprotein transports cortisol, aldosterone, and dexamethasone, but not progesterone. *J Biol Chem* 1992;**267**:24248–24252.
143. Kim, W.Y., Benet, L.Z. P-glycoprotein (P-gp/MDR1)-mediated efflux of sex-steroid hormones and modulation of P-gp expression in vitro. *Pharm Res* 2004;**21**:1284–1293.
144. Oka, A., Oda, M., Saitoh, H., Nakayama, A., Takada, M., Aungst, B.J. Secretory transport of methylprednisolone possibly mediated by P-glycoprotein in Caco-2 cells. *Biol Pharm Bull* 2002;**25**:393–396.
145. Karlsson, J., Kuo, S.M., Ziemniak, J., Artursson, P. Transport of celiprolol across human intestinal epithelial (Caco-2) cells: mediation of secretion by multiple transporters including P-glycoprotein. *Br J Pharmacol* 1993;**110**:1009–1016.
146. Soldner, A., Benet, L.Z., Mutschler, E., Christians, U. Active transport of the angiotensin-II antagonist losartan and its main metabolite EXP 3174; across MDCK-MDR1 and caco-2 cell monolayers. *Br J Pharmacol* 2000;**129**:1235–1243.
147. Saeki, T., Ueda, K., Tanigawara, Y., Hori, R., Komano, T. P-glycoprotein-mediated transcellular transport of MDR-reversing agents. *FEBS Lett* 1993;**324**:99–102.

148. Lee, K., Ng, C., Brouwer, K.L.R., Thakker, D.R. Secretory transport of ranitidine and famotidine across Caco-2 cell monolayers. *J Pharmacol Exp Ther* 2002;**303**:574–580.
149. Lentz, K.A., Polli, J.W., Wring, S.A., Humphreys, J.E., Polli, J.E. Influence of passive permeability on apparent p-glycoprotein kinetics. *Pharm Res* 2000;**17**:1456–1460.
150. Soares-Da-Silva, P., Serrao, M.P. Outward transfer of dopamine precursor L-3,4-dihydroxyphenylalanine (L-dopa) by native and human P-glycoprotein in LLC-PK(1) and LLC-GA5 col300 renal cells. *J Pharmacol Exp Ther* 2000;**293**:697–704.
151. Borgnia, M.J., Eytan, G.D., Assaraf, Y.G. Competition of hydrophobic peptides, cytotoxic drugs, and chemosensitizers on a common P-glycoprotein pharmacophore as revealed by its ATPase activity. *J Biol Chem* 1996;**2716**:3163–3171.
152. Lang, V.B., Langguth, P., Ottiger, C., Wunderli-Allenspach, H., Rognan, D., Rothen-Rutishauser, B., et al. Structure-permeation relations of met-enkephalin peptide analogues on absorption and secretion mechanisms in Caco-2 monolayers. *J Pharm Sci* 1997;**86**:846–853.
153. Faassen, F., Vogel, G., Spanings, H., Vromans, H. Caco-2 permeability, P-glycoprotein transport ratios and brain penetration of heterocyclic drugs. *Int J Pharm* 2003;**263**:113–122.
154. Schinkel, A.H., Wagenaar, E., van Deemter, L., Mol, C.A., Borst, P. Absence of the mdr1a P-Glycoprotein in mice affects tissue distribution and pharmacokinetics of dexamethasone, digoxin, and cyclosporin, A. *J Clin Invest* 1995;**96**:1698–1705.
155. Kim, R.B., Wandel, C., Leake, B., Cvetkovic, M., Fromm, M.F., Dempsey, P.J., et al. Interrelationship between substrates and inhibitors of human CYP3A and P-glycoprotein. *Pharm Res* 1999;**16**:408–414.
156. Smit, J.W., Weert, B., Schinkel, A.H., Meijer, D.K. Heterologous expression of various P-glycoproteins in polarized epithelial cells induces directional transport of small (type 1) and bulky (type 2) cationic drugs. *J Pharmacol Exp Ther* 1998;**286**:321–327.
157. van der Sandt, I.C., Smolders, R., Nabulsi, L., Zuideveld, K.P., de Boer, A.G., Breimer, D.D. Active efflux of the 5-HT(1A) receptor agonist flesinoxan via P-glycoprotein at the blood-brain barrier. *Eur J Pharm Sci* 2001;**14**:81–86.
158. Crowe, A. The influence of P-glycoprotein on morphine transport in Caco-2 cells. Comparison with paclitaxel. *Eur J Pharmacol* 2002;**440**:7–16.
159. Kharasch, E.D., Hoffer, C., Whittington, D., Sheffels, P. Role of P-glycoprotein in the intestinal absorption and clinical effects of morphine. *Clin Pharmacol Ther* 2003;**74**:543–554.
160. Eichelbaum, M., Fromm, M.F., Schwab, M. Clinical aspects of the MDR1 (ABCB1) gene polymorphism. *Ther Drug Monit* 2004;**26**:180–185.
161. Hidalgo, I.J., Raub, T.J., Borchardt, R.T. Characterization of the human colon carcinoma cell line (Caco-2) as a model system for intestinal epithelial permeability. *Gastroenterology* 1989;**96**:736–749.
162. Hunter, J., Jepson, M.A., Tsuruo, T., Simmons, N.L., Hirst, B.H. Functional expression of P-glycoprotein in apical membranes of human intestinal Caco-2 cells. Kinetics of vinblastine secretion and interaction with modulators. *J Biol Chem* 1993;**268**:14991–14997.
163. Collett, A., Tanianis-Hughes, J., Carlson, G.L., Harwood, M.D., Warhurst, G. Comparison of P-glycoprotein-mediated drug-digoxin interactions in Caco-2 with human and rodent intestine: relevance to in vivo prediction. *Eur J Pharm Sci* 2005;**26**:386–393.
164. Maki, N., Hafkemeyer, P., Dey, S. Allosteric modulation of human P-glycoprotein. Inhibition of transport by preventing substrate translocation and dissociation. *J Biol Chem* 2003;**278**: 18132–18139.
165. Fojo, T., Bates, S. Strategies for reversing drug resistance. *Oncogene* 2003;**22**:7512–7523.
166. Mamot, C., Drummond, D.C., Hong, K., Kirpotin, D.B., Park, J.W. Liposome-based approaches to overcome anticancer drug resistance. *Drug Resist Update* 2003;**6**:271–279.
167. Limtrakul, P., Khantamat, O., Pintha, K. Inhibition of P-glycoprotein function and expression by kaempferol and quercetin. *J Chemother* 2005;**17**:86–95.
168. Zhang, S., Morris, M.E. Effects of flavonoids biochanin, A., morin, phloretin, and silymarin on P-glycoprotein-mediated transport. *J Pharmacol Exp Ther* 2003;**304**:1258–1267.
169. Zhou, S., Lim, L.Y., Chowbay, B. Herbal modulation of P-glycoprotein. *Drug Metab Rev* 2004;**36**:57–104.
170. Morris, M.E., Zhang, S. Flavonoid-drug interactions: Effects of flavonoids on ABC transporters. *Life Sci* 2006;**78**:2116–2130.

171. Lown, K.S., Bailey, D.G., Fontana, R.J., Janardan, S.K., Adair, C.H., Fortlage, L.A., et al. Grapefruit juice increases felodipine oral availability in humans by decreasing intestinal protein expression. *J Clin Invest* 1997;**99**:2545–2553.
172. He, K., Iyer, K.R., Hayes, R.N., Sinz, M.W., Woolf, T.F., Hollenberg, P.F. Inactivation of cytochrome P450 3A4 by bergamottin, a component of grapefruit juice. *Chem Res Toxicol* 1998;**11**:252–259.
173. Eagling, V.A., Profit, L., Back, D.J. Inhibition of the CYP3A4-mediated metabolism and P-glycoprotein-mediated transport of the HIV-1 protease inhibitor saquinavir by grapefruit juice components. *Br J Clin Pharmacol* 1999;**48**:543–552.
174. Becquemont, L., Verstuyft, C., Kerb, R., Brinkman, U., Lebot, M., Jaillon, P., Funch-Brentano, C. Effect of grapefruit juice on digoxin pharmacokinetics in humans. *Clin Pharmacol Ther* 2000;**70**:311–316.
175. Greenblatt, D.J., von Moltke, L.L., Harmatz, H.S., Chen, G., Weemhoff, J.L., Jen, C., et al. Time course of recovery of cytochrome P450 3A function after single doses of grapefruit juice. *Clin Pharmacol Ther* 2003;**74**:121–129.
176. Borst, P., Evers, R., Kool, M., Wijnholds, J. A family of drug transporters: the multidrug resistance-associated proteins. *J Natl Cancer Inst* 2000;**92**:1295–1302.
177. Keppler, D., Cui, Y., Konig, J., Leier, I., Nies, A. Export pumps for anionic conjugates encoded by MRP genes. *Adv Enzyme Regul* 1999;**39**:237–246.
178. Taipalensuu, J., Tornblom, H., Lindberg, G., Einarsson, C., Sjoqvist, F., Melhus, H., et al. Correlation of gene expression of ten drug efflux proteins of the ATP-binding cassette transporter family in normal human jejunum and in human intestinal epithelial Caco-2 cell monolayers. *J Pharmacol Exp Ther* 2001;**299**:164–170.
179. Büchler, M., Konig, J., Brom, M., Kartenbeck, J., Spring, H., Horie, T., Keppler, D. cDNA cloning of the hepatocyte canalicular isoform of the multidrug resistance protein, cMrp, reveals a novel conjugate export pump deficient in hyperbilirubinemic mutant rats. *J Biol Chem* 1996;**271**: 15091–15098.
180. Jedlitschky, G., Leier, I., Buchholz, U., Hummel-Eisenbeiss, J., Burchell, B., Keppler, D. ATP-dependent transport of bilirubin glucuronides by the multidrug resistance protein MRP1 and its hepatocyte canalicular isoform MRP2. *Biochem J* 1997;**327**:305–310.
181. Kartenbeck, J., Leuschner, U., Mayer, R., Keppler, D. Absence of the canalicular isoform of the MRP gene-encoded conjugate export pump from the hepatocytes in Dubin-Johnson syndrome. *Hepatology* 1996;**23**:1061–1066.
182. Paulusma, C.C., Kool, M., Bosma, P.J., Scheffer, G.L., ter Borg, F., Scheper, R.J., et al. A mutation in the human canalicular multispecific organic anion transporter gene causes the Dubin-Johnson syndrome. *Hepatology* 1997;**25**:1539–1542.
183. Huisman, M.T., Smit, J.W., Crommentuyn, K.M., Zelcer, N., Wiltshire, H.R., Beijnen, J.H., Schinkel, A.H. Multidrug resistance protein 2 (MRP2) transports HIV protease inhibitors, and transport can be enhanced by other drugs. *AIDS* 2002;**16**:2295–2301.
184. Huisman, M.T., Chhatta, A.A., van Tellingen, O., Beijnen, J.H., Schinkel, A.H. MRP2 (ABCC2) transports taxanes and confers paclitaxel resistance and both processes are stimulated by probenecid. *Int J Cancer* 2005;**116**:824–829.
185. Cui, Y., Konig, J., Buchholz, J.K., Spring, H., Leier, I., Keppler, D. Drug resistance and ATP-dependent conjugate transport mediated by the apical multidrug resistance protein, MRP2, permanently expressed in human and canine cells. *Mol Pharmacol* 1999;**55**:929–937.
186. Evers, R., de Haas, M., Sparidans, R., Beijnen, J., Wielinga, P.R., Lankelma, J., Borst, P. Vinblastine and sulfinpyrazone export by the multidrug resistance protein MRP2 is associated with glutathione export. *Br J Cancer* 2000;**83**:375–383.
187. Sasaki, M., Suzuki, H., Ito, K., Abe, T., Sugiyama, Y. Transcellular transport of organic anions across a double-transfected Madin-Darby canine kidney II cell monolayer expressing both human organic anion-transporting polypeptide (OATP2/SLC21A6) and Multidrug resistance-associated protein 2 (MRP2/ABCC2). *J Biol Chem* 2002;**277**:6497–6503.
188. Matsushima, S., Maeda, K., Kondo, C., Hirano, M., Sasaki, M., Suzuki, H., Sugiyama, Y. Identification of the hepatic efflux transporters of organic anions using double-transfected Madin-Darby canine kidney II cells expressing human organic anion-transporting polypeptide 1B1 (OATP1B1)/multidrug resistance-associated protein 2, OATP1B1/multidrug resistance 1, and OATP1B1/breast cancer resistance protein. *J Pharmacol Exp Ther* 2005;**314**:1059–1067.

189. Lowes, S., Simmons, N.L. Multiple pathways for fluoroquinolone secretion by human intestinal epithelial (Caco-2) cells. *Br J Pharmacol* 2002;**135**:1263–1275.
190. Bakos, E., Evers, R., Sinko, E., Varadi, A., Borst, P., Sarkadi, B. Interactions of the human multidrug resistance proteins MRP1 and MRP2 with organic anions. *Mol Pharmacol* 2000;**57**:760–768.
191. Kim, A.E., Dintaman, J.M., Waddell, D.S., Silverman, J.A. Saquinavir, an HIV protease inhibitor, is transported by P-glycoprotein. *J Pharmacol Exp Ther* 1998;**286**:1439–1445.
192. Liu, L., Cui, Y., Chung, A.Y., Shitara, Y., Sugiyama, Y., Keppler, D., Pang, K.S. (Vectorial Transport of enalapril by Oatp1a1/Mrp2 and OATP1B1 and OATP1B3/MRP2 in rat and human livers. *J Pharmacol Exp Ther* 2006 (Epub ahead of print).
193. Spears, K.J., Ross, J., Stenhouse, A., Ward, C.J., Goh, L.B., Wolf, C.R., et al. Directional trans-epithelial transport of organic anions in porcine LLC-PK1 cells that co-express human OATP1B1 (OATP-C) and MRP2. *Biochem Pharmacol* 2005;**69**:415–423.
194. Guo, A., Marinaro, W., Hu, P., Sinko, P.J. Delineating the contribution of secretory transporters in the efflux of etoposide using Madin-Darby canine kidney (MDCK) cells overexpressing P-glycoprotein (Pgp), multidrug resistance-associated protein (MRP1), and canalicular multispecific organic anion transporter (cMOAT). *Drug Metab Dispos* 2002;**30**:457–463.
195. Kiuchi, Y., Suzuki, H., Hirohashi, T., Tyson, C.A., Sugiyama, Y. cDNA cloning and inducible expression of human multidrug resistance associated protein 3 (MRP3). *FEBS Lett* 1998;**433**:149–152.
196. Uchiumi, T., Hinoshita, E., Haga, S., Nakamura, T., Tanaka, T., Toh, S., et al. Isolation of a novel human canalicular multispecific organic anion transporter, cMOAT2/MRP3, and its expression in cisplatin-resistant cancer cells with decreased ATP-dependent drug transport. *Biochem Biophys Res Commun* 1998;**252**:103–110.
197. Zeng, H., Liu, G., Rea, P.A., Kruh, G.D. Transport of amphipathic anions by human multidrug resistance protein 3. *Cancer Res* 2000;**60**:4779–4784.
198. Inokuchi, A., Hinoshita, E., Iwamoto, Y., Kohno, K., Kuwano, M., Uchiumi, T. Enhanced expression of the human multidrug resistance protein 3 by bile salt in human enterocytes. A transcriptional control of a plausible bile acid transporter. *J Biol Chem* 2001;**276**:46822–46829.
199. Belinsky, M.G., Dawson, P.A., Shchaveleva, I., Bain, L.J., Wang, R., Ling, V., et al. Analysis of the in vivo functions of Mrp3. *Mol Pharmacol* 2005;**68**:160–168.
200. Zelcer, N., Saeki, T., Reid, G., Beijnen, J.H., Borst, P. Characterization of drug transport by the human multidrug resistance protein 3 (ABCC3). *J Biol Chem* 2001;**276**:46400–46407.
201. Krishnamurthy, P., Schuetz, J.D. Role of ABCG2/BCRP in biology and medicine. *Annu Rev Pharmacol Toxicol* 2006;**46**:381–410.
202. Ifergan, I., Shafran, A., Jansen, G., Hooijberg, J.H., Scheffer, G.L., Assaraf, Y.G. Folate deprivation results in the loss of breast cancer resistance protein (BCRP/ABCG2) expression. A role for BCRP in cellular folate homeostasis. *J Biol Chem* 2004;**279**:25527–25534.
203. Sparreboom, A., Loos, W.J., Burger, H., Sissung, T.M., Verweij, J., Figg, W.D., et al. Effect of ABCG2 genotype on the oral bioavailability of topotecan. *Cancer Biol Ther* 2005;**4**:650–658.
204. van Herwaarden, A.E., Schinkel, A.H. The function of breast cancer resistance protein in epithelial barriers, stem cells and milk secretion of drugs and xenotoxins. *Trends Pharmacol Sci* 2006;**27**:10–16.
205. Volk, E.L., Rohde, K., Rhee, M., McGuire, J.J., Doyle, L.A., Ross, D.D., Schneider, E. Methotrexate cross-resistance in a mitoxantrone-selected multidrug-resistant MCF7 breast cancer cell line is attributable to enhanced energy-dependent drug efflux. *Cancer Res* 2000;**60**:3514–3521.
206. Chen, Z.S., Robey, R.W., Belinsky, M.G., Shchaveleva, I., Ren, X.Q., Sugimoto, Y., et al. Transport of methotrexate, methotrexate polyglutamates, and 17beta-estradiol 17-(beta-D-glucuronide) by ABCG2: effects of acquired mutations at R482 on methotrexate transport. *Cancer Res* 2003;**63**:4048–4054.
207. Imai, Y., Asada, S., Tsukahara, S., Ishikawa, E., Tsuruo, T., Sugimoto, Y. Breast cancer resistance protein exports sulfated estrogens but not free estrogens. *Mol Pharmacol* 2003;**64**:610–618.
208. Candeil, L., Gourdier, I., Peyron, D., Vezzio, N., Copois, V., Bibeau, F., et al. ABCG2 overexpression in colon cancer cells resistant to SN38 and in irinotecan-treated metastases. *Int J Cancer* 2004;**109**:848–854.

209. Merino, G., Jonker, J.W., Wagenaar, E., van Herwaarden, A.E., Schinkel, A.H. The breast cancer resistance protein (BCRP/ABCG2) affects pharmacokinetics, hepatobiliary excretion, and milk secretion of the antibiotic nitrofurantoin. *Mol Pharmacol* 2005;**67**:1758–1764.
210. Merino, G., Alvarez, A.I., Pulido, M.M., Molina, A.J., Schinkel, A.H., Prieto, J.G. Breast cancer resistance protein (BCRP/ABCG2) transports fluoroquinolone antibiotics and affects their oral availability, pharmacokinetics, and milk secretion. *Drug Metab Dispos* 2006;**34**:690–695.
211. Huang, L., Wang, Y., Grimm, S. ATP-dependent transport of rosuvastatin in membrane vesicles expressing breast cancer resistance protein. *Drug Metab Dispos* 2006;**34**:738–742.
212. Pavek, P., Merino, G., Wagenaar, E., Bolscher, E., Novotna, M., Jonker, J.W., Schinkel, A.H. Human breast cancer resistance protein: interactions with steroid drugs, hormones, the dietary carcinogen 2-amino-1-methyl-6-phenylimidazo(4,5-b)pyridine, and transport of cimetidine. *J Pharmacol Exp Ther* 2005;**312**:144–152.
213. Merino, G., Jonker, J.W., Wagenaar, E., Pulido, M.M., Molina, A.J., Alvarez, A.I., Schinkel, A.H. Transport of anthelmintic benzimidazole drugs by breast cancer resistance protein (BCRP/ABCG2). *Drug Metab Dispos* 2005;**33**:614–618.
214. Adachi, Y., Suzuki, H., Schinkel, A.H., Sugiyama, Y. Role of breast cancer resistance protein (Bcrp1/Abcg2) in the extrusion of glucuronide and sulfate conjugates from enterocytes to intestinal lumen. *Mol Pharmacol* 2005;**67**:923–928.
215. Zamber, C.P., Lamba, J.K., Yasuda, K., Farnum, J., Thummel, K., Schuetz, J.D., Schuetz, E.G. Natural allelic variants of breast cancer resistance protein (BCRP) and their relationship to BCRP expression in human intestine. *Pharmacogenetics* 2003;**13**:19–28.
216. Sparreboom, A., Gelderblom, H., Marsh, S., Ahluwalia, R., Obach, R., Principe, P., et al. Diflomotecan pharmacokinetics in relation to ABCG2 421C. A genotype. *Clin Pharmacol Ther* 2004;**76**:38–44.
217. Henrich, C.J., Bokesch, H.R., Dean, M., Bates, S.E., Robey, R.W., Goncharova, E.I., et al. A high-throughput cell-based assay for inhibitors of ABCG2 activity. *J Biomol Screen* 2006;**11**:176–183.
218. Ji, Y., Morris, M.E. Effect of organic isothiocyanates on breast cancer resistance protein (ABCG2)-mediated transport. *Pharm Res* 2004;**21**:2261–2269.
219. Zhang, S., Wang, X., Sagawa, K., Morris, M.E. Flavonoids chrysin and benzoflavone, potent breast cancer resistance protein inhibitors, have no significant effect on topotecan pharmacokinetics in rats or mdr1a/1b$^{(-/-)}$ mice. *Drug Metab Dispos* 2005;**33**:341–348.
220. Zhang, S., Yang, X., Morris, M.E. Flavonoids are inhibitors of breast cancer resistance protein (ABCG2)-mediated transport. *Mol Pharmacol* 2004;**65**:1208–1216.
221. Gupta, A., Zhang, Y., Unadkat, J.D., Mao, Q. HIV protease inhibitors are inhibitors but not substrates of the human breast cancer resistance protein (BCRP/ABCG2). *J Pharmacol Exp Ther* 2004;**310**:334–341.
222. Shiozawa, K., Oka, M., Soda, H., Yoshikawa, M., Ikegami, Y., Tsurutani, J., et al. Reversal of breast cancer resistance protein (BCRP/ABCG2)-mediated drug resistance by novobiocin, a coumermycin antibiotic. *Int J Cancer* 2004;**108**:146–151.
223. Litman, T., Brangi, M., Hudson, E., Fetsch, P., Abati, A., Ross, D.D., et al. The multidrug-resistant phenotype associated with overexpression of the new ABC half-transporter, MXR (ABCG2). *J Cell Sci* 2000;**113**:2011–2021.
224. Maliepaard, M., van Gastelen, M.A., Tohgo, A., Hausheer, F.H., van Waardenburg, R.C., de Jong, L.A., et al. Circumvention of breast cancer resistance protein (BCRP)-mediated resistance to camptothecins in vitro using non-substrate drugs or the BCRP inhibitor GF120918. *Clin Cancer Res* 2001;**7**:935–941.
225. Burger, H., van Tol, H., Boersma, A.W., Brok, M., Wiemer, E.A., Stoter, G., Nooter, K. Imatinib mesylate (STI571) is a substrate for the breast cancer resistance protein (BCRP)/ABCG2 drug pump. *Blood* 2004;**104**:2940–2942.
226. Nakanishi, T., Doyle, L.A., Hassel, B., Wei, Y., Bauer, K.S., Wu, S., et al. Functional characterization of human breast cancer resistance protein (BCRP, ABCG2) expressed in the oocytes of Xenopus laevis. *Mol Pharmacol* 2003;**64**:1452–1462.
227. Volk, E.L. Schneider, E. Wild-type breast cancer resistance protein (BCRP/ABCG2) is a methotrexate polyglutamate transporter. *Cancer Res* 2003;**63**:5538–5543.

228. Schmitz, G., Langmann, T., Heimerl, S. Role of ABCG1 and other ABCG family members in lipid metabolism. *J Lipid Res* 2001;**42**:1513–1520.
229. Klett, E.L., Lee, M.H., Adams, D.B., Chavin, K.D., Patel, S.B. Localization of ABCG5 and ABCG8 proteins in human liver, gall bladder and intestine. *BMC Gastroenterology* 2004;**21**:4:21.
230. Repa, J.J., Berge, K.E., Pomajzl, C., Richardson, J.A., Hobbs, H., Mangelsdorf, D.J. Regulation of ATP-binding cassette sterol transporters ABCG5 and ABCG8 by the liver X receptors alpha and beta. *J Biol Chem* 2002;**277**:18793–18800.
231. Repa, J.J., Dietschy, J.M., Turley, S.D. Inhibition of cholesterol absorption by SCH 58053 in the mouse is not mediated via changes in the expression of mRNA for ABCA1, ABCG5, or ABCG8 in the enterocyte. *J Lipid Res* 2002;**43**:1864–1874.
232. Dieter, M.Z., Maher, J.M., Cheng, X., Klaassen, C.D. Expression and regulation of the sterol half-transporter genes ABCG5 and ABCG8 in rats. *Comp Biochem Physiol C Toxicol Pharmacol* 2004;**139**:209–221.
233. Zanlungo, S., Nervi, F. Discovery of the hepatic canalicular and intestinal cholesterol transporters. New targets for treatment of hypercholesterolemia. *Eur Rev Med* Pharmacol Sci 2003;**7**:33–39.
234. Geuken, E., Visser, D.S., Leuvenink, H.G., de Jong, K.P., Peeters, P.M., Slooff, M.J., et al. Hepatic expression of ABC transporters G5 and G8 does not correlate with biliary cholesterol secretion in liver transplant patients. *Hepatology* 2005;**42**:1166–1174.
235. Drobnik, W., Lindenthal, B., Lieser, B., Ritter, M., Christiansen Weber, T., Liebisch, G., et al. ATP-binding cassette transporter A1 (ABCA1) affects total body sterol metabolism. *Gastroenterology* 2001;**120**:1203–1211.
236. Oram, J.F., Heinecke, J.W. ATP-binding cassette transporter A1: a cell cholesterol exporter that protects against cardiovascular disease. Physiol Rev 2005;**85**:1343–1372.
237. Kaminski, W.E., Piehler, A., Wenzel, J.J. ABC A-subfamily transporters: Structure, function and disease. *Biochim Biophys Acta* 2006;**1762**:510–524.
238. Goldiner, I., van der Velde, A.E., Vandenberghe, K.E., van Wijland, M.A., Halpern, Z., Gilat, T., et al. ABCA1-dependent but ApoA-I-independent cholesterol efflux mediated by fatty acid-bile acid conjugates (FABACs). *Biochem J* 2006 [Epub ahead of print].
239. Tam, S.P., Mok, L., Chimini, G., Vasa, M., Deeley, R.G. ABCA1 mediates high-affinity uptake of 25-hydroxycholesterol by membrane vesicles and rapid efflux of the oxysterol by intact cells. *Am J Physiol Cell Physiol* 2006;Apr 12 [Epub ahead of print].
240. Takahashi, K., Kimura, Y., Kioka, N., Matsuo, M., Ueda, K. Purification and ATPase Activity of Human ABCA1. *J Biol Chem* 2006;**281**:10760–10768.
241. Frikke-Schmidt, R., Nordestgaard, B.G., Schnohr, P., Steffensen, R., Tybjaerg-Hansen, A. Mutation in ABCA1 predicted risk of ischemic heart disease in the Copenhagen City Heart Study Population. *J Am Coll Cardiol* 2005;**46**:1516–1520.
242. Charman, W. Lipids, Lipophilic Drugs, and Oral Drug Delivery – Some Emerging Concepts. *J Pharm Sci* 2000;**89**:967–978.
243. Khoo, S.M., Shackleford, D.M., Porter, C.J., Edwards, G.A., Charman, W.N. Intestinal lymphatic transport of halofantrine occurs after oral administration of a unit-dose lipid-based formulation to fasted dogs. *Pharm Res* 2003;**20**:1460–1465.
244. Lespine, A., Chanoit, G., Bousquet-Melou, A., Lallemand, E., Bassissi, F.M., Alvinerie, M., Toutain, P.L. Contribution of lymphatic transport to the systemic exposure of orally administered moxidectin in conscious lymph duct-cannulated dogs. *Eur J Pharm Sci* 2006;**27**:37–43.
245. Hugger, E.D., Novak, B.L., Burton, P.S., Audus, K.L., Borchardt, R.T. A comparison of commonly used polyethoxylated pharmaceutical excipients on their ability to inhibit P-glycoprotein activity in vitro. *J Pharm Sci* 2002;**91**:1991–2002.
246. Rege, B.D., Kao, J.P., Polli, J.E. Effects of nonionic surfactants on membrane transporters in Caco-2 cell monolayers. *Eur J Pharm Sci* 2002;**16**:237–246.
247. Lo, Y.L. Relationships between the hydrophilic–lipophilic balance values of pharmaceutical excipients and their multidrug resistance modulating effect in Caco-2 cells and rat intestines. *J Control Release* 2003;**90**:37–48.
248. Cornaire, G., Woodley, J.F., Saivin, S., Legendre, J.Y., Decourt, S., Cloarec, A., Houin, G. Effect of polyoxyl 35 castor oil and Polysorbate 80 on the intestinal absorption of digoxin in vitro. *Arzneimittelforschung* 2000;**50**:576–579.

249. Shono, Y., Nishihara, H., Matsuda, Y., Furukawa, S., Okada, N., Fujita, T., Yamamoto, A. Modulation of intestinal P-glycoprotein function by Cremophor EL and other surfactants by an in vitro diffusion chamber method using the isolated rat intestinal membranes. *J Pharm Sci* 2004;**93**:877–885.
250. Zhang, H., Yao, M., Morrison, R.A., Chong, S. Commonly used surfactant, Tween 80, improves absorption of P-glycoprotein substrate, digoxin, in rats. *Arch Pharm Res* 2003;**26**:768–772.
251. Seeballuck, F., Ashford, M.B., O'Driscoll, C.M. The effects of pluronics block copolymers and Cremophor EL on intestinal lipoprotein processing and the potential link with P-glycoprotein in Caco-2 cells. *Pharm Res* 2003;**20**:1085–1092.
252. Cornaire, G., Woodley, J., Hermann, P., Cloarec, A., Arellano, C., Houin, G. Impact of excipients on the absorption of P-glycoprotein substrates in vitro and in vivo. *Int J Pharm* 2004;**278**:119–131.
253. Collnot, E.N., Baldes, C., Wempe, M.F., Hyatt, J., Navarro, L., Edgar, K.J., et al. Influence of vitamin E TPGS poly(ethylene glycol) chain length on apical efflux transporters in Caco-2 cell monolayers. *J Control Release* 2006;**111**:35–40.
254. Dintaman, J.M., Silverman, J.A. Inhibition of P-glycoprotein by D-alpha-tocopheryl polyethylene glycol 1000 succinate (TPGS). *Pharm Res* 1999;**16**:1550–1556.
255. Hugger, E.D., Audus, K.L., Borchardt, R.T. Effects of poly(ethylene glycol) on efflux transporter activity in Caco-2 cell monolayers. *J Pharm Sci* 2002;**91**:1980–1990.
256. Batrakova, E.V., Li, S., Miller, D.W., Kabanov, A.V. Pluronic P85 increases permeability of a broad spectrum of drugs in polarized BBMEC and Caco-2 cell monolayers. *Pharm Res* 1999;**16**: 1366–1372.
257. Johnson, B.M., Charman, W.N., Porter, C.J. An in vitro examination of the impact of polyethylene glycol 400, Pluronic P85, and vitamin E D-alpha-tocopheryl polyethylene glycol 1000 succinate on P-glycoprotein efflux and enterocyte-based metabolism in excised rat intestine. *AAPS Pharm Sci* 2002;**4**:E40.
258. Regev, R., Assaraf, Y.G., Eytan, G.D. Membrane fluidization by ether, other anesthetics, and certain agents abolishes P-glycoprotein ATPase activity and modulates efflux from multidrug-resistant cells. *Eur J Biochem* 1999;**259**:18–24.
259. Lo, Y.L. Phospholipids as multidrug resistance modulators of the transport of epirubicin in human intestinal epithelial Caco-2 cell layers and everted gut sacs of rats. *Biochem Pharmacol* 2000;**60**:1381–1390.
260. Risovic, V., Sachs-Barrable, K., Boyd, M., Wasan, K.M. Potential mechanisms by which Peceol increases the gastrointestinal absorption of amphotericin, B. *Drug Dev Ind Pharm* 2004; **30**:767–774.
261. Miller, D.W., Batrakova, E.V., Kabanov, A.V. Inhibition of multidrug resistance-associated protein (MRP) functional activity with pluronic block copolymers. *Pharm Res* 1999;**16**:396–401.
262. Kabanov, A.V., Batrakova, E.V., Alakhov, V.Y. An essential relationship between ATP depletion and chemosensitizing activity of Pluronic block copolymers. *J Control Release* 2003;**91**:75–83.
263. Batrakova, E.V., Li, S., Li, Y., Alakhov, V.Y., Kabanov, A.V. Effect of pluronic P85 on ATPase activity of drug efflux transporters. *Pharm Res* 2004;**21**:2226–2233.
264. Minko, T., Batrakova, E.V., Li, S., Li, Y., Pakunlu, R.I., Alakhov, V.Y., Kabanov, A.V. Pluronic block copolymers alter apoptotic signal transduction of doxorubicin in drug-resistant cancer cells. *J Control Release* 2005;**105**:269–278.
265. Kabanov, A.V., Batrakova, E.V., Alakhov, V.Y. Pluronic block copolymers for overcoming drug resistance in cancer. *Adv Drug Deliv Rev* 2002;**54**:759–779.
266. Yu, L., Bridgers, A., Polli, J., Vickers, A., Long, S., Roy, A., et al. Vitamin E-TPGS increases absorption flux of an HIV protease inhibitor by enhancing its solubility and permeability. *Pharm Res* 1999;**16**:812–1817.
267. Wong, H.L., Bendayan, R., Rauth, A.M., Xue, H.Y., Babakhanian, K., Wu, X.Y. A Mechanistic Study of enhanced doxorubicin uptake and retention in multidrug resistant breast cancer cells using a polymer-lipid hybrid nanoparticle (PLN) system. *J Pharmacol Exp Ther* 2006;Mar 17 [Epub ahead of print].

CHAPTER 2

FORMULATION ISSUES AROUND LIPID-BASED ORAL AND PARENTERAL DELIVERY SYSTEMS

Seong Hoon Jeong, Jae Hyung Park, and Kinam Park

2.1 INTRODUCTION

2.2 LIPIDS FOR DRUG DELIVERY APPLICATIONS

2.3 MECHANISMS OF IMPROVED ORAL BIOAVAILABILITY

2.4 LIPID-BASED DRUG DELIVERY FORMULATIONS

2.5 CONCLUSION

2.1 INTRODUCTION

It is widely known that ingested lipids in food enhance the absorption and bioavailability of poorly soluble drugs [1]. Therefore, the food effect, specifically that of dietary lipids, has been regarded as one of the promising ways of designing oral drug delivery formulations of poorly soluble drugs. To be absorbed through the gastrointestinal (GI) membranes, any drug compounds must be in a dissolved or solubilized state. Lipids can induce the dissolved state of the poorly soluble drugs for better absorption. However, the dissolved state may not be achieved directly from the administered lipids, but more likely from the intraluminal processing to which lipids are subjected before absorption [2], so it is important to understand lipid digestion and the behavior of endogenous lipids for drug solubilization.

The intravenous injection of the poorly soluble drug may cause embolization of blood vessels by the insoluble drug aggregates and may involve local toxicity resulting from the high drug concentration at the site of deposition [3]. For parenteral application, lipid-based drug delivery systems, in the form of fat emulsions, micellar systems, and liposomes, have been used. Recently, many scientists have been

Role of Lipid Excipients in Modifying Oral and Parenteral Drug Delivery, Edited by Kishor M. Wasan

trying to design new lipid-based drug delivery systems to solubilize drugs with sufficient stability, low preparation cost, and easy processing in a sterile form. Examples of the new classes of parenteral lipid formulations are solid lipid nanoparticles and nanosuspensions, lipid-based micelles, lipid microbubbles, and lipoprotein drug carriers.

Lipids are versatile tools for drug administration because they can be formulated into many types of preparations, such as solutions, suspensions, emulations, self-emulsifying systems, and microemulsions. However, there are still challenges for such applications: the complexity of the interfacial and physical chemistry of lipids, the stability and manufacturing problems associated with their commercial production, the limited solubility of some poorly soluble drugs in lipid solvents, the pre-absorptive GI processing that is required of many lipids, the lack of knowledge about what actually happens to the co-administered drugs/lipids, and the lack of predictive in vitro and in vivo testing methodologies [2]. Despite these limitations, lipids have the potential to enhance drug absorption and bioavailability with many formulation opportunities still to be thoroughly investigated.

2.2 LIPIDS FOR DRUG DELIVERY APPLICATIONS

For the purpose of drug delivery formulation, lipids as drug delivery vehicles need to be chosen carefully. For oral drug administration, lipids can be divided roughly into digestible and nondigestible in the GI tract [2]. The digestible lipids are composed of dietary lipids such as glycerides, fatty acids, phospholipids, cholesterol, and cholesterol esters, as well as various synthetic derivatives, e.g. triglyceride lipids can be digested or hydrolyzed into diglycerides and fatty acids by the lingual and gastric lipases in the stomach [4, 5]. The pancreatic lipase hydrolyzes the triglyceride lipids further, producing 2-monoglyceride and two fatty acids. The presence of the hydrolyzed products induces secretion of biliary and pancreatic fluids, causing a substantial change in the luminal environment. Moreover, the digested products are more water-soluble than the parent lipids, and they can be solubilized within bile salt mixed micelles (Figure 2.1). On the other hand, nondigestible lipids include mineral oil (liquid paraffin) and sucrose polyesters. When administered, they remain in the lumen and can decrease drug absorption by holding a fraction of the co-administered drug [6, 7]. Lipids can also be classified depending on raw materials (plant or animal), sources (natural, synthetic, or semi-synthetic), physical states (solid, semisolid, or liquid), polarity (polar or nonpolar) and molecular structures (triglycerides, hydrocarbons, or polyalcohols).

Generally, it is not so easy to tell which lipid can enhance the bioavailability of poorly water-soluble drugs and this depends on each formulation, e.g. when examining the effects of digestible and nondigestible lipids on the bioavailability of poorly soluble drugs (acetyl sulfisoxazole and griseofulvin) in rats, the bioavailability of griseofulvin decreased in all lipids [8]. However, in case of acetyl sulfisoxazole, when digestible lipids were used, bioavailability increased, unlike with nondigestible lipids. Moreover, the maximum serum concentration level of medium-chain triglyceride formulation was four-fold higher than that of the nondigestible formulation.

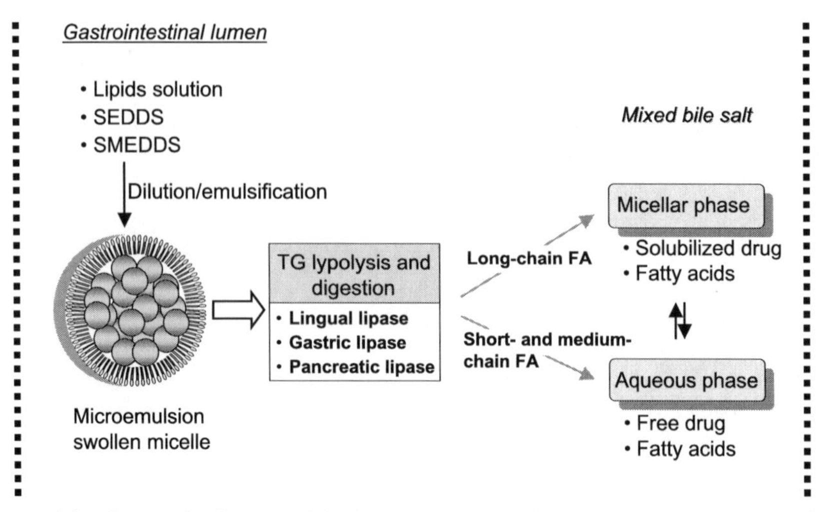

Figure 2.1 Schematic diagram of the in vivo lipid digestion process from lipid-based formulations. SEDDS, self-emulsifying drug delivery systems; SMEDDS, self-emulsifying microemulsion drug delivery systems; TG, triglyceride; FA, fatty acid. (Adapted from Myers and Stella [9].)

As shown in Figure 2.2, when trioctanoin was used for a 10% oil-in-water emulsion, higher bioavailability was observed. This might be the result of the rapid digestion of the medium-chain triglyceride. Low bioavailability from the long-chain triglycerides was a result of the slower and incomplete digestion. The nondigestible mineral oil had lower bioavailability [9]. It has also been implied that the bioavailability of an oleic acid or peanut oil suspension was higher than that of a triolein suspension [10].

2.3 MECHANISMS OF IMPROVED ORAL BIOAVAILABILITY

Lipids may improve the oral bioavailability of poorly water-soluble drugs by a number of possible mechanisms [11]. Lipids can increase effective drug solubility in the GI tract and blood vessels, e.g. the presence of lipids in the GI tract stimulates the secretion of bile salts and endogenous biliary lipids such as phospholipid and cholesterol, forming intestinal mixed micelles, and then increases the solubilization capacity of the GI tract. Moreover, as a result of the intercalation of administered lipids into these bile salt structures, either directly or by secondary digestion, lipids cause the swelling of micelle structures and a further increase in solubilization capacity.

Lipids can also increase the gastric retention time, resulting in slow delivery to the absorption site and increased time available for the absorption [12]. Even though passive intestinal permeability is not considered to be a major limitation to the bioavailability of most poorly water-soluble drugs, various combinations of

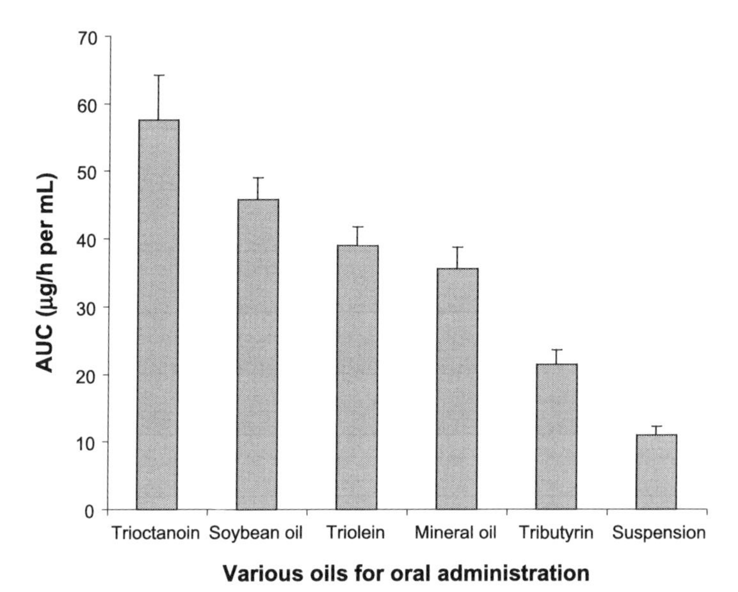

Figure 2.2 Mean plasma AUC (area under the curve) values of penclomedine (5 mg, 16.7 mg/kg) in various 10% oil-in-water emulsions and an aqueous suspension. (Adapted from Myers and Stella [9].)

lipids, lipid digestion products, and surfactants have been shown to have permeability-enhancing properties [13]. They can affect the physical and biochemical barrier function of the GI tract. Lipids and surfactants may reduce the activity of intestinal efflux transporters and the P-glycoprotein efflux pump, and also the extent of enterocyte-based metabolism [14, 15]. For highly lipophilic drugs, lipids may enhance the extent of lymphatic transport and improve bioavailability by reducing the first-pass metabolism [16].

2.4 LIPID-BASED DRUG DELIVERY FORMULATIONS

It is widely known that drug candidates with high biological activities often have poor solubility in water because most of their chemical structures include liphophilic groups that show an affinity toward the target receptor [17, 18]. Low solubility in water has limited the therapeutic application of the drug because it is associated with poor absorption and bioavailability after oral administration [18].

Intravenous injection of poorly soluble drugs may cause embolization of blood vessels by aggregates of the insoluble drugs and may involve local toxicity as a result of the high drug concentration at the site of deposition [3]. To overcome the poor water solubility of drugs, excipients, including ethanol, Cremophor EL (polyethoxylated castor oil), and surfactants, have been used for parenteral application [19–21]. Such components, however, need special manufacturing and packaging conditions because they may extract toxic substances (e.g. plasticizers) from the devices, such as infusion tubing and containers. Furthermore, the drugs dissolved in the presence of excipients may precipitate on dilution with physiological fluids. Surfactants may not retain the solubilized drugs at concentrations lower than their critical micelle concentration (CMC) values that are typically high for low-molecular-mass surfactants. In addition, administration of co-solvents or surfactants may cause undesirable side effects such as venous irritation and respiratory distress [22–24]. When considering the current issues associated with the use of excipients, there is a need for novel formulations that are cost-efficient, biocompatible, and stable in an aqueous condition. These formulations may provide an acceptable shelf-life under storage conditions, allow use of a variety of poorly soluble drugs, and prevent inclusion of toxic solvents such as ethanol and Cremophor EL.

The recent approaches for solubilization of poorly soluble drugs may include the use of emulsions [25, 26], solid–lipid nanoparticles, liposomes [27, 28], and polymeric micelles.

2.4.1 Self-emulsifying and Self-microemulsifying Drug Delivery Systems

Self-emulsifying drug delivery systems (SEDDSs) are isotropic mixtures of oil, surfactant, and drug, and they can form fine oil-in-water (o/w) emulsions when exposed to aqueous phases such as GI environment under gentle agitation. Poorly soluble drugs can be dissolved in SEDDSs for oral administration. These systems can be incorporated into soft gelatin capsules. When the formulations are administered via the lumen, they disperse and form fine emulsions so that the drug can remain in solution in the intestine, avoiding the dissolution step that usually limits the rate of absorption of poorly soluble drugs from the crystalline state [29]. Good examples of SEDDS formulation are Sandimmune Neoral (cyclosporine), Norvir (ritonavir), and Fortovase (saquinavir), and they raised strong interests in such promising emulsion-based drug delivery systems [30]. In cases of self-microemulsifying drug delivery systems (SMEDDSs), they are similar to SEDDSs except that they form finer microemulsions when exposed to the aqueous phase. Generally, microemulsion is considered to be a thermodynamically stable and optically clear dispersion of lipids and amphiphiles.

2.4.2 Dry Emulsions

Dry emulsions are powder-state and lipid-based formulations. They can be easily reconstituted to an o/w emulsion in vivo or when exposed to an aqueous environment. Dry emulsions can be prepared by drying liquid o/w emulsions containing a soluble or insoluble solid carrier in an aqueous medium. Soluble carriers can be gelatin, glycine, lactose, maltodextrin, mannitol, and sucrose, whereas an insoluble carrier can be colloidal silica. There are mainly three drying methods: spray drying, lyophilization, and rotary evaporation. With drying, the aqueous phase can be removed, causing the solid carrier to encapsulate the dispersed lipid phase. The resulting dry powder can fill capsules or be compressed into tablets. The solid carrier may undergo partial or complete transformation into an amorphous state by the drying process, and it may cause physical stability issues of solid phase transformation.

2.4.3 Solid Dispersions

Solid dispersion is a solid matrix containing dispersed or dissolved drug in excess of its solubility in a polymer or lipid matrix. Solid dispersions can include eutectics, solid solutions, glass solutions, glass suspensions, amorphous precipitation in a crystalline carrier, and compound and complex formation [31]. The solid dispersions can be prepared by solvent evaporation or melt formation. A drug can be either dissolved or suspended in the matrix. Especially when suspended, the drug may be present as a crystalline or amorphous form, depending on the composition of the matrix and the preparation method. Therefore, it is important to maintain the physical stability especially when drug and excipients are present in a metastable or amorphous form. The applications of the solid dispersions focus mainly on enhanced dissolution of poorly soluble drugs, sustained release, and improved stability in the GI tract.

2.4.4 Fat Emulsions

Fat emulations have been used mainly for parenteral nutrition. Major components of the fat emulsion are vegetable oils (mainly triglycerides) and phospholipids (emulsion stabilizer from egg or soybean). As many drugs can be dissolved in vegetable oils, drug-loaded emulsions has been used successfully as drug carriers. A good example of the system is the intravenous anesthetic propofol (Diprivan) which has been considered as one of the most successful fat emulsions applied to drug delivery systems so far [32]: it causes less pain on injection than solvent-based or solubilized formulations. Drug emulsions are normally formulated to be isotonic, and the amount of dissolved components can be minimized because of the drug in the lipid phase. Propofol readily dissolves in soybean oil and gives a low concentration of the drug in the aqueous phase, which is another benefit of the system because the pain on injection depends on the concentration. A decrease in toxicity, together with an increase in the therapeutic window, is another potential advantage. If the solubility properties of the lipids need to be adjusted to improve drug loading, other triglycerides can be used: more polar triglycerides (diacetylated monoglycerides and medium-chain triglycerides) with long-chain triglycerides (soybean oil).

As the route of administration of the fat emulsions is parenteral, the systems have to be prepared with strict specifications. For sterility, heat sterilization can be applied and is a preferred way because of the advantages with respect to easy manufacturing and product safety. Therefore, the fat emulsions have to be stable during the process. Alternatively, sterile filtration can be a feasible method for emulsions with a small mean droplet size. A high-pressure homogenization is the major preparation method for small-scale and large-scale manufacturing.

2.4.5 Solid Lipid Nanoparticles

As drug mobility in solid lipids would be much lower compared with liquid oils, the use of solid lipids appeals as a very interesting idea to achieve specific drug delivery purposes. Solid lipids have been used in the form of pellets to achieve a sustained drug delivery. Solid lipid microparticles (by spray drying) and 'Nanopellets' are good examples [33]. Although the primary application of the system has been parenteral, it also can be applied to sustained-release device for oral administration.

Solid lipid nanoparticles (SLNs) can be prepared by dispersing melted lipids with high-speed mixers or ultrasound. The resulting products often contain relatively high amounts of microparticles. This might not be a serious problem for oral administration, but it can cause problems in an intravenous injection. Higher concentrations of the emulsifier may result in a reduction of the particle size and can also increase the risk of side effects. It has been shown that high-pressure homogenization is more efficient for the production of submicrometer-sized dispersions of solid lipids than high shear mixers or ultrasound [33]. Dispersions obtained by this method are called SLNs. Most SLN dispersions produced by high-pressure homogenization are characterized by an average particle size below 500 nm and low microparticle contents.

2.4.6 Liposomes

As one end of phospholipid molecules is water soluble, whereas the opposite end is water insoluble, liposomes, vesicles with spherical shape, can be prepared when the phospholipids are hydrated in an aqueous medium, forming a lipid bilayer (a single bilayer or multiple layers). The liposome, first introduced by Bangham et al. in the 1960s, is a representative example that has been successfully developed as lipid-based drug delivery vehicles [34, 35]. Liposomes can be formed easily by a simple thin-film hydration method [34]. The resulting vesicles, heterogeneous multi-layered vesicles of up to 5 μm in diameter, can be further sonicated or extruded through filters to form smaller, single-layered vesicles [36]. There is a size limitation for liposome formulation, which is approximately 20–25 nm, because the edge energy of a circular bilayer must be sacrificed to provide the bending energy to form a spherical vesicle [37].

Water-soluble drugs can be trapped inside the aqueous compartment of liposomes. However, water-insoluble ones can be incorporated into the bilayer membrane [38]. For hydrophobic drugs, the lipid bilayer can have a function as a solubilizing matrix or a drug delivery carrier depending on the properties of the composition. Water-soluble drugs can remain in the internal aqueous compartment for a long period as a result of the thermodynamically unfavorable environment. The spherical lipid envelop formed by the bilayer membranes constitutes a very effective permeability barrier, limiting the movement of the water-soluble molecules between the internal and external aqueous compartment.

Liposomes have been designed to reduce side effects of incorporated drugs and to enhance the therapeutic efficacy. However, the early formulations indicated limitations in controlling the release rate of the drug as a result of its rapid diffusion into the surrounding medium. By adjusting the liposome structure and incorporating some excipients, a suitable release rate was later achieved [39, 40]. Another problem in extending applications of liposomes was the rapid removal from the circulation because they are recognized by the reticuloendothelial system (RES) as foreign entities. The organs associated with the RES are the liver, lung, and spleen. Large and charged liposomes were removed quickly by the liver and spleen, with a half-life of less than 1 hour, e.g. liposomes with a size larger than 500 nm can be cleared from the blood far faster than smaller ones (100 nm) [41], and liposomes with charged surfaces are cleared from the circulation more rapidly than neutral liposomes [42].

The surface modification of the liposome with polyethyleneglycol (PEG) has increased the residence time in the bloodstream and decreased recognition of the immune system (Figure 2.3a) [43–45]. Clearance of the liposome was two to five times slower than the conventional liposome, which might result from the formation of steric stabilization on the liposome surface reducing protein binding and cellular recognition [46, 47]. Several liposomal formulations have been on the market (e.g. Ambisome, DaunoXome, and Doxil) since the early 1990s. Recent studies demonstrated that the repeated dosage of the PEGylated liposome can result in rapid clearance from the blood, accumulation in the liver, and acute hypersensitivity [48, 49]. Moreover, the chemical and physical stabilities of liposomal formulations in an aqueous environment need to be improved for successful applications [35, 50–52].

For further applications, liposomes with the targeting ability can be developed based on the ligand–acceptor interaction and other mechanisms (active targeting) (Figure 2.3). Tumor site-selective liposomes can be prepared using a membrane system with permeability properties that are sensitive to physicochemical properties, which can be manipulated in sites of tumor growth from extraneous sources. Liposome can be designed as pH sensitive or fusogenic using additional components [53, 54]. The development of thermosensitive and photosensitive liposomes is also possible [55, 56].

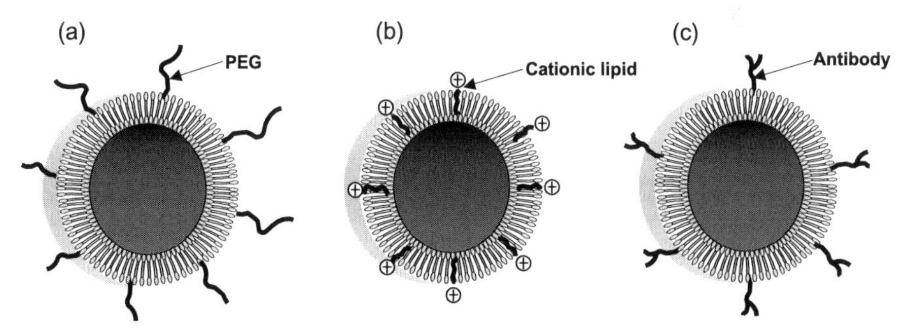

Figure 2.3 Typical examples of liposomes surface-modified with (a) polyethyleneglycol (PEG), (b) cation, and (c) antibody conjugation for the improvement of bioavailability and targeting.

2.4.7 Lipid-based Micelles

The micelles prepared from amphiphilic polymers have received increasing attention because of their potential biotechnological and pharmaceutical applications [57–59]. On contact with an aqueous environment, polymeric amphiphiles undergo intra- and/or intermolecular associations between hydrophobic moieties, resulting in the formation of micelles or micelle-like aggregates (Figure 2.4). Depending on the chemical structures and physical properties of hydrophilic/ hydrophobic constituents, polymeric micelles exhibit unique characteristics such as unusual rheological feature, small hydrodynamic radius (less than microsize) with core-shell structure, and thermodynamic stability [58–60]. Such nanoparticulates have been recognized as a promising drug carrier, because their hydrophobic domain, surrounded by a hydrophilic outer shell, can play a role as a container for various poorly soluble drugs [59, 61].

Polymeric micelles are thermodynamically more stable than micelles prepared from conventional low-molecular-mass surfactants, as demonstrated by the lower CMC values of polymeric amphiphiles [62, 63]. In addition to the thermodynamic stability, polymeric micelles have superior kinetic stability to low-molecular-mass surfactants because of the presence of the multiple sites capable of hydrophobic interaction, leading to the retarded disintegration at concentrations lower than that of the CMCs [64]. The high stability of polymeric micelles enables them to retain the solubilized drug in the hydrophobic core, extend the circulation time in the blood, and lower the toxicity of a drug. In this regard, polymeric micelles are very useful for improving the therapeutic effects of the poorly soluble drugs and reducing their side effects. Many efforts have been made to develop polymeric amphiphiles capable of forming the compact micellar structure in an aqueous media, such as amphiphilic block co-polymers [57, 61] and hydrophobically modified water-soluble polymers [65, 66].

Numerous chemical structures of polymers are available to endow them with the amphiphilicity necessary for self-assemble to form the nano-sized polymeric micelles. For the hydrophilic part, PEG has been the most widely used polymer because it is one of the nontoxic materials for biomedical applications. As a result of the presence of the ether oxygen linkage, which can play a role as a hydrogen bond acceptor, PEG is highly hydrated in aqueous solutions and thus has a high degree of segmental flexibility and large excluded volume. These properties enable

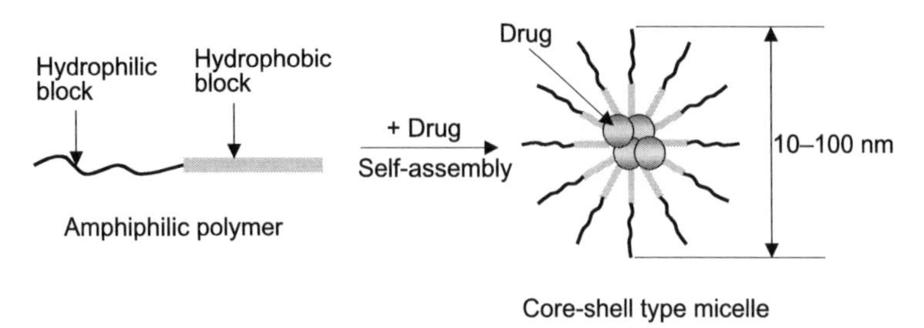

Figure 2.4 Schematics of micelle formation.

PEG to behave as an efficient steric protector of various biological macromolecules (e.g. proteins) and particulate drug delivery systems. Other hydrophilic polymers are also available as micellar corona-forming blocks in an aqueous condition if they are biocompatible and provide effective surface properties. A few examples of the polymers studied as the hydrophilic part include poly(2-ethyl-2-oxazoline) [67], poly(vinyl pyrrolidone) [68], and polysaccharides [69, 70].

Lipids have recently been used as hydrophobic parts of the polymeric amphiphiles [71–73]. Among various lipids, phosphatidylethanolamine (PEA) has been widely studied as a hydrophobic moiety of the polymeric micellar systems because the use of PEA provides high stability for the micelles as a result of the very strong hydrophobic interactions between double acyl chains of the phospholipids residues (Figure 2.5) [3, 72, 73]. In the early 1990s, the PEG–PEA conjugate was studied for preparing PEGylated liposomes to prolong the blood circulation time and prevent rapid uptake by the RES [74, 75]. By simply mixing the PEG–PEA

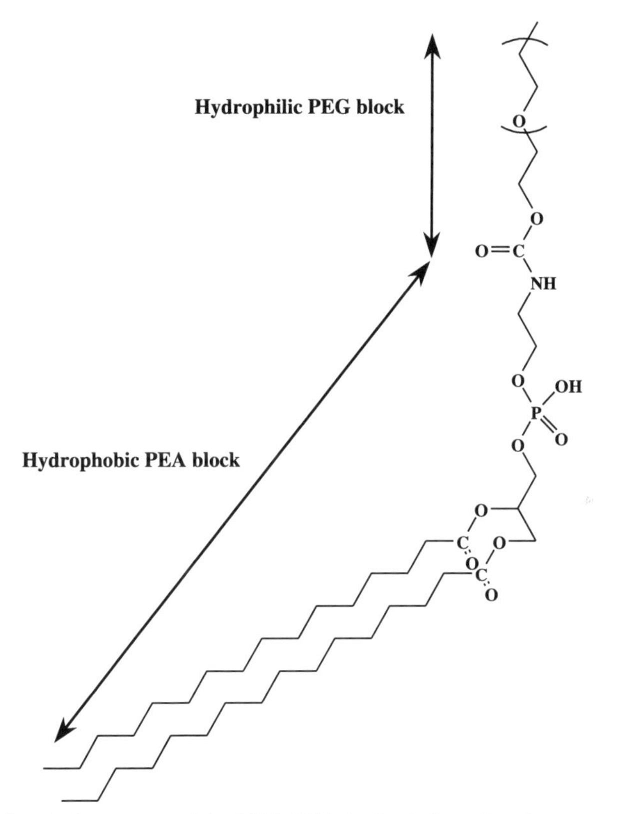

Figure 2.5 Chemical structure of the PEG–PEA (polyethyleneglycol–phosphatidylethanolamine) conjugate. This conjugate can form a stable polymer micelle, because the diacyl phospholipid part (PEA) is very hydrophobic and thus forms a compact lipid core in an aqueous environment. (Adapted from Lukyanov and Torchilin [3].)
Need publisher for this book

conjugate with other lipids, PEG can be incorporated in the liposomal surface, because the lipid part of the conjugate interacts with the lipid bilayer. On the other hand, the mixture forms the micelles instead of liposomes when the PEG–PEA content exceeds certain critical limits [76, 77]. The ability of the PEG–lipid conjugates to form micelles in an aqueous environment was also found in 1994 by Torchilin et al. [78]. As the PEG–lipid conjugates have proved to be nontoxic, they are currently under clinical tests as a component of Doxil [79].

Similar to conventional detergents, lipid-based amphiphiles (LBAs) spontaneously form micelles in an aqueous media, primarily to decrease the Gibbs free energy by minimizing the exposure of hydrophobic segments to the aqueous environment. The simple way to prepare the drug-loaded micelles is to disperse a mixture of LBAs and drug to the aqueous buffer. In detail, the LBAs and drug are dissolved in the volatile organic solvent, and the LBL–drug film is obtained by evaporating the organic solvent under vacuum. After being hydrated in an aqueous buffer, the film is mechanically agitated to form nano-sized micelles. Alternatively, the drug can be loaded inside the micelle by the dialysis method: i.e. the organic solvent is removed from the mixture solution of LBAs and drug by dialysis, through which the LBA micelles are constructed via the intra- and intermolecular association of the hydrophobic parts, and the drug is incorporated inside the lipid core of the micelles [62, 68]. Based on these techniques, a variety of poorly soluble drugs, such as paclitaxel [80, 81], dequalinium [82], and tamoxifen [80], has been successfully incorporated into the LBA micelles. Generally, the incorporation of a drug into LBA micelles reflects a small effect on the micellar size, and the loading efficiency varies from 1.5 to 50 wt%. The size of LBA micelles is in the range of 5–35 nm, depending on the balance between hydrophilicity and hydrophobicity. For PEG–PEA conjugates, the use of PEG blocks with higher molecular mass led to the formation of larger micelles, indicating that the micelle size can be controlled by varying the block length of PEG [3].

The in vitro stability of micelles has been estimated from their CMC values using the pyrene incorporation method [62, 83]. When exposed to polymeric micellar solutions, pyrene molecules preferably locate inside or close to the hydrophobic microdomains of micelles rather than the aqueous phase, resulting in different photophysical characteristics. Therefore, the formation of micelles is associated with the appearance of a hydrophobic core capable of solublizing pyrene molecules. The results have demonstrated that the CMC values of LBA micelles are in a 10^{-5} mol/L range, which is at least 100-fold lower than those of conventional detergents [3]. On the other hand, it has been suggested that the pyrene method does not have sufficient sensitivity for the measurement of the actual CMC values of the amphiphilic compounds bearing the very hydrophobic moieties such as double acyl chains. In fact, a recent study by Lukyanov et al. [71] confirmed that the micelles prepared from PEG–distearoyl PEA could retain their micellar structure at the concentration of 2.5×10^{-6} mol/L, which is fourfold lower than the CMC value determined by the pyrene method. This implies that LBA micelles can maintain their integrity in extremely diluted solutions such as in the blood.

The loading capacity of LBA micelles for the poorly soluble drugs can be significantly augmented by inserting additional compounds that are capable of forming micelles, e.g. Krishnadas et al. [84] prepared mixed micelles, composed of PEG–PEA and egg phosphatidylcholine (egg PC), as a potential carrier of a poorly soluble drug, paclitaxel. They confirmed that the mixed micelles could solubilize 1.5 times more paclitaxel than plain PEG–PEA micelles, and the amount of solubilized paclitaxel increased linearly with an increase in a lipid concentration. Compared with PEG–PEA, the egg PC does not have a bulky hydrophilic PEG chain. Therefore, the addition of the egg PC may increase the content of the hydrophobic inner core in the micelles, which provides a large space for hydrophobic interaction with paclitaxel.

The polymeric micelles have been demonstrated to provide opportunities for the site-specific delivery of drugs because they can solubilize various hydrophobic drugs, increase bioavailabilty, and stay unrecognized during blood circulation [85–87]. In particular, it has become apparent that, when administrated systemically, the biocompatible micelles preferentially accumulate in solid tumors by the enhanced permeability and retention (EPR) effect [88, 89], attributed to leaky tumor vessels and lack of an effective lymphatic drainage system. Similar to most polymeric micelles, it is expected that LBA micelles show a long circulation time in the blood and improved targeting efficiency of the drug by the EPR effect. Lukyanov et al. [71, 90] evaluated the PEG–PEA micelles as the drug carrier in vivo using tumor-bearing mice. They confirmed that the micelles show a prolonged circulation time so that the drug can reach the target site. In particular, the micelles with a larger PEG block had a longer circulation time, indicating that the molecular mass of PEG is associated with the ability to protect the micelles from uptake by the RES.

2.5 CONCLUSION

Lipid-based drug delivery systems have attracted strong interest in the formulation for oral and parenteral administrations. Lipid-based formulations are very attractive ways of increasing drug solubility, improving bioavailability, reducing side effects, and targeting specific tissues. On the other hand, they still present disadvantages: potential drug stability problems, high production cost (third party manufacturing), and migration of excipients into the shell of the dosage form or injection device. The lipid-based drug delivery systems have been designed to have modified surfaces with PEG, ligand, and environment-sensitive moieties, and they will encourage further research to understand the functionality and performance of the systems. Lipid-based drug delivery formulations with novel properties will continue to play an important role in the drug delivery applications.

REFERENCES

1. Charman, W.N., Porter, C.J.H., Mithani, S., Dressman, J.B. Physicochemical and Physiological Mechanisms for the Effects of Food on Drug Absorption: The Role of Lipids and pH. *J Pharm Sci* 1997;**86**:269–282.
2. Humberstone, A.J., Charman, W.N. Lipid-based vehicles for the oral delivery of poorly water soluble drugs. *Adv Drug Deliv Rev* 1997;**25**:103–128.

3. Lukyanov, A.N., Torchilin, V.P. Micelles from lipid derivatives of water-soluble polymers as delivery systems for poorly soluble drugs. *Adv Drug Deliv Rev* 2004;**56**:1273–1289.
4. Liao, T.H., Hamosh, P., Hamosh, M. Fat digestion by lingual lipase: mechanism of lipolysis in the stomach and upper small intestine. *Pediatr Res* 1984;**18**:402–409.
5. Hamosh, M., Scanlon, J.W., Ganot, D., Likel, M., Scanlon, K.B., Hamosh, P. Fat digestion in the newborn. Characterization of lipase in gastric aspirates of premature and term infants. *J Clin Invest* 1981;**67**:838–846.
6. Volpenhein, R.A., Webb, D.R., Jandacek, R.J. Effect of a nonabsorbable lipid, sucrose polyester, on the absorption of DDT by the rat. *J Toxicol Env Health* 1980;**6**:679–683.
7. Rozman, K., Ballhorn, L., Rozman, T. Mineral oil in the diet enhances fecal excretion of DDT in the rhesus monkey. *Drug Chem Toxicol* 1983;**6**:311–316.
8. Bloedow, D.C., Hayton, W.L. Effects of lipids on bioavailability of sulfisoxazole acetyl, dicumarol, and griseofulvin in rats. *J Pharm Sci* 1976;**65**:328–334.
9. Myers, R.A., Stella, V.J. Systemic bioavailability of penclomedine (NSC-338720) from oil-in-water emulsions administered intraduodenally to rats. *Int J Pharm* 1992;**78**:217–226.
10. Yamahira, Y., Noguchi, T., Takenaka, H., Maeda, T. Biopharmaceutical studies of lipid-containing oral dosage forms: relationship between drug absorption rate and digestibility of vehicles. *Int J Pharm* 1979;**3**:23–31.
11. Porter, C.J.H., Charman, W.N. In vitro assessment of oral lipid based formulations. *Adv Drug Deliv Rev* 2001;**50**:S127–S147.
12. Hunt, J.N., Knox, M.T. Relation between the chain length of fatty acids and the slowing of gastric emptying. *J Physiol* 1968;**194**:327–336.
13. Aungst, B.J. Intestinal permeation enhancers. *J Pharm Sci* 2000;**89**:429–442.
14. Dintaman, J.M., Silverman, J.A. Inhibition of P-glycoprotein by D-a-tocopheryl polyethylene glycol 1000 succinate (TPGS). *Pharm Res* 1999;**16**:1550–1556.
15. Nerurkar, M.M., Burton, P.S., Borchardt, R.T. The use of surfactants to enhance the permeability of peptides through Caco-2 cells by inhibition of an apically polarized efflux system. *Pharm Res* 1996;**13**:528–534.
16. Porter, C.J.H., Charman, W.N. Intestinal lymphatic drug transport: an update. *Adv Drug Deliv Rev* 2001;**50**:61–80.
17. Lipinski, C.A. Drug-like properties and the causes of poor solubility and poor permeability. *J Pharmacol Toxicol Methods* 2000;**44**:235–249.
18. Lipinski, C.A., Lombardo, F., Dominy, B.W., Feeney, P.J. Experimental and computational approaches to estimate solubility and permeability in drug discovery and development settings. *Adv Drug Deliv Rev* 2001;**46**:3–26.
19. Alkan-Onyuksel, H., Ramakrishnan, S., Chai, H.B., Pezzuto, J.M. A mixed micellar formulation suitable for the parenteral administration of taxol. *Pharm Res* 1994;**11**:206–212.
20. Kawakami, K., Miyoshi, K., Ida, Y. Solubilization behavior of poorly soluble drugs with combined use of Gelucire 44/14 and cosolvent. *J Pharm Sci* 2004;**93**:1471–1479.
21. Chambin, O., Jannin, V. Interest of multifunctional lipid excipients: case of Gelucire 44/14. *Drug Dev Ind Pharm* 2005;**31**:527–534.
22. Dye, D., Watkins, J. Suspected anaphylactic reaction to Cremophor EL. *BMJ* 1980;**280**:1353.
23. Hassan, S., Dhar, S., Sandstrom, M., Arsenau, D., Budnikova, M., Lokot, I., et al. Cytotoxic activity of a new paclitaxel formulation, Pacliex, in vitro and in vivo. *Cancer Chemother Pharmacol* 2005;**55**:47–54.
24. Constantinides, P.P., Tustian, A., Kessler, D.R. Tocol emulsions for drug solubilization and parenteral delivery. *Adv Drug Deliv Rev* 2004;**56**:1243–1255.
25. Park, K.M., Lee, M.K., Hwang, K.J., Kim, C.K. Phospholipid-based microemulsions of flurbiprofen by the spontaneous emulsification process. *Int J Pharm* 1999;**183**:145–154.
26. Constantinides, P.P., Lambert, K.J., Tustian, A.K., Schneider, B., Lalji, S., Ma, W., et al. Formulation development and antitumor activity of a filter-sterilizable emulsion of paclitaxel. *Pharm Res* 2000;**17**:175–182.
27. Straubinger, R.M., Arnold, R.D., Zhou, R., Mazurchuk, R., Slack, J.E. Antivascular and antitumor activities of liposome-associated drugs. *Anticancer Res* 2004;**24**:397–404.
28. Straubinger, R.M., Balasubramanian, S.V. Preparation and characterization of taxane-containing liposomes. *Methods Enzymol* 2005;**391**:97–117.

29. Pouton, C.W. Formulation of self-emulsifying drug delivery systems. *Adv Drug Deliv Rev* 1997;**25**:47–58.
30. Neslihan Gursoy, R., Benita, S. Self-emulsifying drug delivery systems (SEDDS) for improved oral delivery of lipophilic drugs. *Biomed Pharmacother* 2004;**58**:173–182.
31. Breitenbach, J. Melt extrusion: from process to drug delivery technology. *Eur J Pharm Biopharm* 2002;**54**:107–117.
32. Davis, S.S. Coming of age of lipid-based drug delivery systems. *Adv Drug Deliv Rev* 2004;**56**:1241–1242.
33. Mehnert, W., Mader, K. Solid lipid nanoparticles. Production, characterization and applications. *Adv Drug Deliv Rev* 2001;**47**:165–196.
34. Bangham, A.D., Horne, R.W. Negative staining of phospholipids and their structural modification by surface-active agents as observed in the electron microscope. *J Mol Biol* 1964;**12**:660–668.
35. Wissing, S.A., Kayser, O., Muller, R.H. Solid lipid nanoparticles for parenteral drug delivery. *Adv Drug Deliv Rev* 2004;**56**:1257–1272.
36. Feng, S.-S., Chien, S. Chemotherapeutic engineering: application and further development of chemical engineering principles for chemotherapy of cancer and other diseases. *Chem Eng Sci* 2003; **58**:4087–4114.
37. Helfrich, W. Size of bilayer vesicles generated by sonication. *Phys Lett A* 1974;**50A**:115–116.
38. Mayer, L.D., Krishna, R., Bally, M.B. Liposomes for cancer therapy applications. In: *Polymeric Biomaterials,* 2nd edn. 2002, pp. 823–841.
39. Fielding, R.M., Abra, R.M. Factors affecting the release rate of terbutaline from liposome formulations after intratracheal instillation in the guinea pig. *Pharm Res* 1992;**9**:220–223.
40. Harashima, H., Kiwada, H. Liposomal targeting and drug delivery: kinetic consideration. *Adv Drug Deliv Rev* 1996;**19**:424–444.
41. Abra, R.M., Hunt, C.A. Liposome disposition in vivo. III. Dose and vesicle-size effects. *BBA-Lipid Lipid Met* 1981;**666**:493–503.
42. Senior, J., Gregoriadis, G. Stability of small unilamellar liposomes in serum and clearance from the circulation: the effect of the phospholipid and cholesterol components. *Life Sci* 1982;**30**:2123–2136.
43. Klibanov, A.L., Maruyama, K., Torchilin, V.P., Huang, L. Amphipathic polyethyleneglycols effectively prolong the circulation time of liposomes. *FEBS Lett* 1990;**268**:235–237.
44. Papahadjopoulos, D., Allen, T.M., Gabizon, A., Mayhew, E., Matthay, K., Huang, S.K., et al. Sterically stabilized liposomes: improvements in pharmacokinetics and antitumor therapeutic efficacy. *Proc Natl Acad Sci USA* 1991;**88**:11460–11464.
45. Simoes, S., Moreira, J.N., Fonseca, C., Duzgunes, N., de Lima, M.C. On the formulation of pH-sensitive liposomes with long circulation times. *Adv Drug Deliv Rev* 2004;**56**:947–965.
46. Guo, X., Szoka, F.C., Jr. Steric stabilization of fusogenic liposomes by a low-pH sensitive PEG-diortho ester-lipid conjugate. *Bioconjug Chem* 2001;**12**:291–300.
47. MacKay, J.A., Deen, D.F., Szoka, F.C. Distribution in brain of liposomes after convection enhanced delivery; modulation by particle charge, particle diameter, and presence of steric coating. *Brain Res* 2005;**1035**:139–153.
48. Ishida, T., Harada, M., Wang, X.Y., Ichihara, M., Irimura, K., Kiwada, H. Accelerated blood clearance of PEGylated liposomes following preceding liposome injection: effects of lipid dose and PEG surface-density and chain length of the first-dose liposomes. *J Control Release* 2005;**105**:305–317.
49. Judge, A., McClintock, K., Phelps, J.R., Maclachlan, I. Hypersensitivity and loss of disease site targeting caused by antibody responses to PEGylated liposomes. *Mol Ther* 2006;**13**:328–337.
50. Sarbolouki, M.N., Toliat, T. Storage stability of stabilized MLV and REV liposomes containing sodium methotrexate (aqueous and lyophilized). *PDA J Pharm Sci Technol* 1998;**52**:23–27.
51. Brandl, M. Liposomes as drug carriers: a technological approach. *Biotechnol Annu Rev* 2001;**7**:59–85.
52. Glavas-Dodov, M., Fredro-Kumbaradzi, E., Goracinova, K., Simonoska, M., Calis, S., Trajkovic-Jolevska, S., Hincal, A.A. The effects of lyophilization on the stability of liposomes containing 5-FU. *Int J Pharm* 2005;**291**:79–86.
53. Yamada, Y., Shinohara, Y., Kakudo, T., Chaki, S., Futaki, S., Kamiya, H., Harashima, H. Mitochondrial delivery of mastoparan with transferrin liposomes equipped with a pH-sensitive fusogenic peptide for selective cancer therapy. *Int J Pharm* 2005;**303**:1–7.

54. Mizoue, T., Horibe, T., Maruyama, K., Takizawa, T., Iwatsuru, M., Kono, K., et al. Targetability and intracellular delivery of anti-BCG antibody-modified, pH-sensitive fusogenic immunoliposomes to tumor cells. *Int J Pharm* 2002;**237**:129–137.
55. Gaver, M.H., Wu, N.Z., Hong, K., Huang, S.K., Dewhirst, M.W., Papahadjopoulos, D. Thermosensitive liposomes: extravasation and release of contents in tumor microvascular networks. *Int J Radiat Oncol* 1996;**36**:1177–1187.
56. Namiki, Y., Namiki, T., Date, M., Yanagihara, K., Yashiro, M., Takahashi, H. Enhanced photodynamic antitumor effect on gastric cancer by a novel photosensitive stealth liposome. *Pharmacol Res* 2004;**50**:65–76.
57. Gref, R., Minamitake, Y., Peracchia, M.T., Trubetskoy, V., Torchilin, V., Langer, R. Biodegradable long-circulating polymeric nanospheres. *Science* 1994;**263**:1600–1603.
58. Tobio, M., Gref, R., Sanchez, A., Langer, R., Alonso, M.J. Stealth PLA-PEG nanoparticles as protein carriers for nasal administration. *Pharm Res* 1998;**15**:270–275.
59. Kataoka, K., Harada, A., Nagasaki, Y. Block copolymer micelles for drug delivery: design, characterization and biological significance. *Adv Drug Deliv Rev* 2001;**47**:113–131.
60. Cudd, A., Bhogal, M., O'Mullane, J., Goddard, P. Formation of cage-like particles by poly(amino acid)-based block copolymers in aqueous solution. *Proc Natl Acad Sci USA* 1991;**88**:10855–10859.
61. Nakanishi, T., Fukushima, S., Okamoto, K., Suzuki, M., Matsumura, Y., Yokoyama, M., et al. Development of the polymer micelle carrier system for doxorubicin. *J Control Release* 2001;**74**:295–302.
62. La, S.B., Okano, T., Kataoka, K. Preparation and characterization of the micelle-forming polymeric drug indomethacin-incorporated poly(ethylene oxide)-poly(beta-benzyl L-aspartate) block copolymer micelles. *J Pharm Sci* 1996;**85**:85–90.
63. Kabanov, A.V., Batrakova, E.V., Alakhov, V.Y. Pluronic block copolymers as novel polymer therapeutics for drug and gene delivery. *J Control Release* 2002;**82**:189–212.
64. Yokoyama, M., Sugiyama, T., Okano, T., Sakurai, Y., Naito, M., Kataoka, K. Analysis of micelle formation of an Adriamycin-conjugated poly(ethylene glycol)-poly(aspartic acid) block copolymer by gel permeation chromatography. *Pharm Res* 1993;**10**:895–899.
65. Park, J.H., Kwon, S., Nam, J.O., Park, R.W., Chung, H., Seo, S.B., et al. Self-assembled nanoparticles based on glycol chitosan bearing 5beta-cholanic acid for RGD peptide delivery. *J Control Release* 2004;**95**:579–588.
66. Kim, K., Kwon, S., Park, J.H., Chung, H., Jeong, S.Y., Kwon, I.C., Kim, I.S. Physicochemical characterizations of self-assembled nanoparticles of glycol chitosan-deoxycholic acid conjugates. *Biomacromolecules* 2005;**6**:1154–1158.
67. Lee, S.C., Kim, C., Kwon, I.C., Chung, H., Jeong, S.Y. Polymeric micelles of poly(2-ethyl-2-oxazoline)-block-poly(epsilon-caprolactone) copolymer as a carrier for paclitaxel. *J Control Release* 2003;**89**:437–446.
68. Benahmed, A., Ranger, M., Leroux, J.C. Novel polymeric micelles based on the amphiphilic diblock copolymer poly(*N*-vinyl-2-pyrrolidone)-block-poly(D,L-lactide). *Pharm Res* 2001;**18**:323–328.
69. Son, Y.J., Jang, J.S., Cho, Y.W., Chung, H., Park, R.W., Kwon, I.C., et al. Biodistribution and antitumor efficacy of doxorubicin loaded glycol-chitosan nanoaggregates by EPR effect. *J Control Release* 2003;**91**:135–145.
70. Na, K., Lee, K.H., Bae, Y.H. pH-sensitivity and pH-dependent interior structural change of self-assembled hydrogel nanoparticles of pullulan acetate/oligo-sulfonamide conjugate. *J Control Release* 2004;**97**:513–525.
71. Lukyanov, A.N., Gao, Z., Mazzola, L., Torchilin, V.P. Polyethylene glycol-diacyllipid micelles demonstrate increased accumulation in subcutaneous tumors in mice. *Pharm Res* 2002;**19**:1424–1429.
72. Torchilin, V.P., Lukyanov, A.N., Gao, Z., Papahadjopoulos-Sternberg, B. Immunomicelles: targeted pharmaceutical carriers for poorly soluble drugs. *Proc Natl Acad Sci USA* 2003;**100**:6039–6044.
73. Vakil, R., Kwon, G.S. PEG-phospholipid micelles for the delivery of amphotericin B. *J Control Release* 2005;**101**:386–389.
74. Senior, J., Delgado, C., Fisher, D., Tilcock, C., Gregoriadis, G. Influence of surface hydrophilicity of liposomes on their interaction with plasma protein and clearance from the circulation: studies with poly(ethylene glycol)-coated vesicles. *Biochim Biophys Acta* 1991;**1062**:77–82.

75. Blume, G., Cevc, G. Molecular mechanism of the lipid vesicle longevity in vivo. *Biochim Biophys Acta* 1993;**1146**:157–168.
76. Bedu-Addo, F.K., Tang, P., Xu, Y., Huang, L. Interaction of polyethyleneglycol-phospholipid conjugates with cholesterol-phosphatidylcholine mixtures: sterically stabilized liposome formulations. *Pharm Res* 1996;**13**:718–724.
77. Edwards, K., Johnsson, M., Karlsson, G., Silvander, M. Effect of polyethyleneglycol-phospholipids on aggregate structure in preparations of small unilamellar liposomes. *Biophys J* 1997;**73**:258–266.
78. Torchilin, V.P., Omelyanenko, V.G., Papisov, M.I., Bogdanov, A.A., Jr, Trubetskoy, V.S., Herron, J.N., Gentry, C.A. Poly(ethylene glycol) on the liposome surface: on the mechanism of polymer-coated liposome longevity. *Biochim Biophys Acta* 1994;**1195**:11–20.
79. Gabizon, A.A. Pegylated liposomal doxorubicin: metamorphosis of an old drug into a new form of chemotherapy. *Cancer Invest* 2001;**19**:424–436.
80. Gao, Z., Lukyanov, A.N., Singhal, A., Torchilin, V. Diacyllipid-polymer micelles as nanocarriers for poorly soluble anticancer drugs. *Nano Lett* 2002;**2**:979–982.
81. Gao, Z., Lukyanov, A.N., Chakilam, A.R., Torchilin, V.P. PEG-PE/phosphatidylcholine mixed immunomicelles specifically deliver encapsulated taxol to tumor cells of different origin and promote their efficient killing. *J Drug Target* 2003;**11**:87–92.
82. Lizano, C., Weissig, V., Torchilin, V.P., Sancho, P., Garcia-Perez, A.I., Pinilla, M. In vivo biodistribution of erythrocytes and polyethyleneglycol-phosphatidylethanolamine micelles carrying the antitumour agent dequalinium. *Eur J Pharm Biopharm* 2003;**56**:153–157.
83. Kwon, S., Park, J.H., Chung, H., Kwon, I.C., Jeong, S.Y. Physicochemical characteristics of self-assembled nanoparticles based on glycol chitosan bearing 5beta-cholanic acid. *Langmuir* 2003;**19**:10188–10193.
84. Krishnadas, A., Rubinstein, I., Onyuksel, H. Sterically stabilized phospholipid mixed micelles: in vitro evaluation as a novel carrier for water-insoluble drugs. *Pharm Res* 2003;**20**:297–302.
85. Kataoka, K., Matsumoto, T., Yokoyama, M., Okano, T., Sakurai, Y., Fukushima, S., et al. Doxorubicin-loaded poly(ethylene glycol)-poly(beta-benzyl-L-aspartate) copolymer micelles: their pharmaceutical characteristics and biological significance. *J Control Release* 2000;**64**:143–153.
86. Lavasanifar, A., Samuel, J., Kwon, G.S. Poly(ethylene oxide)-block-poly(L-amino acid) micelles for drug delivery. *Adv Drug Deliv Rev* 2002;**54**:169–190.
87. Hubbell, J.A. Materials science. Enhancing drug function. *Science* 2003;**300**:595–596.
88. Matsumura, Y., Maeda, H. A new concept for macromolecular therapeutics in cancer chemotherapy: mechanism of tumoritropic accumulation of proteins and the antitumor agent smancs. *Cancer Res* 1986;**46**:6387–6392.
89. Duncan, R. Polymer conjugates for tumour targeting and intracytoplasmic delivery. The EPR effect as a common gateway? *Pharm Sci Technol Today* 1999;**2**:441–449.
90. Lukyanov, A.N., Gao, Z., Torchilin, V.P. Micelles from polyethylene glycol/phosphatidylethanolamine conjugates for tumor drug delivery. *J Control Release* 2003;**91**:97–102.

CHAPTER 3

LIPID-BASED PARENTERAL DRUG DELIVERY SYSTEMS: BIOLOGICAL IMPLICATIONS

Vladimir P. Torchilin

3.1 INTRODUCTION

3.2 LIPOSOMES IN DRUG DELIVERY

3.3 MODIFICATION OF THE BIOLOGICAL PROPERTIES OF LIPOSOMES: LIPOSOME TARGETING

3.4 INFLUENCE OF LIPOSOMES ON BIOLOGICAL PROPERTIES OF INCORPORATED DRUGS

3.5 LIPID-CORE MICELLES AS PHARMACEUTICAL CARRIERS FOR POORLY SOLUBLE DRUGS

3.6 CONCLUSION

3.1 INTRODUCTION

Among a variety of lipid-based pharmaceutical carriers for parenteral use aimed to modify and improve significantly the biological properties of the carrier-loaded drugs and enhance their therapeutic activity, most of the studies have been performed with liposomes and lipid-core micelles. Microreservoir-type systems, such as liposomes (mainly, for water-soluble drugs) and micelles (mainly, for water-insoluble drugs), have certain advantages over other delivery systems, e.g., a maximal load of the drug, a small necessary quantity of a targeting component, because just a few targeting moieties can carry multiple drug moieties loaded into the reservoir, and a possibility of easy control of composition, size, and in vivo stability of a microreservoir.

Liposomes are artificial phospholipid vesicles (spherical self-closed structures formed by one or several concentric lipid bilayers with aqueous phase both inside

Role of Lipid Excipients in Modifying Oral and Parenteral Drug Delivery, Edited by Kishor M. Wasan

and between those bilayers) with the size varying from 50 to 1000 nm and even more, which can be loaded with a variety of water-soluble drugs (into their inner aqueous compartment) and water-insoluble drugs (into the hydrophobic compartment of the phospholipid bilayer); they have been considered promising drug carriers for well over two decades [1]. The size, charge, and surface properties of liposomes can easily be changed simply by adding new ingredients to the lipid mixture before liposome preparation and/or by variation of preparation methods.

Micelles, including polymeric micelles [2], represent another promising type of pharmaceutical carrier. Micelles are colloidal dispersions with a particle size within range 5–100 nm. An important property of micelles is their ability to increase the solubility and bioavailability of poorly soluble pharmaceuticals. The use of certain special amphiphilic molecules as micelle building blocks can also introduce the property of micelle-extended blood half-life on intravenous administration.

As a result of their small size, micelles, as well as small-size liposomes, demonstrate a spontaneous penetration into the interstitium in the body compartments with the leaky vasculature (tumors and infarcts) by the enhanced permeability and retention (EPR) effect – a form of selective targeted delivery termed 'passive targeting' [3–5]. This effect is based on certain specificities of pathological tissues, such as tumors. The first specificity is that the tumor vasculature, unlike the vasculature of healthy tissues, is permeable to macromolecules with a molecular mass of 50 kDa or even higher. This allows macromolecules to enter into the interstitial tumor space. Another specificity is that the lymphatic system, responsible for the drainage of macromolecules from normal tissues, virtually does not work in the case of many tumors as a result of the disease [6]. Consequently, macromolecules and nanoparticulates that have entered tumor tissues are retained there for a prolonged time. Unlike macromolecules or nanoparticles, low-molecular-mass conventional pharmaceutics are not retained in tumors because of their ability to return to the circulation by diffusion [6]. Diffusion and accumulation parameters for drug carriers in tumors have recently been shown to be strongly dependent on the cutoff size of tumor blood vessel wall, and the cutoff size varies for different tumors [7–8]. The details on the mechanism of the EPR effect and the list of preparations for which this effect has been observed can be found in several review articles [6, 9].

In this chapter, we consider the biological implications of drug loading into liposomes and lipid-core micelles. Some other lipid-based nanocarriers, such as solid lipid nanoparticles, as well as the use of lipid-based nanocarriers for gene delivery, are outside the scope of the chapter.

3.2 LIPOSOMES IN DRUG DELIVERY

3.2.1 Liposomes

For many years, liposomes have been considered promising carriers for biologically active substances and some liposomal drugs have already found their way into clinical practice [10, 11] (Table 3.1). The initial success with many of these drugs has fueled further clinical investigations. To mention just the most recent ones, the

TABLE 3.1 Some liposomal drugs (approved or under clinical evaluation)

Active drug (and product name for liposomal preparation)	Indications
All-*trans*-retinoic acid (Altragen)	Acute promyelocytic leukemia; non-Hodgkin's lymphoma; renal cell carcinoma; Kaposi's sarcoma
Amphotericin B (AmBisome)	Fungal infections
Annamycin	Doxorubicin-resistant tumors
Cytarabine (DepoCyt)	Lymphomatous meningitis
Daunorubicin (DaunoXome)	Kaposi's sarcoma
DNA plasmid encoding HLA-B7 and β_2-microglobulin (Allovectin-7)	Metastatic melanoma
Doxorubicin (Mycet)	Combinational therapy of recurrent breast cancer
Doxorubicin in PEG liposomes (Doxil, Caelyx)	Refractory Kaposi's sarcoma, ovarian cancer, recurrent breast cancer
E1A gene	Various tumors
LipoMASC (liposomes for various drugs and diagnostic agents)	Broad applications
Lurtotecan (NX211)	Ovarian cancer
Nystatin (Nyotran)	Topical anti-fungal agent
Platinum compounds (Platar)	Solid tumors
Vincristine (Onco TCS)	Non-Hodgkin's lymphoma

research on the use of doxorubicin in PEG (polyethyleneglycol) liposomes for the treatment of solid tumors concentrated on the selective delivery of the drug in patients with breast carcinoma metastases and improvement in survival [12–14]. The same set of indications was targeted by the combination therapy involving liposomal doxorubicin and paclitaxel [15] or Caelyx (doxorubicin in PEG liposomes) and carboplatin [16]. Caelyx is also in a phase II study for patients with squamous cell cancer of the head and neck [17] and ovarian cancer [18]. Patient research showed the impressive effect of doxorubicin in PEG liposomes against unresectable hepatocellular carcinoma [19], cutaneous T-cell lymphoma [20], and sarcoma [21]. Liposomal lurtotecan was found to be effective in patients with topotecan-resistant ovarian cancer [22]. Among other indications, one may notice the use of the liposomal amphotericin B for the treatment of visceral leishmaniasis [23] and long-acting analgesia with liposomal bupivacaine in healthy volunteers [24].

Initially, 'plain' phospholipid liposomes with nonmodified surface have been thought to be effective and easy to prepare pharmaceutical carriers for favorable modification of the biological properties of various drugs; however, their fast elimination from the blood and the capture of the liposomal preparations by the cells of the reticuloendothelial system (RES), primarily in the liver as a result of rapid opsonization of the liposomes [25], hindered or limited their practical application.

3.2.2 Immunoliposomes

To increase liposome accumulation in the desired areas and the efficacy of liposomal pharmaceutical agents, the use of targeted liposomes with surface-attached ligands capable of recognition and binding to tissues or cells of interest has been suggested. Immunoglobulins, primarily of the IgG class, and their Fab fragments are the most widely used targeting moieties for liposomes, although many other ligands have also been suggested and successfully tried (see more below). Routine methods for attaching specific ligands to liposomes without affecting the liposome integrity and antibody properties include a covalent binding of an antibody to a reactive group on the liposome surface, and hydrophobic insertion of proteins modified with hydrophobic residues into the liposomal membrane [26]. Still, despite improvements in the targeting efficacy, most immunoliposomes ended in the liver as a consequence of insufficient time for interaction with the target. This is especially true in cases when a target of choice has diminished the blood supply (ischemic or necrotic areas) or the concentration of the target antigen is low. In both cases better accumulation can be expected if liposomes can stay in the circulation for prolonged periods of time.

3.2.3 Long-circulating Liposomes

Different methods have been suggested to achieve long circulation of liposomes in vivo, including coating the liposome surface with inert, biocompatible polymers, such as PEG, which form a protective layer over the liposome surface and slows down the liposome recognition by opsonins and subsequent clearance [27, 28]. Long-circulating liposomes are now investigated in detail and widely used in biomedical in vitro and in vivo studies; they have also found their way into clinical practice [29, 30]. The longevity phenomenon is based on the decreased rate of plasma protein adsorption on the hydrophilic surface of pegylated liposomes [31]. An important feature of protective polymers is their flexibility, which allows a relatively small number of their surface-grafted molecules to create dense protective ‘clouds’ over the liposome surface [32, 33]. As a result, long-circulating liposomes demonstrate dose-independent, nonsaturable, log-linear kinetics, and increased bioavailability as a result of the inhibition of liposome destruction and their capture by the RES [34].

Although PEG remains the gold standard in liposome steric protection, attempts continue to identify other polymers that could be used to prepare long-circulating liposomes. Earlier studies with various water-soluble flexible polymers have been summarized in [33, 35]. More recent papers describe long-circulating liposomes prepared using poly[*N*-(2-hydroxypropyl)methacrylamide] [36], poly-*N*-vinylpyrrolidones [37], L-amino acid-based, biodegradable, polymer–lipid conjugates [38], and polyvinyl alcohol [39]. The relative role of the liposome charge and protective polymer molecular size was investigated [40], showing that opsonins with different molecular sizes may be involved in the clearance of liposomes containing different charged lipids. Interactions of plasma proteins with liposomes of different composition may be shielded differently by PEG [41].

3.2.4 Long-circulating Immunoliposomes

The further development of the concept of long-circulating liposomes involves the combination of the properties of long-circulating liposomes and immunoliposomes in one preparation [42–44]. To achieve better selective targeting of PEG-coated liposomes, the targeting ligand is not attached directly to the particles, but via a PEG spacer arm, so that the ligand is extended outside the dense PEG layer, excluding steric hindrances for its binding to the target receptors. To achieve this, PEG has to be modified with a phospholipid on one side and the targeting ligand on the other. Currently, various advanced technologies have been used, and the targeting moiety is usually attached above the protecting polymer layer, by coupling it with the distal water-exposed terminus of activated liposome-grafted polymer molecule [43, 45]. A recently described new technique of a single step attachment of specific ligands, including monoclonal antibodies, to pegylated liposomes via *p*-nitrophenylcarbonyl-terminated PEG-phosphatidylethanolamine (pNP-PEG-PE) provides new opportunities for preparing targeted long-circulating liposomes [45].

3.3 MODIFICATION OF THE BIOLOGICAL PROPERTIES OF LIPOSOMES: LIPOSOME TARGETING

3.3.1 Immunotargeting

Specific liposome targeting can dramatically change biodistribution and target accumulation of the liposome-incorporated drugs. Multiple experiments with immunotargeted liposomes, especially for tumor targeting, have been reviewed on many occasions [46–48]. Still, the research in this area continues with some interesting results obtained recently. Thus, it was established that internalizing antibodies are needed to achieve a really improved therapeutic efficacy of antibody-targeted liposomal drugs, as was shown using B-lymphoma cells and internalizable epitopes (CD19) [49]. HER2-overexpressing tumors have been successfully targeted with anti-HER2 liposomes [50]. Nucleosome-specific antibodies capable of recognition of various tumor cells via tumor cell surface-bound nucleosomes improved Doxil targeting to tumor cells and increased its cytotoxicity [51]. GD2-targeted immunoliposomes with the novel antitumoral drug, fenretinide, inducing apoptosis in neuroblastoma and melanoma cell lines, demonstrated strong anti-neuroblastoma activity both in vitro and in vivo in mice [52]. Antibody CC52 against rat colon adenocarcinoma CC531 attached to pegylated liposomes provided specific accumulation of liposomes in a rat model of metastatic CC531 [53]. Combining immunoliposome and endosome-disruptive peptide improves cytosolic delivery of the liposomal drug, increases cytotoxicity, and opens new approaches to constructing targeted liposomal systems, as shown with the diphtheria toxin A chain incorporated together with pH-dependent fusogenic peptide, diINF-7, into liposomes specific towards ovarian carcinoma [54].

3.3.2 Transferrin-mediated Targeting

In addition to antibodies, a variety of low-molecular-mass ligands can also target liposomes to certain cells and tissues. Thus, as transferrin (Tf) receptors (TfRs) are overexpressed on the surface of many tumor cells, antibodies against TfR as well as Tf itself are among the popular ligands for liposome targeting to tumors and inside tumor cells [55]. Recent studies involve the coupling of Tf to PEG on pegylated liposomes in order to combine longevity and the ability to be targeted for drug delivery into solid tumors [56]. Similar approach was applied both for delivering, into tumors, agents for photodynamic therapy including hypericin [57, 58], and for intracellular delivery of cisplatin into gastric cancer [59]. Tf-coupled doxorubicin-loaded liposomes demonstrate increased binding and toxicity against C6 glioma [60]. Interestingly, the increase in the expression of the TfR was also discovered in post-ischemic cerebral endothelium, which was used to deliver Tf-modified PEG liposomes to post-ischemic brains in rats [61]. Tf [62], as well as anti-TfR, antibodies [63, 64] were also used to facilitate gene delivery into cells by cationic liposomes. Tf-mediated liposome delivery was successfully used for brain targeting, and immunoliposomes with the OX26 monoclonal antibody to the rat TfR were found to concentrate on brain microvascular endothelium [65].

3.3.3 Folate-mediated Targeting

Targeting tumors with folate-bearing liposomes now represents a very popular approach, because folate receptor (FR) expression is frequently overexpressed in many tumor cells. After early studies demonstrated the possibility of delivering macromolecules [66] and then liposomes [67] into living cells using FR endocytosis, which could bypass multidrug resistance, the interest to folate-targeted drug delivery by liposomes grew fast (see important reviews in the literature [68, 69]). Liposomal daunorubicin [70] and doxorubicin [71] demonstrated increased cytotoxicity when delivered into various tumor cells via folate receptor. Recent studies have suggested more complex schemes for using folate liposomes. Thus, the application of folate-modified doxorubicin-loaded liposomes for the treatment of acute myelogenous leukemia was combined with the induction of FR using all-*trans*-retinoic acid [72]. Folate-targeted liposomes have been suggested as delivery vehicles for boron neutron capture therapy [73] and also been used for targeting tumors with haptens for tumor immunotherapy [74]. Within the frame of gene therapy, folate-targeted liposomes were used for both gene targeting to tumor cells [75] and targeting tumors with antisense oligonucleotides [76].

3.3.4 Other Targeting Ligands

The search for new ligands for liposome targeting concentrates around specific receptors overexpressed on target cells and certain specific components characteristic of pathologic cells. Thus, liposome targeting to tumors has been achieved by using vitamin and growth factor receptors [77]. Vasoactive intestinal peptide (VIP) was used to target PEG liposomes with radionuclides to VIP receptors of the tumor, which resulted in an enhanced breast cancer inhibition in rats [78]. PEG liposomes were targeted by arginine-glycine-aspartic acid (RGD) – containing peptides to integrins of tumor vasculature and, being loaded with doxorubicin, demonstrated increased efficiency against C26 colon carcinoma in a murine model [79]. RGD peptide was also used for targeting liposomes to integrins on activated platelets and, thus, could be used for specific cardiovascular targeting [80]. Similar angiogenic homing peptide was used for targeted delivery to vascular endothelium of drug-loaded liposomes in experimental treatment of tumors in mice [81].

Epidermal growth factor receptor (EGFR)-targeted immunoliposomes were specifically delivered to variety of tumor cells overexpressing EGFR [82]. Mitomycin C in long-circulating hyaluronan-targeted liposomes increases its activity against tumors overexpressing hyaluronan receptors [83]. Galactosylated liposomes target drugs to the liver for therapy of liver tumors or metastases [84]. The ability of galactosylated liposomes to concentrate in parenchymal cells was applied for gene delivery in these cells (see Hashida et al. [85] for a review). As the increased level of chondroitin sulfate expression was found in many tumor cells, cisplatin-loaded cationic non-pegylated and pegylated liposomes specifically binding chondroitin sulfate were used for the suppression of tumor growth and metastases in vivo [86].

3.3.5 Virosomes

Virosomes represent one more line of evolution for liposomes when, to enhance tissue targeting, the liposome surface was modified with fusogenic viral envelope proteins [87]. Initially, virosomes were intended for intracellular delivery of drugs and DNA [88, 89]. Later, virosomes became a cornerstone for new vaccine development. Delivery of protein antigens to the immune system by fusion-acting virosomes was found to be very effective [90], in particular into dendritic cells [91]. As a result, a whole set of virosome-based vaccines have been developed for application in humans and animals. Special attention was paid to influenza vaccine using virosomes containing the spike proteins of influenza virus [92], because it elicits high titers of influenza-specific antibodies. Trials of virosome influenza vaccine in children showed that it is highly immunogenic and well tolerated [93]. A similar approach was used to prepare virosomal hepatitis A vaccine that elicited high antibody titers after primary and booster immunization of infants and young children [94]; the data have been confirmed in healthy adults [95] and elderly patients [96]. Combination of influenza protein-based virosomes with other antigens may be used to prepare other vaccines [97]. In general, virosomes can provide an excellent opportunity for efficient delivery of both various antigens and many drugs (nucleic acids, cytotoxic drugs, toxoids) [97, 98], although they might represent certain problems associated with their stability and immunogenicity.

3.3.6 The pH-sensitive Liposomes

In some cases, liposome targeting inside cells is based on physical principles rather than biological interactions, and different methods of liposomal content delivery into the cytoplasm have been elaborated [99]. According to one of these methods, the liposome is made of pH-sensitive components and, after being endocytosed in the intact form, it fuses with the endovacuolar membrane under the action of lowered pH inside the endosome, releasing its contents into the cytoplasm [100, 101]. The optimized approach combines pH sensitivity of liposomes with their longevity and ligand-mediated targeting. Thus, long-circulating, pegylated, pH-sensitive liposomes, although demonstrating a decreased pH sensitivity, still effectively deliver their contents into cytoplasm (recent review by Simoes et al. [102]). Antisense oligonucleotides are delivered into cells by anionic, pH-sensitive, phosphatidylethanolamine (PE)-containing liposomes, which are stable in the blood; however, they undergo phase transition at acidic endosomal pH and facilitate the release of the incorporated oligonucleotides into the cell cytoplasm (recent review by Fattal et al. [103]). New pH-sensitive liposomal additives were recently described including oleyl alcohol [104] and pH-sensitive morpholine lipids (monostearoyl derivatives of morpholine) [105]. Combination of liposome pH sensitivity and specific ligand targeting for cytosolic drug delivery using decreased endosomal pH values was described for both folate and Tf-targeted liposomes [106–109].

3.3.7 Intracellular Targeting

Even the successful delivery of various to their targets (such as tumors) by means of (immuno)liposomes does, however, solve only part of a general drug efficiency problem. For many drugs, the next task is to achieve their intracellular delivery, because many targets for anticancer drugs are located inside cells.

The very nature of cell membranes prevents entry of the peptide or protein unless there is an active transport mechanism, which is usually the case for very short peptides [110]. Vector molecules promote the delivery of associated drug carriers inside the cells via receptor-mediated endocytosis [50]. This process involves attachment of the vector molecule and an associated drug carrier to specific ligands on target cell membranes, followed by endosome formation. An efficient cellular uptake via endocytosis is generally observed, but the delivery of intact peptides and proteins is compromised by an insufficient endosomal escape and subsequent lysosomal degradation. Enhanced endosomal escape can be achieved through the use of, for example, lytic peptides [54, 111, 112], pH-sensitive polymers [113] or swellable dendritic polymers [114].

An approach has recently emerged that allows for a much more straightforward and efficient way for delivery of peptides and proteins to the cytoplasm. This approach is based on the phenomenon called transduction, and uses the ability of certain peptides to ferry conjugated macromolecules, such as proteins [115] and DNA [116] and even nanoparticles [117–120], across cell membranes directly into cytoplasm. Peptides that cause transduction (protein transduction domains, PTDs, or cell-penetrating peptides, CPPs) were derived from proteins of viruses and *DrosophilaAntennapedia* transcription factor and can be as short as polymers with 10–16 units [115, 121, 122]. The details of CPP action on cell membranes are not completely elucidated, but the recent data assume more than one mechanism for CPPs and CPP-mediated intracellular delivery of various molecules and particles. TAT-mediated intracellular delivery of large molecules and nanoparticles was been proved to proceed via the energy-dependent macropinocytosis with subsequent enhanced escape from endosomes into the cell cytoplasm [123–125], whereas individual CPPs or CPP-conjugated small molecules penetrate cells via electrostatic interactions and hydrogen bonding and do not seem to depend on the energy [124, 126, 127]. As traversal through cellular membranes represents a major barrier for efficient delivery of macromolecules into cells, CPPs, whatever their mechanism of action, may serve to transport various drugs and even drug-loaded pharmaceutical carriers into mammalian cells in vitro and in vivo.

Several proteins, including those involved in oncogenesis, cancer-related signal transduction and cell proliferation pathways, have been delivered in active form into various human cells in vitro using fused PTD peptides [128–131]. It has also been shown that TAT PTD allows delivery of biologically active proteins into various cells in vivo [132]. The recent review of the current status and perspectives of the PTD-mediated delivery of proteins and peptides into cancer cells can be found in Gupta et al. [124]. These results open new avenues into the development of peptide- and protein-based anticancer therapeutics with intracellular molecular targets. The use of PTD-modified liposomes should provide a combination of the advantages of microreservoir carriers, enhanced accumulation in tumors via the EPR effect and cytoplasmic drug delivery. It was demonstrated that liposomes, could be delivered into various cells by multiple TAT-peptide or other CPP molecules attached to the liposome surface [119, 120, 133]. Complexes of TAT-peptide–liposomes with a plasmid (plasmid pEGFP-N1 encoding for the green fluorescence protein, GFP) were used for successful in vitro transfection of various tumor and normal cells as well as for in vivo transfection of tumor cells in mice bearing Lewis lung carcinoma [134]. The covalent coupling of PTDs to microparticulate drug carriers may provide an efficient tool for cytosolic delivery of various drugs, including proteins and peptides, in vitro and even in vivo in certain protocols of the local treatment. This is the next level in modifying drug biological behavior by liposomal carriers.

3.4 INFLUENCE OF LIPOSOMES ON BIOLOGICAL PROPERTIES OF INCORPORATED DRUGS

3.4.1 Protein and Peptide Drugs

The most dramatic influence of lipid-based pharmaceutical carriers on biological properties of incorporated drugs could be seen in the case of protein and peptide drugs. Many proteins and peptides possess biological activity that makes them potent therapeutics. Enzymes represent an important, and probably the best investigated, group of protein drugs. Their clinical use already has a rather long history [135–137]. Certain diseases (usually inherited) connected with the deficiency of some lysosomal enzymes (so-called storage diseases) can be treated only by the administration of exogenous enzymes [138, 139]. Therapeutic enzymes include antitumor enzymes acting by destroying certain amino acids needed for tumor growth, enzymes for replacement therapy (usually digestive enzymes) for the correction of various insufficiencies of the digestive tract, enzymes for the treatment of lysosomal storage diseases, enzymes for thrombolytic therapy, antibacterial and antiviral enzymes, and hydrolytic and antiinflammatory enzymes.

Among the antitumor enzymes, the most frequently used L-asparaginase [140] hydrolyzes asparagine via deamination of the amino acid with the formation of aspartic acid, which inhibits the growth of asparagine-dependent tumors. As a result, L-asparaginase became a standard tool in the treatment of leukemia [141]. Other enzymes of interest include glutaminase, cysteine desulfatase, cysteine aminotransferase, cysteine oxidase, arginase, arginine deaminase, arginine decarboxylase [142], folate-degrading enzymes [143], ribonucleases (RNAases), and exonucleases [144]. Enzymes for the therapy of lysosmal storage diseases include glucocerebrosidase, various glucosidases, phenylalanine ammonia lyase, and some others [138, 139]. Therapy with blood clotting factor VIII can be useful for the treatment of the hemophilia A [145].

Enzyme therapy includes also a thrombolytic therapy that uses many different enzymes – fibrinolysin (plasmin), streptokinase, urokinase (pro-urokinase), tissue plasminogen activator [146–148]. The application of enzymes detoxifying free radical oxygen derivatives (superoxide dismutase or SOD, and catalase) is recommended for different pathological conditions, including the damage caused by post-ischemic reperfusion [149]. Other enzymes considered as promising therapeutic agents include: elastase, which can be used in the treatment of arthritis [150]; pronase for the treatment of spleen and liver diseases [151]; collagenase and its mixture with hyaluronidase for the treatment of spleen and liver diseases, as well as for the treatment of postoperative scars [152]; mixtures of proteases with RNAase and deoxyribonuclease (DNAase) showed good results in the prevention and treatment of postoperative pneumonia; lysozyme is highly effective against viral diseases, including hepatitis, and in stomatology for caries treatment [135–137].

Peptide hormones, first of all insulin, are among the most broadly used drugs. More recently, peptides such as somatostatin analogs (octretide, lanreotide, vapreotide) have become available in the clinic for the treatment of pituitary and gastrointestinal tumors [153]. Peptide inhibitors of angiogenesis, including endostatin, are currently in different stages of clinical trials and show a great promise for cancer treatment [154, 155]. Research on depsipeptides has also revealed a set of potential anticancer agents [156].

The use of protein and peptides as therapeutic agents is, however, hampered by their low stability at physiological pH values and temperatures. Different processes leading to the inactivation of various biologically active proteins and peptides in vivo include: conformational intramolecular protein transformation into inactive conformation from the effect of temperature, pH, high salt concentration, or detergents; the dissociation of protein subunits or, in the case of enzymes, enzyme–cofactor complexes, and the association of protein or peptide molecules with the formation of inactive associates; noncovalent complexation with ions or low-molecular-mass and high-molecular-mass compounds, affecting the native structure of the protein or peptide; proteolytic degradation under the action of endogenous proteases; chemical modification by different compounds in solution (e.g., oxidation of -SH groups in sulfhydryl enzymes and Fe (II) atoms in heme-containing proteins by oxygen; and thiol–disulfide exchange, and destruction of labile side groups such as tryptophan and methionine). A very important point is the immune response of the macroorganism to foreign proteins containing different antigenic determinants.

One of the most popular and well-elaborated technologies to stabilize protein and peptide drugs and improve their pharmacological properties is their incorporation into liposomes. The improved properties of the liposomal proteins and peptides include: (1) increased stability and prolonged activity in the body; (2) decreased immunogenicity and affinity to specific antibodies which permits to administer repeated therapy; (3) decreased affinity to natural inhibitors; (4) the possibility of administering the whole therapeutic dose of a protein/peptide drug in a single injection; and (5) a decrease in the total quantity of a protein/peptide drug needed for the treatment, which makes the treatment more economical.

3.4.2 Liposomal Enzymes and Proteins

From the clinical point of view, the potential ability of liposome-encapsulated enzymes to enter the cytoplasm or lysosomes of live cells is of primary importance for the treatment of inherited diseases caused by the abnormal functioning of some intracellular enzymes and subsequent accumulation of non-degraded substrates in the lysosomes or cytoplasm of affected cells. Despite relative rarity of these diseases, together they pose a serious medical problem [139, 157]. The attempts to treat these diseases by the direct infusion of the deficient or malfunctioning enzyme were unsuccessful. Intravenous administration of purified hexose aminidase has been tried in the treatment of Tay–Sachs disease [158]. The enzyme was thought to enter the affected cells of the central nervous system via endocytosis and to destroy the ganglioside GM_2 that accumulated in cell lysosomes. It was found that endocytosis could not provide sufficient amounts of the enzyme inside the cells; furthermore, predominant accumulation of hexose aminidase was discovered in the liver instead of in target central nervous system (CNS) cells. Similar problems are linked to unsuccessful attempts with purified glucocerebrosidase and ceramide trihexosidase [159, 160].

The use of liposome-immobilized enzymes instead of their native precursors opens new opportunities for enzyme therapy [161, 162], especially in the treatment of diseases localized in liver cells that are natural targets for liposomes. Thus, the biodistribution of liposomes made of phosphatidylcholine, phosphatidic acid, and cholesterol (in 7 : 1 : 2 molar ratio) and containing β-fructofuranosidase has been studied [163]. It was shown that, within an hour, 50% of the administered enzyme can still be found in the circulation, but, after 6 hours, 45% of the enzyme activity accumulated in the liver. The enzyme preserves its activity for a long time – 25% of the administered activity can still be found in the liver after 48 hours. Even after 100 hours, 5% of the active enzyme is still present. It is very important that up to 50% of intracellular enzyme activity be localized in the lysosomal fraction. Similar data have been obtained for intravenously administered liposome-encapsulated α-mannosidase [164] and neuraminidase [165].

β-Glucuronidase, immobilized into charged liposomes composed mainly from phosphatidylcholine dipalmitoyl, also demonstrated fast accumulation in the liver of experimental mice. The enzyme remained active for more than a week, associated with the lysosomes of liver cells [166]. After lysosome overloading with dextran administered intravenously into mice, the dextran content in the liver of control mice remained unchanged for 6 days, whereas rats treated with liposome-immobilized dextranase (intravenous injection) demonstrated the decrease in the dextran content by 70% in 2 days [167]. Glucocerebroside β-glucosidase of human origin was encapsulated into neutral liposomes made of egg lecithin [168]. The efficiency of enzyme entrapment into liposomes is very high, the latency of the encapsulated enzyme being more than 95%. The ability of liposome-immobilized β-galactosidase to degrade GM_1 ganglioside in lysosomes of feline fibroblasts, with pathological accumulation of this substrate, has been demonstrated [169]. β-Galactosidase-containing liposomes obtained from a mixture of lecithin, cholesterol, and sulfatide (in 7:2:1 molar ratio) are readily incorporated into the liver and spleen of mice hosting the model of globoid cell leukodystrophy, The liposome-encapsulated therapeutic enzyme, administered as a single injection after preliminary injection of liposomes with galactocerebroside into experimental mice, causes the breakdown of 70–80% of intracellular galactocerebroside [170].

Animal experiments have also clearly demonstrated the suitability of liposomes for immobilization of enzymes used for the therapy of cancer and some other diseases. L-Asparaginase used for the treatment of asparagine-dependent tumors is a good example. Thus, the increase in the circulation half-life of the liposomal L-asparaginase and the decrease in its antigenicity and susceptibility towards the proteolytic degradation, together with the increase in the efficacy of experimental tumor therapy in mice, have been shown [171]. The longevity of the liposomal L-asparaginase (in non-long-circulating liposomes) also depends on the liposome size: in large liposomes (about 1200 nm) the circulation time was decreased, whereas in small liposomes (about 170 nm) it was prolonged tenfold compared with free enzyme [172]. The use of the liposome-encapsulated asparaginase improves the survival of animals with P1534 tumors compared with free enzyme. It is also important that the encapsulation into liposomes prevents the production of anti-asparaginase antibodies. Palmitoyl-L-asparaginase was also incorporated into liposomes and demonstrated prolongation of blood life by almost tenfold, a decrease in acute toxicity and improved antitumor activity in vivo [173].

Superoxide dismutase encapsulated into liposomes demonstrated improved pharmacokinetics including longer plasma half-life; it did not provoke acute or delayed toxic effects [174]. Liposome-entrapped SOD reduces ischemia–reperfusion oxidative stress in gerbil brains on intraperitoneal bolus injection by increasing enzyme activity and decreasing membrane peroxidation in various regions of the brain [175]. Liposomes can also be used for the transmembrane intracellular delivery of SOD and catalase [176, 177]. This is extremely important for the elimination of oxygen-derived free radicals inside cells, because their increased generation causes the toxic effect of ischemia on endothelial and many other cells. Using liposomes, intracellular SOD activity can be increased by 15-fold. Improved delivery of SOD to pulmonary epithelium via pH-sensitive liposomes [178] was demonstrated in vitro using cultured cells, and receptor-mediated endocytosis was shown to be at least partially responsible.

Experimental thrombolytic therapy with the liposome-incorporated tissue-type plasminogen activator in rabbits with jugular vein thrombosis clearly demonstrated the benefits of the liposomal enzyme over the native one: four times lower dose of the liposomal enzyme was required to provide the same degree of the lysis [179].

The use of liposomes for the transfer of therapeutic enzymes through the 'blood–brain' barrier, which allows delivery of these enzymes into cells of the CNS, also seems very attractive. It has already been shown that horse radish peroxidase, encapsulated into liposomes made of phosphatidylcholine, cholesterol, and phosphatidic acid [7:2:1 molar ratio), acquires the ability to cross the hematoencephalic barrier, whereas the native enzyme cannot. The presence of peroxidase in brain cells has been proved by histochemical methods [180].

From the practical point of view, new artificial oxygen-transporting systems capable of prolonged activity in the circulation are of special interest. The topic has been extensively elaborated on; we mention only several examples: where natural hemoglobin was incorporated into liposomes of different composition (so-called hemosomes). Thus, it was shown that the maximal quantity of hemoglobin obtained from lysed erythrocytes incorporates into negatively charged liposomes [181]. Kato and Tanaka [182] have stabilized hemosomes with carboxymethylchitin. Stabilized hemosomes bind oxygen in the same way as human blood hemolysates. The acute toxicity of hemosomes was moderate – in mice, the LD_{50} was 13.8 mL hemosomes/kg weight. To make long-circulating hemosomes, the technology was developed of PEG post-insertion, and the resulting liposomes do not lose any hemoglobin and circulate longer in rabbits [183]. Pegylated liposomal hemoglobin was found to be stable on storage for 1 year even at room temperature [184] and to circulate for a long time in rabbits when labeled with ^{99m}Tc (half-clearance time of 48 h) [185]. A good microvascular perfusion was achieved with the liposomal hemoglobin in hamsters [186]. Hemoglobin vesicles suspended in recombinant human serum albumin helped to treat hemorrhagic shock in rats [187]. However, some side effects were found for PEG hemosomes. Thus, they were phagocytosed by human peripheral blood monocytes and macrophages via the opsonin-independent pathway [188]. In addition, some studies show complement activation upon administration of pegylated Hb liposomes [189].

Eventually, it was found that the encapsulation into liposomes could affect certain properties of proteins or even cause certain undesirable side effects. Thus, liposomes can cause undesirable changes in hemodynamics, including immediate hypersensitivity and cardiopulmonary distress [190] – this was shown in pigs receiving the liposomal hemoglobin (complement activation-related pseudoallergy). The liposomal DNAase I can provoke neoplastic transformation in embryonal Syrian hamster cells in culture [191]. It was also shown that the liposomal membrane can influence catalytic properties of the liposome-immobilized β-D-galactoside α2→3-sialyltransferase [192]. The probability of similar effects should always be taken into account with liposome-immobilized enzymes (especially membrane enzymes). Special attention has also been paid to the immunological properties of liposome-immobilized enzymes. The ability of liposomes to demonstrate adjuvant properties in some cases is already well known. Hudson et al. [193] demonstrated the possibility of enhanced immune response development in mice receiving intravenous liposomes with immobilized bovine β-glucuronidase, despite the expected protection of the intraliposomal enzyme from contact with immunocompetent cells.

3.4.3 Liposomal Peptides

Liposomes have also been repeatedly used favorably to modify biological activity of various peptides. Thus, earlier studies on the incorporation of insulin into liposomes were reviewed in Sprangler [194]. Later, the liposomal insulin was used for intratracheal administration [195]. It was shown that insulin incorporation into liposomes made of dipalmitoylphosphatidyl choline and cholesterol [7:2] resulted in an improved pulmonary uptake of insulin in rats and enhanced the hypoglycemic effect. The attempt to improve the bioavailability of the oral liposomal insulin by coating insulin-containing liposomes with chitosan for better mucoadhesion in the gastrointestinal (GI) tract turned out to be successful in rats and resulted in an efficient and long-lasting lowering of glucose level [196]. Similar results have been also obtained with insulin-containing liposomes coated with PEG or mucin [197]. It was explained by better interaction of polymer-coated liposomes with the mucus layer and better retention of insulin under aggressive conditions of the stomach and GI tract.

The efficiency of the oral administration of the liposomal insulin in liposomes of different phospholipid composition was also confirmed in Kisel et al. [198]. However, high variability of effects on the oral administration of the liposomal insulin still represents a challenge. Buccal delivery of the liposomal insulin, which showed encouraging results in rabbit experiments [199], might represent an interesting alternative. The pharmacodynamics of insulin in PEG-coated liposomes on intravenous administration was studied in rats [200]. Pegylated liposomes provided the strongest and longest decrease in the glucose level, supporting the hypothesis on slow release of the hormone from liposomes in the blood.

Cytokines were also frequent candidates for liposomal dosage forms that have been expected to extend their lifetime in the body. Thus, the incorporation of recombinant interleukin-2 into liposomes increased its blood circulation time by eightfold [201]. Asialofetuin liposomes were shown to deliver efficiently human recombinant interferon-γ into hepatocytes in vitro [202]. Liposomal preparations of granulocyte–macrophage colony-stimulating factor (GM-CSF) and tumor necrosis factor α (TNF-α) demonstrated improved pharmacokinetics and biological activity on the background of reduced toxicity in experiment in mice [203]. Liposomal muramyl tripeptide was successfully used in patients with relapsed osteosarcoma [204]. Mannosylated liposomes with muramyl dipeptide significantly inhibited liver metastases in tumor-bearing mice [205]. PEG-coated liposomes have also been proposed for the oral delivery of recombinant human EGF [206]. Liposomal delivery of the peptide inhibitor of the transcription factor nuclear factor-κB (NF-κB) was shown to inhibit the proliferation of vascular smooth muscle cells significantly [207]. Liposomal recombinant human TNF strongly suppressed parasitemia and protects against *Plasmodium berghei* k173-induced experimental cerebral malaria in mice [208].

The possibility of the topical delivery of the liposomal interferon was considered [209] and the details of the dermal penetration of the liposomal interferon-γ as the key role of the transfollicular route were investigated [210]. Topical delivery of growth hormone-releasing peptide was suggested in mice [211]; such liposomes for peptide delivery may be further improved by modification with hyaluronic acid, which increases their bioadhesion [212]. Topical delivery of the liposomal enkephalin was demonstrated in Vutla et al. [213], confirming the earlier finding that nonionic liposomes facilitate topical delivery of peptide drugs [214]. The incorporation of ciclosporin into liposomes of various compositions was shown to minimize the toxic side effects associated with traditional intravenous formulations of ciclosporin, and to maintain good drug activity in dogs [215]. Topical delivery of the liposomal ciclosporin in murine model was also described [216].

Leupeptide (the tripeptide inhibitor of proteolytic enzymes) can be delivered into the brain by means of liposomes obtained by reverse phase evaporation from a mixture of lecithin, cholesterol, and sulfatide [4:5:1 molar ratio) [217]. The inhibitor was used for the treatment of experimental allergic encephalomyelitis in guinea-pigs. The therapeutic efficiency of the liposomal inhibitor was very high as estimated by the histopathology data and survival of experimental animals. Liposome-encapsulated inhibitors of aldose reductase (quercitine, quercitrine, AU22–284, and sorbinyl) have been successfully delivered into the ocular lens [218, 219].

3.4.4 Liposomal ATP

An interesting example of how the ability of the liposome to change the biological properties of an entrapped drug relates to liposomal ATP. As, in a free form, ATP is a subject of immediate hydrolysis by a variety of free and cell-bound nucleases, there is a certain interest in liposomal forms of ATP, and some encouraging results with ATP-loaded liposomes in various in vitro and in vivo models have been reported. Thus, ATP liposomes were shown to protect human endothelial cells from the energy failure in a cell culture model of sepsis [220]. In a brain ischemia model, the use of the liposomal ATP increased the number of ischemic episodes tolerated before there was brain electrical silence and death [221].

In a hypovolemic shock–reperfusion model in rats, the administration of ATP liposomes provided an effective protection to the liver [222]. ATP liposomes also improved rat liver energy state and metabolism during cold storage preservation [223]. Similar properties were also demonstrated for the liposomal coenzyme Q10 [224, 225]. Interestingly, biodistribution studies with ATP liposomes demonstrated their significant accumulation in the damaged myocardium [226]. Recently, ATP-loaded liposomes were shown effectively to preserve the mechanical properties of the heart under ischemic conditions in an isolated rat heart model [227, 228] and in rabbits with experimental myocardial infarction [229]. As shown in these studies, the main role of the liposomal carrier is the protection of the liposomal ATP from hydrolysis by the biological surroundings and improved intracellular delivery of ATP into hypoxic cells. ATP-loaded immunoliposomes were also prepared possessing specific affinity towards myosin, i.e. capable of specific recognition of hypoxic cells [230].

3.4.5 Liposomes as Adjuvants

Liposomes have also been shown to be effective immunological adjuvants for protein and peptide antigens (see summary of numerous studies in Gregoriadis [231] and Friede [232]. They are capable of inducing both humoral and cellular immune responses towards the liposomal antigens. Liposomes with encapsulated protein or peptide antigen are phagocytosed by macrophages and eventually end up in lysosomes. There, proteins and peptides are degraded by the lysosomal enzymes, and their fragments are then presented on the macrophage surface, being associated with major histocompatibility complex (MHC) II. This results in the stimulation of specific T-helper cells and, via lymphokine secretion and interaction of T cells with B cells that captured free antigen, stimulation of specific B cells and subsequent secretion of antibodies [232]. In some cases, however, a fraction of the liposomal antigen can escape from endosomes into the cytoplasm (e.g., when pH-sensitive liposomes are used) and in this case the liberated antigen is processed and presented being associated with MHC I, thus inducing cytotoxic T lymphocytes (CTL response). The ability to induce the CTL response provides liposomes with certain benefits when compared with traditional adjuvants (such as Freund's adjuvant) that do not induce any significant CTL response.

The variety of protein antigens has been incorporated into liposomes (e.g., diphtheria toxoid [233], hepatitis B antigens [234, 235], influenza virus antigens [236, 237], tumor-associated antigens [238], and many others – see quite a few examples in Gregoriadis [231]. A pronounced immunoadjuvant effect of liposomes can also be seen when proteins (enzymes) or other immunogens are bound to the outer surface of liposomal membranes [239].

Liposomal antigens have been used to enhance the mucosal immune response. Thus, the colonic/rectal IgA response to liposomal ferritin was significantly enhanced over the response to free antigen when cholera toxin was used as adjuvant [240]. The protective efficiency of the 30-kDa secretory protein of *Mycobacteriun tuberculosis* H37Ra against TB in mice was significantly enhanced by incorporating this protein into liposomes serving as adjuvant [241]. Synthetic human MUC1 peptides, which are considered to be candidates for therapeutic cancer vaccines, were incorporated into liposomes or attached to the surface of liposomes; in both cases they elicited strong antigen-specific T responses [242]. Formaldehyde-inactivated ricin toxoid in liposomes was used for intrapulmonary immunization to create the protection against inhaled ricin with good results [243]. Liposomal composition incorporating an antennapedia homeodomain fused to a poorly immunogenic CTL epitope increased the immunogenicity of the construct and improved immune response (activation of CD8+ T cells), evidently because of protection of the antigen by liposomes [244]. Cytokine-containing liposomes have been used as vaccine adjuvants [245].

Oral delivery of antigens in liposomes (ovalbumin was used as a model antigen) effectively induced oral tolerance [246]. Liposomes with the surface-attached recombinant B subunit of cholera toxin were shown to be an effective oral antigen delivery system [247]. In recent developments, liposomes were successfully used for the delivery of peptide vaccines and CTL epitopes to dendritic cells, improving immune response [248–250]. Various approaches to deliver liposomal proteins to the cytoplasm and Golgi of antigen-presenting cells have recently been reviewed [251].

3.5 LIPID-CORE MICELLES AS PHARMACEUTICAL CARRIERS FOR POORLY SOLUBLE DRUGS

Another interesting example of the favorable modification of drug properties with lipid-based pharmaceutical nanocarriers relates to the use of lipid-core micelles.

3.5.1 Poorly Soluble Drugs

The availability of biocompatible and biodegradable drug carriers possessing small particle size, high loading capacity, extended circulation time, and ability to accumulate in required pathological sites in the body, and capable of carrying poorly soluble pharmaceuticals is especially important based on the fact that therapeutic application of hydrophobic, poorly water-soluble agents is associated with various serious problems. Low water solubility results in poor absorption and low bioavailability, especially on the oral administration [252]. In addition, the aggregation of poorly soluble drugs on intravenous administration might lead to various complications, including embolism which result in side effects as severe as respiratory system failure [253, 254], and can also lead to high local drug concentrations at the sites of aggregate deposition; these could be associated with local toxic effects of the drug and its diminished systemic bioavailability [255]. On the other hand, the hydrophobicity and low solubility in water seem to be intrinsic properties of many drugs (including anticancer agents, many of which are bulky polycyclic compounds, such as camptothecin, paclitaxel, or tamoxifen) [256], because it helps a drug molecule to penetrate a cell membrane and reach important intracellular targets [257, 258]. It was also observed that a drug or, in a more general case, a biologically active molecule may need a lipophilic group to acquire a sufficient affinity toward the appropriate target receptor [252, 259].

To overcome the poor solubility of some drugs certain clinically acceptable organic solvents are used in formulations, such as ethanol or Cremophor EL (polyethoxylated castor oil) [255]. More recent approaches include the use of liposomes [260] and cyclodextrins [261]. The administration of many co-solvents or surfactants causes toxic or other undesirable side effects [262]. The use of liposomes and cyclodextrins demonstrated some promising results with certain poorly soluble drugs, although the capacity of the liposomal membrane or cyclodextrin inner cavity for water-insoluble molecules is rather limited. Another option is to use certain micelle-forming surfactants in formulations of insoluble drugs [255].

3.5.2 Micelles and Micellization

Micelles represent so-called colloidal dispersions that belong to a large family of dispersed systems consisting of particulate matter or dispersed phase, distributed within a continuous phase or dispersion medium. In terms of size, colloidal dispersions occupy a position between molecular dispersions with particle size under 1 nm and coarse dispersions with particle size greater than 0.5 µm. More specifically, micelles normally have particle size within range of 5–100 nm. The concentration of a monomeric amphiphile at which micelles appear is called the critical micelle concentration (CMC). Hydrophobic fragments of amphiphilic molecules form the core of a micelle, which can solubilize poorly soluble pharmaceuticals [263]. This solubilization phenomenon was extensively investigated and reviewed in many publications [264, 265]. In aqueous systems, nonpolar molecules will be solubilized within the micelle core, polar molecules will be adsorbed on the micelle surface, and substances with intermediate polarity will be distributed along surfactant molecules in certain intermediate positions.

As drug carriers, micelles provide a set of clear advantages [266–269]. The solubilization of drugs using micelle-forming surfactants (which results in formation of mixed micelles) causes an increased water solubility of sparingly soluble drug and its improved bioavailability, reduction of toxicity and other adverse effects, enhanced permeability across the physiological barriers, and substantial changes in drug biodistribution. The use of certain special amphiphilic molecules as surfactants can also introduce the property of micelle-extended blood half-life on intravenous administration. Besides, micelles may be become targets by chemical attachment of a targeting moiety to their surface. On the other hand, being in a micellar form en route to the target organ or tissue, the drug is well protected from possible inactivation under the effect of biological surroundings, and does not itself provoke undesirable side effects on nontarget organs and tissues. A very important property of micelles is their size, which normally varies between 5 and 50–100 nm and fills the gap between such drug carriers as individual macromolecules (antibodies, albumin, dextran) with the size below 5 nm, and nanoparticulates (liposomes, microcapsules) with a size of about 50 nm and above. According to the analysis of the available literature, the most usual size of a pharmaceutical micelle is between 10 and 80 nm, optimal CMC value being expected to be in a low millimolar region or even lower, and the loading efficacy towards a hydrophobic drug should be between 5 and 25% wt.

3.5.3 Polymeric Micelles

Polymeric micelles represent a class of micelles and are formed from block copolymers consisting of hydrophilic and hydrophobic monomer units. It has been repeatedly shown that amphiphilic block AB-type copolymers, with the length of a hydrophilic block exceeding to some extent that of a hydrophobic one, can form spherical micelles in aqueous solutions. The particulates are composed of the core of the hydrophobic blocks stabilized by the corona of hydrophilic polymeric chains. If the length of a hydrophilic block is too high, copolymers exist in water as unimers (individual molecules), whereas molecules with very long hydrophobic block form a structure with non-micellar morphology, such as rods and lamellae [32d].

The core compartment of the pharmaceutical polymeric micelle should demonstrate high loading capacity, controlled-release profile for the incorporated drug, and good compatibility between the core-forming block and incorporated drug. The micelle corona should provide an effective steric protection for the micelle. It should also determine, for the micelle, hydrophilicity, charge, the length and surface density of hydrophilic blocks, and the presence of reactive groups suitable for further micelle derivatization, such as an attachment of targeting moieties [271–275]. These properties control important biological characteristics of a micellar carrier, such as its pharmacokinetics, biodistribution, biocompatibility, longevity, surface adsorption of biomacromolecules, adhesion to biosurfaces, and targetability [271–274, 276–278]. In most cases the structure of amphiphilic unimers follows some simple regulations: PEG blocks with a molecular mass from 1 to 15 kDa are the usual corona-forming blocks, and the length of a hydrophobic core-forming block is close or somewhat lower than that of a hydrophilic block [279]. Although some other hydrophilic polymers may be used to make corona blocks [289, 281], PEG still remains the hydrophilic block of choice. At the same time, a variety of polymers may be used to build hydrophobic core-forming blocks: propylene oxide [273, 282], L-lysine [283, 284], aspartic acid [285, 286], β-benzoyl-L-aspartate [287, 288], γ-benzyl-L-glutamate [289], caprolactone [290, 291], D,L-lactic acid [274, 292], and many others.

3.5.4 Lipid-core Micelles

In some cases, phospholipid residues – short, however, and extremely hydrophobic as a result of the presence of two long-chain fatty acyl groups – can also be successfully used as hydrophobic core-forming groups [293]. The use of lipid moieties as hydrophobic blocks capping hydrophilic polymer (such as PEG) chains can provide additional advantages for particle stability when compared with conventional amphiphilic polymer micelles, as a result of the existence of two fatty acid acyls that might contribute considerably to an increase in the hydrophobic interactions between the polymeric chains in the micelle's core. Similar to other PEG-containing amphiphilic block copolymers, diacyl-lipid–PEG conjugates (such as PEG–PE) were found to form very stable micelles in an aqueous environment [27, 294]. Their CMC values can be as low as 10^{-6} mol/L [267, 269], which is at least 100-fold lower than those of conventional detergents [262], so that micelles prepared from these polymers will maintain their integrity even on strong dilution (e.g., in the blood during a therapeutic application). The high stability of polymeric micelles allows for good retention of encapsulated drugs in the solubilized form on parenteral administration.

Chemically, PEG–PE molecules consist of hydrophilic, flexible PEG fragments conjugated with a very hydrophobic diacyl phospholipid parts. The ability of PEG–PE molecules to form micelles in an aqueous environment was observed as early as 1994 [32], since, on attempts to prepare long-circulating liposomes, it has been noticed that phospholipid/PEG–PE mixtures of certain compositions form micelles instead of liposomes if the PEG–PE content exceeds a certain critical limit [295, 296]. This phenomenon was considered as a nuisance until it was realized that PEG–PE and similar micelles have a substantial potential as a particulate carrier for the delivery of therapeutic and diagnostic agents [297].

The micelles made of such lipid-containing conjugates can be loaded with various poorly soluble drugs (paclitaxel, camptothecin, etc.) and demonstrate good stability, longevity, and the ability to accumulate in areas with damaged vasculature (enhanced permeability and retention, EPR, effect in leaky areas, such as infarcts and tumors) [293, 298, 299]. Mixed micelles made of PEG–PE and other micelle-forming components are described that provide even better solubilization of certain poorly soluble drugs as a result of the increase in the capacity of the hydrophobic core [300–302]. Certain PEG–PE-based mixed micelles may allow for an increased intracellular delivery of micelle-incorporated drugs [300].

Micelle preparation of lipid–polymer conjugates is a simple process, because, similar to conventional detergents, such polymers, including PEG–PEs with a molecular mass of PEG chains from 750 to 5000 Da, form micelles spontaneously on shaking a dry polymer lipid film in the presence of an aqueous medium. All versions of PEG–PE conjugates form micelles with the size 7–35 nm. Micelles formed from conjugates with polymer (PEG) blocks of higher molecular mass have a slightly larger size, indicating that the micelle size may be tailored for a particular application by varying the length of PEG. Preparation of lipid-based micelles by a detergent or water-miscible solvent removal method results in formation of particles with very similar diameters.

3.5.5 Drug-loaded Micelles

The simplest and most convenient technique for the preparation of drug-loaded PEG–PE micelles involves simple dispersing a dry PEG–PE/drug mixture in an aqueous buffer. A typical protocol for the preparation of drug-loaded micelles includes the following steps.

Solutions of PEG–PE and a drug of interest in miscible volatile organic solvents are mixed, and organic solvents are evaporated to form a PEG–PE/drug film. The film obtained is then hydrated in the presence of an aqueous buffer and the micelles are formed by intensive shaking. If the amount of a drug exceeds the solubilization capacity of micelles, the excess drug precipitates in a crystalline form and the precipitated crystals are removed by filtration. The partial list of poorly soluble therapeutic agents incorporated into PEG–PE micelles is presented in Table 3.2. The loading efficiency for different compounds varies from 1.5 to 50% by weight. This value apparently correlates with the hydrophobicity of a drug. As already mentioned, in some cases, to improve drug solubilization, additional mixed micelle-forming compounds may be added to PEG–PE micelles. Thus, to increase the encapsulation efficiency of paclitaxel, egg phosphatidylcholine (PC) was added to the micelle composition, which approximately doubled the paclitaxel encapsulation efficiency (from 15 to 33 mg drug/g micelle-forming material [302–304]. This may be explained by the fact that egg PC, unlike PEG–PE, does not have a large hydrophilic PEG domain, and its addition into micelle composition results in particles with higher hydrophobic content [305]. Paclitaxel in mixed PEG–PE/egg PC micelles demonstrated high cytotoxic activity against MCF-7 human mammary adenocarcinoma cells [304].

TABLE 3.2 Some lipid-containing co-polymers used to prepare lipid-core micelles loaded with various pharmaceuticals

Block co-polymer	Micelle-incorporated pharmaceuticals
PEG-PE	Dequalinium
	Soybean trypsin inhibitor
	Paclitaxel
	Camptothecin
	Tamoxifen
	Porphyrine
	Vitamin K_3
	^{111}In (via DTPA-PE, diagnostic)
	Gd (via DTPA-PE, diagnostic)
PEG–PE/egg phosphatidylcholine (mixed micelles)	Paclitaxel
	Camptothecin
Various polymer–lipid conjugates	Corticosteroids
	Sulfonylbenzoylpiperazine

PE, phosphatidylethanolamine; PEG, polyethyleneglycol.

A drug incorporated into lipid-core polymeric micelles is associated with micelles: when PEG–PE micelles loaded with several drugs were dialyzed against aqueous buffer at sink conditions, all tested preparations retained more than 90% of encapsulated drug within first the 7 h of incubation. The micelles retain 95%, 75%, and 87% of initially incorporated chlorine 6-trimethyl ester, tamoxifen, and paclitaxel, respectively, even after 48 h incubation [303].

3.5.6 'Passive' Micelle Targeting

Targeting micelles to pathological organs or tissues can further increase pharmaceutical efficiency of a micelle-encapsulated drug. Thus, for example, in some cases, targeting could proceed spontaneously via the enhanced permeability and retention (EPR) effect [306]. Direct correlations between the longevity of a particulate drug carrier in the circulation and its ability to reach its target site have been observed on many occasions [5, 6]. The prolonged circulation provides a drug with a better chance to reach and/or interact with its target [5]. The results of the blood clearance study of various PEG–PE micelles clearly demonstrated their longevity: the micelle formulations studied had circulation half-lives from in mice, rats, and rabbits from 1.2 to 2.0 h depending on the molecular size of the PEG block [298]. The increase in the size of a PEG block increases the micelle circulation time in the blood, probably by providing a better steric protection against opsonin penetration to the hydrophobic micelle core. Still, circulation times for PEG–PE micelles are somewhat shorter compared with those for PEG-coated long-circulating liposomes [27], which could be explained in part by their more rapid extravasation of the micelles from the vasculature associated with their considerably smaller size compared with liposomes [307]. Slow dissociation of micelles under physiological conditions as a result of continuous clearance of unimers with a micelle–unimer equilibrium being shifted towards the unimer formation [284] can also play its role.

As with long-circulating liposomes [308–310], micelles formed by PEG_{750}–PE, PEG_{2000}–PE, and PEG_{5000}–PE accumulate efficiently in tumors via the EPR effect [54]. It is worth mentioning that micelles prepared from several different PEG–PE conjugates studied demonstrated much higher accumulation in tumors compared with non-target tissue (muscle) even in the case of an experimental Lewis lung carcinoma (LLC) in mice, known to have the relatively small vasculature cutoff size [307, 311]. In other words, because of their smaller size, micelles may have additional advantages as a tumor drug delivery system, which utilizes the EPR effect compared with particulate carriers with larger individual particle size. Thus, the micelle-incorporated model protein (soybean trypsin inhibitor or STI, M_r 21.5 kDa) accumulates to a higher extent in subcutaneously established murine LLC than the same protein in larger liposomes [307].

The accumulation pattern of PEG–PE micelles prepared from all versions of PEG–PE conjugates is characterized by the peak tumor accumulation times of about 3–5 h. The largest total tumor uptake of the injected dose 5 h post-injection (as AUC) was found for micelles formed by the unimers with relatively large PEG block (PEG_{5000}–PE). This may be explained by the fact that these micelles have the longest circulation time and a lesser extravasation into the normal tissue compared with micelles prepared from the smaller PEG–PE conjugates. Micelles prepared from PEG–PE conjugates with shorter versions of PEG might, however, be more efficient carriers of poorly soluble drugs because they have a greater hydrophobic : hydrophilic phase ratio and can be loaded with drug more efficiently on a weight : weight basis. Similar results have been obtained with another murine tumor model, EL4 T cell lymphoma [298]. Some other recent data also clearly indicate spontaneous targeting of PEG–PE-based micelles into other experimental tumors [301] in mice, as well as into the damaged heart areas in rabbits with experimental myocardial infarction [299].

Another targeting mechanism is based on the fact that many pathological processes in various tissues and organs are accompanied with local temperature increase and/or acidosis [212, 213]. So, the efficiency of the micellar carriers can be further improved by making micelles capable of disintegration under the increased temperature or decreased pH values in pathological sites, i.e. by combining the EPR effect with stimuli responsiveness. For this purpose, micelles are made of thermo- or pH-sensitive components, such as poly(*N*-isopropylacrylamide) and its copolymers with poly(D,L-lactide) and other blocks, and acquire the ability to disintegrate in target areas, releasing the micelle-incorporated drug [279, 314–317]. The pH-responsive polymeric micelles loaded with phthalocyanine seem to be promising carriers for the photodynamic cancer therapy [318], whereas doxorubicin-loaded polymeric micelles containing acid-cleavable linkages provided an enhanced intracellular drug delivery into tumor cells, and thus higher efficiency [319]. Thermoresponsive polymeric micelles were shown to demonstrate an increased drug release on temperature changes [320].

3.5.7 Ligand-targeted Micelles

As with other delivery systems, the drug delivery potential of polymeric micelles may be enhanced still further by attaching targeting ligands to the micelle surface. Among those ligands are various sugar moieties [321], transferrin [322], and folate residues [333], because many target cells, especially cancer cells, overexpress appropriate receptors (such as transferrin and folate receptors) on their surface. Thus, it was shown that galactose- and lactose-modified micelles made of PEG–polylactide co-polymer specifically interact with lectins, thus modeling targeting delivery of the micelles to hepatic sites [321, 324]. Transferrin-modified micelles based on PEG and polyethyleneimine with sizes between 70 and 100 kDa are expected to target tumors with overexpressed transferrin receptors [322]. Mixed micelle-like complexes of pegylated DNA and PEI modified with transferrin [325, 326] were designed for the enhanced DNA delivery into cells overexpressing the same transferrin receptors. A similar targeting approach was successfully tested with folate-modified micelles [327]. Poly(L-histidine)/PEG and poly(L-lactic acid)/PEG block co-polymer micelles, carrying folate residue on their surface, were shown to be efficient for the delivery of doxorubicin to tumor cells in vitro, demonstrating a potential for solid tumor treatment and combined targetability and pH-sensitivity [328].

Among all specific ligands, antibodies provide the broadest opportunities in terms of diversity of targets and specificity of interaction. Several attempts to attach an antibody covalently to a surfactant or polymeric micelles (i.e. to prepare immunomicelles) have been described [269, 273, 301, 322]. Thus, micelles modified with fatty acid-conjugated Fab fragments of antibodies to antigens of brain glial cells (acid gliofibrillar antigen and α_2-glycoprotein) loaded with neuroleptic trifluoperazine increasingly accumulated in the rat brain on intracarotid administration [273, 329].

By adapting the coupling technique developed for attaching specific ligands to liposomes [45], we have prepared PEG–PE-based immunomicelles modified with monoclonal antibodies. The approach uses PEG–PE with the free PEG terminus activated with a *p*-nitrophenylcarbonyl (pNP) group. Diacyl-lipid fragments of such bifunctional PEG derivatives firmly incorporate into the micelle core, whereas the water-exposed pNP group, stable at pH values <6, efficiently interacts with amino groups of various ligands (such as antibodies and their fragments) at pH values >7.5, yielding a stable urethane (carbamate) bond. To prepare immunotargeted micelles, the antibody to be attached was simply incubated with drug-loaded micelles at pH around 8.0. The micelle-attached protein was quantified using fluorescent labels or by SDS-PAGE (sodium docecylsulfate polyacrylamide gel electrophoresis) [301, 304]. It was calculated that 10–20 antibody molecules could be attached to a single micelle. Antibodies attached to the micelle corona preserve their specific binding ability, and immunomicelles specifically recognize their target substrates, as was confirmed by ELISA (enzyme-linked immunosorbent assay) with corresponding substrate monolayers.

To enhance the tumor accumulation specifically of PEG–PE-based micelles, the latter have been modified with tumor-specific monoclonal antibodies [301, 304] Although anticancer antibodies are usually tumor type specific and unable to react with different tumors, earlier we have shown that certain nonpathogenic monoclonal antinuclear autoantibodies with nucleosome-restricted specificity (monoclonal antibody 2C5 or mAb2C5 being among them) recognize the surface of numerous tumor, but not normal, cells via tumor cell surface-bound nucleosomes [330, 331]. As these antibodies bind a broad variety of cancer cells, they may serve as specific ligands for the delivery of drugs and drug carriers into tumors. Fluorescently labeled empty and paclitaxel-loaded 2C5 immunomicelles have been shown to bind effectively to the surface of many unrelated tumor cells lines. Such specific recognition of cancer cells by drug-loaded immunomicelles results in dramatically improved cancer cell killing by such micelles. At paclitaxel concentrations of 40 ng/mL, free drug kills less than 2% of MCF7 cells in culture. The activity of paclitaxel in plain micelles was slightly higher (around 5% cell killing). At the same time, paclitaxel-loaded immunomicelles kill more than 50% of cancer cells.

In vivo experiments with LLC tumor-bearing mice revealed a dramatically enhanced tumor uptake of paclitaxel-loaded, radiolabeled 2C5 immunomicelles compared with nontargeted micelles. An enhanced accumulation of 2C5-targeted micelles over plain micelles in the tumor (up to 30%) was observed both at 30 min and at 2 h after injection, evidence of the specific recognition and tumor binding of 2C5-targeted immunomicelles. These data suggest the possibility that drug-loaded immunomicelles may also be better internalized by tumor cells, similar to antibody-targeted liposome [50]; thus, they deliver more drug inside tumor cells than might be achieved in cases of simple EPR effect-mediated tumor accumulation. By analyzing the absolute quantity of tumor-accumulated paclitaxel delivered by different drug formulations, it was shown that mAb 2C5 immunomicelles were capable of bringing into tumors substantially higher quantities of paclitaxel than in the case of paclitaxel-loaded nontargeted micelles or free drug formulation, which results in much stronger tumor growth inhibition [301].

3.5.8 Lipid-core Micelles for Intracellular Drug Delivery

In addition to using targeting ligands (such as mAb 2C5 described above), one may try further to improve the efficiency of drug-loaded micelles by enhancing their intracellular delivery, thus compensating for an excessive drug degradation in lysosomes as a result of the endocytosis-mediated capture of therapeutic micelles by cells. In some cases this could be achieved by controlling the micelle charge. The net positive charge usually enhances the uptake of various nanoparticles by cells. It is now believed that endocytosis is the major mechanism for the intracellular delivery of DNA-loaded particles made of positively charged lipids. Cationic lipid formulations such as Lipofectin (an equimolar mixture of *N*-[1-(2,3-dioleyloxy)propyl]-*N,N,N*-trimethylammonium chloride or DOTMA, and dioleoyl phosphatidylethanolamine or DOPE), noticeably improve the endocytosis-mediated intracellular delivery of various drugs and DNA entrapped in liposomes and other lipid constructs made of these compositions [332–335]. After endocytosis, drug-loaded particles could escape from the endosomes and enter a cell's cytoplasm through disruptive interactions of the cationic lipid with endosomal membranes [336]. PEG–PE micelles, on the other hand, have been found to carry a net negative charge [299], which might hinder their internalization by cells. The compensation of this negative charge by the addition of positively charged lipids to PEG–PE could improve the uptake by cancer cells of paclitaxel-loaded mixed PEG–PE/positively charged lipid micelles. It is also possible that, after enhanced endocytosis, drug-loaded mixed micelles made of PEG–PE and positively charged lipids could escape from the endosomes and enter the cytoplasm of cancer cells. With this in mind, the attempt was made to increase an intracellular delivery and, thus, the anticancer activity of the micellar paclitaxel by preparing paclitaxel-containing micelles from the mixture of PEG–PE and Lipofectin lipids (LLs).

The cell interaction (BT-20 breast adenocarcinoma cells were used in this case) and intracellular fate of paclitaxel-containing PEG-PE/LL micelles and similar micelles, prepared without the addition of the LLs, were investigated by fluorescence microscopy. Fluorescently labeled PEG–PE and PEG–PE/LL micelles were both endocytosed by BT-20 cells; however, in the case of PEG–PE/LL micelles, endosomes were shown to degrade fast and release drug-loaded micelles into the cell cytoplasm as a result of the destabilizing effect of the LL component on the endosomal membranes [337]. The increased cytoplasmic delivery of paclitaxel-loaded PEG–PE/LL micelles revealed by fluorescence microscopy resulted in their increased in vitro cytotoxicity against BT-20 cells and A2780 cells (human ovarian carcinoma). The in vitro anticancer effects of drug-loaded micelles were significantly improved for paclitaxel-containing PEG–PE/LLs compared with that of free paclitaxel or paclitaxel delivered using noncationic LL-free PEG–PE micelles: in A2780 cancer cells, the IC50 values of free paclitaxel, paclitaxel in PEG–PE micelles, and paclitaxel in PEG-PE/LL micelles were 22.5, 5.8, and 1.2 μmol/L, respectively. In BT-20 cancer cells, the IC_{50} values of the same preparations were 24.3, 9.5, and 6.4 μmol/L, respectively.

3.6 CONCLUSION

Lipid-based pharmaceutical nanocarriers, such as liposomes and lipid-core polymeric micelles can significantly improve biological characteristics of various water-soluble and water-insoluble drugs, including protein and peptide drugs, and increase their accumulation in pathological areas. These nanocarriers can also be made capable of specific targeting disease sites via variety of mechanisms and of intracellular delivery. As a result, therapeutic efficiency of many drugs can be strongly increased.

REFERENCES

1. Lasic, D.D. *Liposomes: From Physics to Applications*. Amsterdam: Elsevier Science Publishers; 1993.
2. Torchilin, V.P., Weissig, V. Polymeric micelles for the delivery of poorly soluble drugs. In: Park, K., Mrsny, R.J. (eds), *Controlled Drug Delivery. Designing technologies for the future American Chemical Society*. Washington, DC: American Chemical Society, 2000: pp 297–313.

3. Palmer, T.N., Caride, V.J., Caldecourt, M.A., Twickler, J., Abdullah, V. The mechanism of liposome accumulation in infarction, *Biochim Biophys Acta* 1984;**797**:363–368.
4. Maeda, H., Wu, J., Sawa, T., Matsumura, Y., Hori, K. Tumor vascular permeability and the EPR effect in macromolecular therapeutics: a review. *J Control Release* 2000;**65**:271–284.
5. Gabizon, A.A. Liposome circulation time and tumor targeting: implications for cancer chemotherapy. *Adv Drug Deliv Rev* 1995;**16**:285–294.
6. Maeda, H., Sawa, H.T., Konno, T. Mechanism of tumor-targeted delivery of macromolecular drugs, including the EPR effect in solid tumor and clinical overview of the prototype polymeric drug SMANCS. *J Control Release* 2001;**74**:47–61.
7. Jain, R.K. Transport of molecules, particles, and cells in solid tumors. *Annu Rev Biomed* Eng. 1999;**1**:241–263.
8. Yuan, F., Dellian, M., Fukumura, M., Leunig, M., Berk, D.A., Torchilin, V.P., Jain, R.K. Vascular permeability in a human tumor xenograft: Molecular size dependence and cutoff size. *Cancer Res* 1995;**55**:3752–3756.
9. Maeda, H. SMANCS and polymer-conjugated macromolecular drugs: advantages in cancer chemotherapy. *Adv Drug Deliv Rev* 2001;**46**:169.
10. Muggia, F.M. Clinical efficacy and prospects for use of pegylated liposomal doxorubicin in the treatment of ovarian and breast cancers. Drugs. 1997;**54**(suppl 4):22–29.
11. Valero, V., Buzdar, A.U., Theriault, R.L., Azarnia, N., Fonseca, G.A., Willey, J., et al. Phase II trial of liposome-encapsulated doxorubicin, cyclophosphamide, and fluorouracil as first-line therapy in patients with metastatic breast cancer. *J Clin Oncol* 1999;**17**:1425–1434.
12. Symon, Z., Peyser, A., Tzemach, D., Lyass, O., Sucher, E., Shezen, E., Gabizon, A. Selective delivery of doxorubicin to patients with breast carcinoma metastases by stealth liposomes. *Cancer* 1999;**86**:72–78.
13. Perez, A.T., Domenech, G.H., Frankel, C., Vogel, C.L. Pegylated liposomal doxorubicin (Doxil) for metastatic breast cancer: the Cancer *Res*earch Network, Inc., experience. *Cancer Invest* 2002;**20**(suppl 2):22–29.
14. O'Shaughnessy, J.A. Pegylated liposomal doxorubicin in the treatment of breast cancer. *Clin Breast Cancer* 2003;**4**:318–328.
15. Schwonzen, M., Kurbacher, C.M., Mallmann, P. Liposomal doxorubicin and weekly paclitaxel in the treatment of metastatic breast cancer. *Anticancer Drugs* 2000;**11**:681–685.
16. Goncalves, A., Braud, A.C., Viret, F., Genre, D., Gravis, G., Tarpin, C., et al. Phase I study of pegylated liposomal doxorubicin (Caelyx) in combination with carboplatin in patients with advanced solid tumors. *Anticancer Res* 2003;**23**:3543–3548.
17. Harrington, K.J., Lewanski, C., Northcote, A.D., Whittaker, J., Peters, A.M., Vile, R.G., Stewart, J.S. Phase II study of pegylated liposomal doxorubicin (Caelyx) as induction chemotherapy for patients with squamous cell cancer of the head and neck. *Eur J Cancer* 2001;**37**:2015–2022.
18. Johnston, S.R., Gore, M.E. Caelyx: phase II studies in ovarian cancer. *Eur J Cancer* 2001;**37**(suppl 9):S8–S14.
19. Schmidinger, M., Wenzel, C., Locker, G.J., Muehlbacher, F., Steininger, R., Gnant, M., et al. Pilot study with pegylated liposomal doxorubicin for advanced or unresectable hepatocellular carcinoma. *Br J Cancer* 2001;**85**:1850–1852.
20. Wollina, U., Dummer, R., Brockmeyer, N.H., Konrad, H., Busch, J.O., Kaatz, M., et al. Multicenter study of pegylated liposomal doxorubicin in patients with cutaneous T-cell lymphoma. *Cancer* 2003;**98**:993–1001.
21. Skubitz, K.M. Phase II trial of pegylated-liposomal doxorubicin (Doxil) in sarcoma. *Cancer Invest* 2003;**21**:167–176.
22. Seiden, M.V., Muggia, F., Astrow, A., Matulonis, U., Campos, S., Roche, M., et al. A phase II study of liposomal lurtotecan (OSI-211) in patients with topotecan resistant ovarian cancer. *Gynecol Oncol* 2004;**93**:229–232.
23. Sundar, S., Jha, T.K., Thakur, C.P., Mishra, M., Singh, V.P., Buffels, R. Single-dose liposomal amphotericin B in the treatment of visceral leishmaniasis in India: a multicenter study. *Clin Infect Dis* 2003;**37**:800–804.

24. Grant, G.J., Barenholz, Y., Bolotin, E.M., Bansinath, M., Turndorf, H., Piskoun, B., Davidson, E.M. A novel liposomal bupivacaine formulation to produce ultralong-acting analgesia. *Anesthesiology* 2004;**101**:133–137.
25. Senior, J.H. Fate and behavior of liposomes in vivo: a review of controlling factors. *CRC Crit Rev Ther Drug Carrier Syst* 1987;**3**:123–193.
26. Torchilin, V.P. Liposomes as targetable drug carriers. *CRC Crit Rev Ther Drug Carrier Syst* 1985;**1**:65–115.
27. Klibanov, A.L., Maruyama, K., Torchilin, V.P., Huang, L. Amphipathic polyethyleneglycols effectively prolong the circulation time of liposomes. *FEBS Lett* 1990;**268**:235–238.
28. Blume, G., Cevc, G. Molecular mechanism of the lipid vesicle longevity in vivo. *Biochim Biophys Acta* 1993;**1146**:157–168.
29. Gabizon, A.A. Pegylated liposomal doxorubicin: metamorphosis of an old drug into a new form of chemotherapy. *Cancer Invest* 2001;**19**:424–436.
30. Martin, F., Lasic, D. (eds). Stealth Liposomes. Boca Raton, FL: CRC Press, 1995.
31. Woodle, M.C., Newman, M.S., Cohen, J.A. Sterically stabilized liposomes: physical and biological properties. *J Drug Target* 1994;**5**:397–403.
32. Torchilin, V.P., Omelyanenko, V.G., Papisov, M.I., Bogdanov, A.A. Jr., Trubetskoy, V.S., Herron, J.N., Gentry, C.A. Poly(ethylene glycol) on the liposome surface: on the mechanism of polymer-coated liposome longevity. *Biochim Biophys Acta* 1994;**1195**:11–20.
33. Torchilin, V.P., Trubetskoy, V.S. Which polymers can make nanoparticulate drug carriers long-circulating? *Adv Drug Deliv Rev* 1995;**16**:141–155.
34. Allen, T.M., Hansen, C. Pharmacokinetics of stealth versus conventional liposomes: effect of dose. *Biochim Biophys Acta* 1991;**1068**:133–141.
35. Woodle, M.C. Controlling liposome blood clearance by surface-grafted polymers. *Adv Drug Deliv Rev* 1998;**32**:139–152.
36. Whiteman, K.R., Subr, V., Ulbrich, K., Torchilin, V.P. Poly(HPMA)-coated liposomes demonstrate prolonged circulation in mice. *J Liposome Res* 2001;**11**:153–164.
37. Torchilin, V.P., Levchenko, T.S., Whiteman, K.R., Yaroslavov, A.A., Tsatsakis, A.M., Rizos, A.K., et al. Amphiphilic poly-*N*-vinylpyrrolidones: synthesis, properties and liposome surface modification. *Biomaterials* 2001;**22**:3035–3044.
38. Metselaar, J.M., Bruin, P., de Boer, L.W., de Vringer, T., Snel, C., Oussoren, C., et al. A novel family of L-amino acid-based biodegradable polymer-lipid conjugates for the development of long-circulating liposomes with effective drug-targeting capacity. *Bioconj Chem* 2003;**14**: 1156–1164.
39. Takeuchi, H., Kojima, H., Yamamoto, H., Kawashima, Y. Evaluation of circulation profiles of liposomes coated with hydrophilic polymers having different molecular weights in rats. *J Control Release* 2001;**75**:83–91.
40. Levchenko, T.S., Rammohan, R., Lukyanov, A.N., Whiteman, K.R., Torchilin, V.P. Liposome clearance in mice: the effect of a separate and combined presence of surface charge and polymer coating. *Int J Pharm* 2002;**240**:95–102.
41. Chiu, G.N., Bally, M.B., Mayer, L.D. Selective protein interactions with phosphatidylserine containing liposomes alter the steric stabilization properties of poly(ethylene glycol). *Biochim Biophys Acta* 2001;**1510**:56–69.
42. Torchilin, V.P., Klibanov, A.L., Huang, L., O'Donnell, S., Nossif, N.D., Khaw, B.A. Targeted accumulation of polyethylene glycol-coated immunoliposomes in infarcted rabbit myocardium. *FASEB J* 1992;**6**:2716–2719.
43. Blume, G., Cevc., G., Crommelin, M.D.J.A., Bakker-Woundenberg, I.A.J.M., Kluft, C., Storm, G. Specific targeting with poly(ethylene glycol)-modified liposomes: coupling of homing devices to the ends of polymeric chains combines effective target binding with long circulation times. *Biochim Biophys Acta* 1993;**1149**:180–184.
44. Abra, R.M., Bankert, R.B., Chen, F., Egilmez, N.K., Huang, K., Saville, R., et al. The next generation of liposome delivery systems: recent experience with tumor-targeted, sterically-stabilized immunoliposomes and active-loading gradients. *J Liposome Res* 2002;**12**:1–3.
45. Torchilin, V.P., Levchenko, T.S., Lukyanov, A.N., Khaw, B.A., Klibanov, A.L., Rammohan, R., et al. *p*-Nitrophenylcarbonyl-PEG-PE-liposomes: fast and simple attachment of specific ligands,

including monoclonal antibodies, to distal ends of PEG chains via *p*-nitrophenylcarbonyl groups. *Biochim Biophys Acta* 2001;**1511**:397–411.
46. Torchilin, V.P. Affinity liposomes in vivo: factors influencing target accumulation. *J Mol Recognit* 1996;**9**(5–6):335–346.
47. Park, J.W., Benz, C.C., Martin, F.J. Future directions of liposome- and immunoliposome-based cancer therapeutics. *Semin Oncol* 2004;**6**(suppl 13):196–205.
48. Sapra, P., Tyagi, P., Allen, T.M. Ligand-targeted liposomes for cancer treatment. *Curr Drug Deliv* 2005;**2**:369–381.
49. Sapra, P., Allen, T.M. Internalizing antibodies are necessary for improved therapeutic efficacy of antibody-targeted liposomal drugs. *Cancer Res* 2002;**62**:7190–7194.
50. Park, J.W., Kirpotin, D.B., Hong, K., Shalaby, R., Shao, Y., Nielsen, U.B., et al. Tumor targeting using anti-her2 immunoliposomes. *J Control Release* 2001;**74**:95–113.
51. Lukyanov, A.N., Elbayoumi, T.A., Chakilam, A.R., Torchilin, V.P. Tumor-targeted liposomes: doxorubicin-loaded long-circulating liposomes modified with anti-cancer antibody. *J Control Release* 2004;**100**(1):135–144.
52. Raffaghello, L., Pagnan, G., Pastorino, F., Cosimo, E., Brignole, C., Marimpietri, D., et al. Immunoliposomal fenretinide: a novel antitumoral drug for human neuroblastoma. *Cancer Lett* 2003;**197**:151–155.
53. Kamps, J.A., Koning, G.A., Velinova, M.J., Morselt, H.W., Wilkens, M., Gorter, A., et al. Uptake of long-circulating immunoliposomes, directed against colon adenocarcinoma cells, by liver metastases of colon cancer. *J Drug Target* 2000;**8**:235–245.
54. Mastrobattista, E., Koning, G.A., van Bloois, L., Filipe, A.C., Jiskoot, W., Storm, G. Functional characterization of an endosome-disruptive peptide and its application in cytosolic delivery of immunoliposome-entrapped proteins. *J Biol Chem* 2002;**277**:27135–27143.
55. Hatakeyama, H., Akita, H., Maruyama, K., Suhara, T., Harashima, H. Factors governing the in vivo tissue uptake of transferrin-coupled polyethylene glycol liposomes in vivo. *Int J Pharm* 2004;**281**:25–33.
56. Ishida, O., Maruyama, K., Tanahashi, H., Iwatsuru, M., Sasaki, K., Eriguchi, M., Yanagie, H. Liposomes bearing polyethyleneglycol-coupled transferrin with intracellular targeting property to the solid tumors in vivo. *Pharm Res* 2001;**18**:1042–1048.
57. Derycke, A.S., De Witte, P.A. Transferrin-mediated targeting of hypericin embedded in sterically stabilized PEG-liposomes. *Int J Oncol* 2002;**20**:181–187.
58. Gijsens, A., Derycke, A., Missiaen, L., De Vos, D., Huwyler, J., Eberle, A., De Witte, P. Targeting of the photocytotoxic compound AlPcS4 to Hela cells by transferring conjugated PEG-liposomes. *Int J Cancer* 2002;**101**:78–85.
59. Iinuma, H., Maruyama, K., Okinaga, K., Sasaki, K., Sekine, T., Ishida, O., et al. Intracellular targeting therapy of cisplatin-encapsulated transferrin-polyethylene glycol liposome on peritoneal dissemination of gastric cancer. *Int J Cancer* 2002;**99**:130–137.
60. Eavarone, D.A., Yu, X., Bellamkonda, R.V. Targeted drug delivery to C6 glioma by transferrin-coupled liposomes. *J Biomed Mater Res* 2000;**51**:10–14.
61. Omori, N., Maruyama, K., Jin, G., Li, F., Wang, S.J., Hamakawa, Y., et al. Targeting of post-ischemic cerebral endothelium in rat by liposomes bearing polyethylene glycol-coupled transferrin. *Neurol Res* 2003;**25**:275–279.
62. Joshee, N., Bastola, D.R., Cheng, P.W. Transferrin-facilitated lipofection gene delivery strategy: characterization of the transfection complexes and intracellular trafficking. *Hum Gene Ther* 2002;**13**:1991–2004.
63. Xu, L., Huang, C.C., Huang, W., Tang, W.H., Rait, A., Yin, Y.Z., et al. Systemic tumor-targeted gene delivery by anti-transferrin receptor scFv-immunoliposomes. *Mol Cancer Ther* 2002; **1**:337–346.
64. Tan, P.H., Manunta, M., Ardjomand, N., Xue, S.A., Larkin, D.F., Haskard, D.O., et al. Antibody targeted gene transfer to endothelium. *J Gene Med* 2003;**5**:311–323.
65. Huwyler, J., Wu, D., Pardridge, W.M. Brain drug delivery of small molecules using immunoliposomes. *Proc Natl Acad Sci USA* 1996;**93**:14164–14169.
66. Leamon, C.P., Low, P.S. Delivery of macromolecules into living cells: a method that exploits folate receptor endocytosis. *Proc Natl Acad Sci USA* 1991;**88**:5572–5576.

67. Lee, R.J., Low, P.S. Delivery of liposomes into cultured KB cells via folate receptor-mediated endocytosis. *J Biol Chem* 1994;**269**:3198–3204.
68. Lu, Y., Low, P.S. Folate-mediated delivery of macromolecular anticancer therapeutic agents. *Adv Drug Deliv Rev* 2002;**54**:675–693.
69. Gabizon, A., Shmeeda, H., Horowitz, A.T., Zalipsky, S. Tumor cell targeting of liposome-entrapped drugs with phospholipid-anchored folic acid-PEG conjugates. *Adv Drug Deliv Rev* 2004;**56**:1177–1192.
70. Ni, S., Stephenson, S.M., Lee, R.J. Folate receptor targeted delivery of liposomal daunorubicin into tumor cells. *Anticancer Res* 2002;**22**:2131–2135.
71. Pan, X.Q., Wang, H., Lee, R.J. Antitumor activity of folate receptor-targeted liposomal doxorubicin in a KB oral carcinoma murine xenograft model. *Pharm Res* 2003;**20**:417–422.
72. Pan, X.Q., Zheng, X., Shi, G., Wang, H., Ratnam, M., Lee, R.J. Strategy for the treatment of acute myelogenous leukemia based on folate receptor beta-targeted liposomal doxorubicin combined with receptor induction using all-trans retinoic acid. *Blood* 2002;**100**:594–602.
73. Stephenson, S.M., Yang, W., Stevens, P.J., Tjarks, W., Barth, R.F., Lee, R.J. Folate receptor-targeted liposomes as possible delivery vehicles for boron neutron capture therapy. *Anticancer Res* 2003;**23**:3341–3345.
74. Lu, Y., Low, P.S. Folate targeting of haptens to cancer cell surfaces mediates immunotherapy of syngeneic murine tumors. *Cancer Immunol Immunother* 2002;**51**:153–162.
75. Reddy, J.A., Abburi, C., Hofland, H., Howard, S.J., Vlahov, I., Wils, P., Leamon, C.P. Folate-targeted, cationic liposome-mediated gene transfer into disseminated peritoneal tumors. *Gene Ther* 2002;**9**:1542–1550.
76. Leamon, C.P., Cooper, S.R., Hardee, G.E. Folate-liposome-mediated antisense oligodeoxynucleotide targeting to cancer cells: evaluation in vitro and in vivo. *Bioconj Chem* 2003;**14**:738–747.
77. Drummond, D.C., Hong, K., Park, J.W., Benz, C.C., Kirpotin, D.B. Liposome targeting to tumors using vitamin and growth factor receptors. *Vitam Horm* 2000;**60**:285–332.
78. Dagar, S., Krishnadas, A., Rubinstein, I., Blend, M.J., Onyuksel, H. VIP grafted sterically stabilized liposomes for targeted imaging of breast cancer: in vivo studies. *J Control Release* 2003;**91**:123–133.
79. Schiffelers, R.M., Koning, G.A., ten Hagen, T.L., Fens, M.H., Schraa, A.J., Janssen, A.P., et al. Anti-tumor efficacy of tumor vasculature-targeted liposomal doxorubicin. *J Control Release* 2003;**91**:115–122.
80. Lestini, B.J., Sagnella, S.M., Xu, Z., Shive, M.S., Richter, N.J., Jayaseharan, J., et al. Surface modification of liposomes for selective cell targeting in cardiovascular drug delivery. *J Control Release* 2002;**78**:235–247.
81. Asai, T., Shimizu, K., Kondo, M., Kuromi, K., Watanabe, K., Ogino, K., et al. Anti-neovascular therapy by liposomal DPP-CNDAC targeted to angiogenic vessels. *FEBS Lett* 2002;**520**:167–170.
82. Mamot, C., Drummond, D.C., Greiser, U., Hong, K., Kirpotin, D.B., Marks, J.D., Park, J.W. Epidermal growth factor receptor (EGFR)-targeted immunoliposomes mediate specific and efficient drug delivery to EGFR- and EGFRvIII-overexpressing tumor cells. *Cancer Res* 2003;**63**:3154–3161.
83. Peer, D., Margalit, R. Loading mitomycin C inside long circulating hyaluronan targeted nano-liposomes increases its antitumor activity in three mice tumor models. *Int J Cancer* 2004;**108**:780–789.
84. Matsuda, I., Konno, H., Tanaka, T., Nakamura, S. Antimetastatic effect of hepatotropic liposomal adriamycin on human metastatic liver tumors. *Surg Today* 2001;**31**:414–420.
85. Hashida, M., Nishikawa, M., Yamashita, F., Takakura, Y. Cell-specific delivery of genes with glycosylated carriers. *Adv Drug Deliv Rev* 2001;**52**:187–196.
86. Lee, C.M., Tanaka, T., Murai, T., Kondo, M., Kimura, J., Su, W., et al. Novel chondroitin sulfate-binding cationic liposomes loaded with cisplatin efficiently suppress the local growth and liver metastasis of tumor cells in vivo. *Cancer Res* 2002;**62**:4282–4288.
87. Kaneda, Y. Virosomes: evolution of the liposome as a targeted drug delivery system. *Adv Drug Deliv Rev* 2000;**43**:197–205.
88. Sarkar, D.P., Ramani, K., Tyagi, S.K. Targeted gene delivery by virosomes. *Methods Mol Biol* 2002;**199**:163–173.

89. Cusi, M.G., Terrosi, C., Savellini, G.G., Di Genova, G., Zurbriggen, R., Correale, P. Efficient delivery of DNA to dendritic cells mediated by influenza virosomes. *Vaccine* 2004;**22**:735–739.
90. Bungener, L., Huckriede, A., Wilschut, J., Daemen, T. Delivery of protein antigens to the immune system by fusion-active virosomes: a comparison with liposomes and ISCOMs. *Biosci Rep* 2002;**22**:323–338.
91. Bungener, L., Serre, K., Bijl, L., Leserman, L., Wilschut, J., Daemen, T., Machy, P. Virosome-mediated delivery of protein antigens to dendritic cells. *Vaccine* 2002;**20**:2287–2295.
92. Huckriede, A., Bungener, L., Daemen, T., Wilschut, J. Influenza virosomes in vaccine development. *Methods Enzymol* 2003;**373**:74–91.
93. Herzog, C., Metcalfe, I.C., Schaad, U.B. Virosome influenza vaccine in children. *Vaccine* 2002;**20**(suppl 5):B24–B28.
94. Usonis, V., Bakasenas, V., Valentelis, R., Katiliene, G., Vidzeniene, D., Herzog, C. Antibody titres after primary and booster vaccination of infants and young children with a virosomal hepatitis A vaccine (Epaxal). *Vaccine* 2003;**21**:4588–4592.
95. Ambrosch, F., Finkel, B., Herzog, C., Koren, A., Kollaritsch, H. Rapid antibody response after vaccination with a virosomal hepatitis a vaccine. *Infection* 2004;**32**:149–152.
96. Ruf, B.R., Colberg, K., Frick, M., Preusche, A. Open, randomized study to compare the immunogenicity and reactogenicity of an influenza split vaccine with an MF59-adjuvanted subunit vaccine and a virosome-based subunit vaccine in elderly. *Infection* 2004;**32**;191–198.
97. Gluck, R., Moser, C., Metcalfe, I.C. Influenza virosomes as an efficient system for adjuvanted vaccine delivery. *Expert Opin Biol Ther* 2004;**4**:1139–1145.
98. Moser, C., Metcalfe, I.C., Viret, J.F. Virosomal adjuvanted antigen delivery systems. *Expert Rev Vaccines* 2003;**2**:189–196.
99. Connor, J., Huang, L. pH-sensitive immunoliposomes as an efficient and target-specific carrier for antitumor drugs. *Cancer Res* 1986;**46**:3431.
100. Drummond, D.C., Zignani, M., Leroux, J. Current status of pH-sensitive liposomes in drug delivery. *Prog Lipid Res* 2000;**39**:409–460.
101. Hong, M.S., Lim, S.J., Oh Y.K., Kim, C.K. pH-sensitive, serum-stable and long-circulating liposomes as a new drug delivery system. *J Pharm Pharmacol* 2002;**54**(1):51–58.
102. Simoes, S., Moreira, J.N., Fonseca, C., Duzgunes, N., de Lima, M.C. On the formulation of pH-sensitive liposomes with long circulation times. *Adv Drug Deliv Rev* 2004;**56**:947–965.
103. Fattal, E., Couvreur, P., Dubernet, C. 'Smart' delivery of antisense oligonucleotides by anionic pH-sensitive liposomes. *Adv Drug Deliv Rev* 2004;**56**:931–946.
104. Sudimack, J.J., Guo, W., Tjarks, W., sLee, R.J. A novel pH-sensitive liposome formulation containing oleyl alcohol. *Biochim Biophys Acta* 2002;**1564**:31–37.
105. Asokan, A., Cho, M.J. Cytosolic delivery of macromolecules. II. Mechanistic studies with pH-sensitive morpholine lipids. *Biochim Biophys Acta* 2003;**1611**:151–160.
106. Reddy, J.A., Low, P.S. Enhanced folate receptor mediated gene therapy using a novel pH-sensitive lipid formulation. *J Control Release* 2000;**64**:27–37.
107. Turk, M.J., Reddy, J.A., Chmielewski, J.A., Low, P.S. Characterization of a novel pH-sensitive peptide that enhances drug release from folate-targeted liposomes at endosomal pHs. *Biochim Biophys Acta* 2002;**1559**:56–68.
108. Kakudo, T., Chaki, S., Futaki, S., Nakase, I., Akaji, K., Kawakami, T., et al. Transferrin-modified liposomes equipped with a pH-sensitive fusogenic peptide: an artificial viral-like delivery system. *Biochemistry* 2004;**43**:5618–5628.
109. Shi, G., Guo, W., Stephenson, S.M., Lee, R.J. Efficient intracellular drug and gene delivery using folate receptor-targeted pH-sensitive liposomes composed of cationic/anionic lipid combinations. *J Control Release* 2002;**80**:309–319.
110. Egleton, R.D., Davis, T.P. Bioavailability and transport of peptides and peptide drugs into the brain. *Peptides* 1997;**18**:1431–1439.
111. Kamata, H., Yagisawa, H., Takahashi, S., Hirata, H. Amphiphilic peptides enhance the efficiency of liposome-mediated DNA transfection. *Nucleic Acids Res* 1994;**22**:536–537.
112. Midoux, P., Kichler, A., Boutin, V., Maurizot, J.C., Monsigny, M. Membrane permeabilization and efficient gene transfer by a peptide containing several histidines. *Bioconj Chem* 1998;**9**: 260–267.

113. Lackey, C.A., Press, O.W., Hoffman, A.S., Stayton, P.S. A biomimetic pH-responsive polymer directs endosomal release and intracellular delivery of an endocytosed antibody complex. *Bioconjug Chem* 2002;**13**:996–1001.
114. Padilla, De Jesus, O.L., Ihre, H.R., Gagne, L., Frechet, J.M., Szoka, F.C. Jr. Polyester dendritic systems for drug delivery applications: in vitro and in vivo evaluation. *Bioconj Chem* 2002;**13**: 453–461.
115. Vives, E., Brodin, P., Lebleu, B. A truncated HIV-1 Tat protein basic domain rapidly translocates through the plasma membrane and accumulates in the cell nucleus. *J Biol Chem* 1997;**272**: 16010–16017.
116. Wagner, E. Application of membrane-active peptides for nonviral gene delivery. *Adv Drug Deliv Rev* 1999;**38**:279–289.
117. Lewin, M., Carlesso, N., Tung, C.H., Tang, X.W., Cory, D., Scadden, D.T., Weissleder, R. Tat peptide-derivatized magnetic nanoparticles allow in vivo tracking and recovery of progenitor cells. *Nat Biotechnol* 2000;**18**:410–414.
118. Bhorade, R., Weissleder, R., Nakakoshi, T., Moore, A., Tung, C.H. Macrocyclic chelators with paramagnetic cations are internalized into mammalian cells via a HIV-tat derived membrane translocation peptide. *Bioconjug Chem* 2000;**11**:301–305.
119. Torchilin, V.P., Rammohan, R., Weissig, V., Levchenko, T.S. TAT peptide on the surface of liposomes affords their efficient intracellular delivery even at low temperature and in the presence of metabolic inhibitors. *Proc Natl Acad Sci USA* 2001;**98**:8786–8791.
120. Tseng, Y.L., Liu, J.J., Hong, R.L. Translocation of liposomes into cancer cells by cell-penetrating peptides penetratin and tat: a kinetic and efficacy study. *Mol Pharmacol* 2002;**62**:864–872.
121. Derossi, D., Joliot, A.H., Chassaing, G., Prochiantz, A. The third helix of the Antennapedia homeodomain translocates through biological membranes. *J Biol Chem* 1994;**269**:10444–10450.
122. Elliott, G., O'Hare, P. Intercellular trafficking and protein delivery by a herpesvirus structural protein. *Cell* 1997;**88**:223–233.
123. Wadia, J.S., Stan, R.V., Dowdy, S.F. Transducable TAT-HA fusogenic peptide enhances escape of TAT fusion proteins after lipid raft macropinocytosis. *Nat Med* 2004;**10**:310–315.
124. Gupta, B., Levchenko, T.S., Torchilin, V.P. Intracellular delivery of large molecules and small particles by cell-penetrating proteins and peptides. *Adv Drug Deliv Rev* 2005;**57**:637–651.
125. Wadia, J.S., Dowdy, S.F. Transmembrane delivery of protein and peptide drugs by TAT-mediated transduction in the treatment of cancer. *Adv Drug Deliv Rev* 2005;**57**:579–596.
126. Rothbard, J.B., Jessop, T.C., Lewis, R.S., Murray, B.A., Wender, P.A. Role of membrane potential and hydrogen bonding in the mechanism of translocation of guanidinium-rich peptides into cells. *J Am Chem Soc* 2004;**126**:9506–9507.
127. Rothbard, J.B., Jessop, T.C., Wender, P.A. Adaptive translocation: the role of hydrogen bonding and membrane potential in the uptake of guanidinium-rich transporters into cells. *Adv Drug Deliv Rev* 2005;**57**:495–504.
128. Vocero-Akbani, A., Lissy, N.A., Dowdy, S.F. Transduction of full-length Tat fusion proteins directly into mammalian cells: analysis of T cell receptor activation-induced cell death. *Methods Enzymol* 2000;**322**:508–521.
129. Soga, N., Namba, N., McAllister, S., Cornelius, L., Teitelbaum, S.L., Dowdy, S.F., et al. Rho family GsTPases regulate VEGF-stimulated endothelial cell motility. *Exp Cell Res* 2001;**269**:73–87.
130. Zezula, J., Casaccia-Bonnefil, P., Ezhevsky, S.A., Osterhout, D.J., Levine, J.M., Dowdy, S.F., et al. p21cip1 is required for the differentiation of oligodendrocytes independently of cell cycle withdrawal. *EMBO Rep* 2001;**2**:27–34.
131. Hsia, C.Y., Cheng, S., Owyang, A.M., Dowdy, S.F., Liou, H.C. c-Rel regulation of the cell cycle in primary mouse B lymphocytes. *Int Immunol* 2002;**14**:905–916.
132. Schwarze, S.R., Ho, A., Vocero-Akbani, A., Dowdy, S.F. In vivo protein transduction: delivery of a biologically active protein into the mouse. *Science* 1999;**285**:1569–1572.
133. Gorodetsky, R., Levdansky, L., Vexler, A., Shimeliovich, I., Kassis, I., Ben-Moshe, M., et al. Liposome transduction into cells enhanced by haptotactic peptides (Haptides) homologous to fibrinogen C-termini. *J Control Release* 2004;**95**:477–488.

134. Torchilin, V.P., Levchenko, T.S., Rammohan, R., Volodina, N., Papahadjopoulos-Sternberg, B., D'Souza, G.G. Cell transfection in vitro and in vivo with nontoxic TAT peptide-liposome-DNA complexes. *Proc Natl Acad Sci USA* 2003;**100**:1972–1977.
135. Wolf, M., Ransberger, K. *Enzyme Therapy*. New York: Vantage Press, 1972.
136. Holcenberg, J.S., Roberts, J. (eds). *Enzyme as Drugs*. New York: John Wiley, 1981.
137. Torchilin, V.P. *Immobilized Enzymes in Medicine*. Berlin: Springer-Verlag, 1991.
138. Tager, J.M., Daems, W.T., Hoodhwinkel, G.J.M. (eds). *Enzyme Therapy in Lysosomal Storage Diseases*. Amsterdam: North Holland, 1974.
139. Grabowsky, G.A., Desnick, R.J. Enzyme replacement in genetic diseases. In: Holcenberg, J.S., Roberts, J. (eds), *Enzymes as Drugs*. New York: John Wiley, 1981: p 167.
140. Capizzi, R.L., Cheng, Y-C. Therapy of neoplasia with asparaginase. In: Holcenberg, J.S., Roberts, J. (eds), *Enzymes as Drugs*. New York: John Wiley, 1981: p 1.
141. Asselin, B.L. The three asparaginases. Comparative pharmacology and optimal use in childhood leukemia. *Adv Exp Med Biol* 1999;**457**:621–629.
142. Roberts, J. Therapy of neoplasia by deprivation of essential amino acids. In: Holcenberg, J.S., Roberts, J. (eds), *Enzymes as Drugs*. New York: John Wiley, 1981: p 63.
143. Kalghatgi, K.K., Bertino, J.R. Folate-degrading enzymes: a review with special emphasis on carboxypeptidase G. In: Holcenberg, J.S., Roberts, J. (eds), *Enzymes as Drugs*. New York: John Wiley, 1981: p 77.
144. Levy, C.C., Karpetsky, T.P. Human ribonucleases. In: Holcenberg, J.S., Roberts, J. (eds), *Enzymes as Drugs*. New York: John Wiley, 1981: p 103.
145. Lazerson, *J*. Clotting factor replacement. In: Holcenberg, J.S., Roberts, J. (eds), *Enzymes as Drugs*. New York: John Wiley, 1981: p 241.
146. Fletcher, A.P., Alkjaersing, N.K. Fibrinolytic and defibrinating enzymes. In: Holcenberg, J.S., Roberts, J. (eds), *Enzymes as Drugs*. New York: John Wiley, 1981: p 209.
147. Lijnen, H.R. Fibrinolysis: molecular mechanism and pathophysiological aspects. *Sangre* 1984;**29**:755–761.
148. Matsuo, O., Bando, H., Okada, K., Tanaka, K., Iga, Y., Arimura, H. Thrombolytic effect of single-chain pro-urokinase in a rabbit jugular vein thrombosis model. *Thromb Res* 1986;**42**: 187–194.
149. Myers, M.L., Bolli, R., Lekich, R.F., Hartley, C.J., Roberts, R. Enhancement of recovery of myocardial function by oxygen Tree-radical scavengers after reversible regional ischemia. *Circulation* 1985;**72**:915–921.
150. Sawada, T. Method for treatment of arthrosis deformations with elastase. US Patent 4446192, 1984.
151. Kujisaki, S., Mitani, M. Pronase used for the treatment of diseases of the liver and kidneys in humans and animals. US Patent 4485095, 1984.
152. Pinnell, S.R. Method for the prevention and treatment of scars with enzymes. US Patent 4524065, 1985.
153. Froidevaux, S., Eberle, A.N. Somatostatin analogs and radiopeptides in cancer therapy. Biopolymers 2002;**66**:161–183.
154. Figg, W.D., Kruger, E.A., Price, D.K., Kim, S., Dahut, W.D. Inhibition of angiogenesis: treatment options for patients with metastatic prostate cancer. *Invest New Drugs* 2002;**20**:183–194.
155. Kerbel, R., Folkman, *J*. Clinical translation of angiogenesis inhibitors. *Nat* Rev *Cancer* 2002;**2**:727–739.
156. Ballard, C.E., Yu, H., Wang, B. Recent developments in depsipeptide research. *Curr Med Chem* 2002;**9**:471–498.
157. Desnick, R.J., Thorpe, S.R., Fidler, M.B. Toward enzyme therapy for lysosomal storage diseases, *Physiol Rev* 1976;**56**:57–99.
158. Johnson, W.G., Desnick, R.J., Long, D.M., Sharp, H.L., Krivit, L., Brady, B., Brady, R.O. Intravenous injection of purified hexosaminidase A into a patient with Tay-Sachs disease. In: Bergsma, D. (ed.), *Enzyme Therapy in Genetic Diseases*, vol 9. Baltimore, M.D.: Williams & Wilkins, 1973: pp 120–124.
159. Brady, R.O., Pentchev, P.G., Gal, A.E., Hibbert, S.R., Dekaban, A.S. Replacement therapy for inherited enzyme deficiency. Use of purified glucocerebrosidase in Gaucher's disease. *N Engl J Med* 1973;**291**:989–993.

160. Brady, R.O., Tallman, J.F., Johnson, W.G., Gal, A.E., Leahy, W.R., Quirk. J.M., Dekaban, A.S. Replacement therapy for inherited enzyme deficiency. Use of purified ceramidetrihexosidase in Fabry's disease. *N Engl J Med* 1973;**289**:9–14.
161. Gregoriadis, G. Liposomes in the therapy of lysosomal storage diseases. *Nature* 1978;**275**:695–696.
162. Gregoriadis, G., Dean, M.F. Enzyme therapy in genetic diseases. *Nature* 1979;**278**:603–604.
163. Gregoriadis, G., Ryman, B.L. Lysosomal localization of β-fructofuranosidase-containing liposomes injected into rats. *Biochem J* 1972;**129**:123–133.
164. Patel, H.M., Ryman, B.E. α-Mannosidase in zinc-deficient rats. Possibility of liposomal therapy in mannosidosis. *Bioch Soc Trans* 1974;**2**:1014–1017.
165. Gregoriadis, G., Putman, D., Louis, L., Neerunjun, D. Comparative effect and fate of non-entrapped and liposome-entrapped neuraminidase injected into rats. *Biochem J* 1974;**140**:323–330.
166. Steger, L.D., Desnick, R.J. Enzyme therapy. VI. Comparative in vivo fates and effects on lysosomal integrity of enzyme entrapped in negatively and positively charged liposomes. *Biochim Biophys Acta* 1977;**464**:530–546.
167. Colley, C.M., Ryman, B.E. The use of a liposomally entrapped enzyme in the treatment of an artificial storage condition. *Biochim Biophys Acta* 1976;**451**:417–425.
168. Braidman, I.P., Gregoriadis, G. Rapid partial purification of placental glucocerebroside β-glucosidase and its entrapment in liposomes. *Biochem J* 1977;**164**:439–445.
169. Reynolds, G.O., Baker, H.J., Reynolds, R.H. Enzyme replacement using liposome carriers in feline GM1 gangliosidosis fibroblasts. *Nature* 1978;**275**:754–755.
170. Umezawa, F., Eto, Y., Tokoro, T., Ito, F., Maekawa, K. Enzyme replacement with liposomes containing beta-galactosidase from *Charonia lumpas* in murine globoid cell leukodystrophy (Twitcher). *Biochem Biophys Res Commun* 1985;**127**:663–667.
171. Fishman, Y., Citri, N. L-asparaginase entrapped in liposomes: preparation and properties. *FEBS Lett* 1975;**60**:17–20.
172. Gaspar, M.M., Perez-Soler, R., Cruz, M.E. Biological characterization of L-asparaginase liposomal formulations. *Cancer Chemother Pharmacol* 1996;**38**:373–377.
173. Jorge, J.C., Perez-Soler, R., Morais, J.G., Cruz, M.E. Liposomal palmitoyl-L-asparaginase: characterization and biological activity. *Cancer Chemother Pharmacol* 1994;**34**:230–234.
174. Jadot, G., Vaille, A., Maldonado, J., Vanelle, P. Clinical pharmacokinetics and delivery of bovine superoxide dismutase, *Clin Pharmacokinet* 1995;**28**:17–25.
175. Stanimirovic, D.B., Markovic, M., Micic, D.V., Spatz, M., Mrsulja, B.B. Liposome-entrapped superoxide dismutase reduces ischemia/reperfusion 'oxidative stress' in gerbil brain. *Neurochem Res* 1994;**19**:1473–1478.
176. Freeman, B.A., Young, S.L., Crapo, J.D. Liposome-mediated augmentation of superoxide dismutase in endothelial cells prevents oxygen injury. *J Biol Chem* 1983;**258**:12534–12542.
177. Freeman, B.A., Turrens, J.F., Mirza, Z., Crapo, J.D., Young, S.L. Modulation of oxidant lung injury by using liposome-entrapped superoxide dismutase and catalase. *Fed Proc* 1985;**44**: 2591–2595.
178. Briscoe, P., Caniggia, I., Graves, A., Benson, B., Huang, L., Tanswell, A.K., Freeman, B.A. Delivery of superoxide dismutase to pulmonary epithelium via pH-sensitive liposomes. *Am J Physiol* 1995;**268**:L374–380.
179. Heeremans, J.L., Prevost, R., Bekkers, M.E., Los, P., Emeis, J.J., Kluft, C., Crommelin, D.J. Thrombolytic treatment with tissue-type plasminogen activator (t-PA) containing liposomes in rabbits: a comparison with free t-PA. *Thromb Haemost* 1995;**73**:488–494.
180. Yagi, N., Naoi, M., Sasaki, H., Abe, H., Konishi, H., Arichi, S. Incorporation of enzyme into the brain by means of liposomes of novel composition. *J Appl Biochem* 1982;**4**:121–125.
181. Szebeni, J., Di Iorio, E.E., Hauser, H., Winterhalter, K.H. Encapsulation of hemoglobin in phospholipid liposomes: characterization and stability. *Biochemistry* 1985;**24**:2827–2832.
182. Kato, A., Tanaka, I. Liposome-type artificial red blood cells stabilized with carboxymethyl chitin, *Biomat Med Dev Art Org* 1985;**13**:61–82.
183. Awasthi, V.D., Garcia, D., Klipper, R., Goins, B.A., Phillips, W.T. Neutral and anionic liposome-encapsulated hemoglobin: effect of postinserted poly(ethylene glycol)-distearoylphosphatidylethanolamine on distribution and circulation kinetics. *J Pharmacol Exp Ther* 2004;**309**: 241–248.

184. Sakai, H., Tomiyama, K.I., Sou, K., Takeoka, S., Tsuchida, E. Poly(ethylene glycol)-conjugation and deoxygenation enable long-term preservation of hemoglobin-vesicles as oxygen carriers in a liquid state. *Bioconj Chem* 2000;**11**:425–432.
185. Phillips, W.T., Klipper, R.W., Awasthi, V.D., Rudolph, A.S., Clif, R., Kwasiborski, V., Goins, B.A. Polyethylene glycol-modified liposome-encapsulated hemoglobin: a long circulating red cell substitute. *J Pharmacol Exp Ther* 1999;**288**:665–670.
186. Sakai, H., Tsai, A.G., Rohlfs, R.J., Hara, H., Takeoka, S., Tsuchida, E., Intaglietta, M. Microvascular responses to hemodilution with Hb vesicles as red blood cell substitutes: influence of O2 affinity. *Am J Physiol* 1999;**276**:H553–H562.
187. Sakai, H., Masada, Y., Horinouchi, H., Yamamoto, M., Ikeda, E., Takeoka, S., et al. Hemoglobin-vesicles suspended in recombinant human serum albumin for resuscitation from hemorrhagic shock in anesthetized rats. *Crit Care Med* 2004;**32**;539–545.
188. Shibuya-Fujiwara, N., Hirayama, F., Ogata, Y., Ikeda, H., Ikebuchi, K. Phagocytosis in vitro of polyethylene glycol-modified liposome-encapsulated hemoglobin by human peripheral blood monocytes plus macrophages through scavenger receptors. *Life Sci* 2001;**70**:291–300.
189. Szebeni, J., Alving, C.R. Complement-mediated acute effects of liposome-encapsulated hemoglobin. *Artif Cells Blood Substit Immobil Biotechnol* 1999;**27**:23–41.
190. Szebeni, J., Fontana, J.L., Wassef, N.M., Mongan, P.D., Morse, D.S., Dobbins, D.E., et al. Hemodynamic changes induced by liposomes and liposome-encapsulated hemoglobin in pigs. *Circulation* 1999;**99**:2302–2309.
191. Zajac-Kaye, M., Ts'o POP. DNAase I encapsulated in liposomes can induce neoplastic transformation of Syrian hamster embryo cells in culture. *Cell* 1984;**39**:427–437.
192. Westcott, K.R., Hill, R.L. Reconstitution of a porcine submaxillary gland β-D-galactoside $\alpha 2 \rightarrow 3$ sialyltransferase into liposomes. *J Biol Chem* 1985;**260**:13116–13121.
193. Hudson, L.D.S., Fiddler, M.B., Desnick, R.J. Enzyme therapy. X. Immune response induced by enzyme and buffer-loaded liposomes in C3H/HeJ Gush mice. *J Pharmacol Exp Ther* 1979;**208**: 507–514.
194. Sprangler, R.S. Insulin administration via liposomes. *Diab Care* 1990;**13**:911–922.
195. Liu, F.Y., Shao, Z., Kildsig, D.O., Mitra, A.K. Pulmonary delivery of free and liposomal insulin. *Pharm Res* 1993;**10**:228–232.
196. Takeuchi, H., Yamamoto, H., Niwa, T., Hino, T., Kawashima, Y. Enteral adsorption of insulin in rats from mucoadhesive chitosan coated liposomes. *Pharm Res* 1996;**13**:896–901.
197. Iwanaga, K., Ono, S., Narioka, K., Kakemi, M., Morimoto, K., Yamashita, S., et al. Application of surface-coated liposomes for oral delivery of peptide: effects of coating the liposome's surface on the GI transport of insulin. *J Pharm Sci* 1999;**88**:248–252.
198. Kisel, M.A., Kulik, L.N., Tsybovsky, I.S., Vlasov, A.P., Vorob'yov, M.S., Kholodova, E.A., Zabarovskaya, Z.V. Liposomes with phosphatidylethanol as a carrier for oral delivery of insulin: studies in rat. *Int J Pharm* 2001;**216**:105–114.
199. Yang, T.Z., Wang, X.T., Yan, X.Y., Zhang, Q. Phospholipid deformable vesicles for buccal delivery of insulin. *Chem Pharm Bull (Tokyo)* 2002;**50**:749–753.
200. Kim, A., Yun, M.O., Oh, Y.K., Ahn, W.S., Kim, C.K. Pharmacodynamics of insulin in polyethylene glycol-coated liposomes. *Int J Pharm* 1999;**180**:75–81.
201. Kanaoka, E., Takahashi, K., Yoshikawa, T., Jizomoto, H., Nishihara, Y., Hirano, K. A novel and simple type of liposome carrier for recombinant interleukin-2. *J Pharm Pharmacol* 2001;**53**: 295–302.
202. Ishihara, H., Hara, T., Aramaki, Y., Tsuchiya, S., Hosoi, K. Preparation of asialofetuin-labeled liposomes with encapsulated human interferon-gamma and their uptake by isolated rat hepatocytes. *Pharm Res* 1990;**7**:542–546.
203. Kedar, E., Palgi, O., Golod, G., Babai, I., Barenholz, Y. Delivery of cytokines by liposomes. III. Liposome-encapsulated GM-CSF and TNF-alpha show improved pharmacokinetics and biological activity and reduced toxicity in mice. *J Immunother* 1997;**20**:180–193.
204. Kleinerman, E.S., Gano, J.B., Johnston, D.A., Benjamin, R.S., Jaffe, N. Efficacy of liposomal muramyl tripeptide (CGP 19835A) in the treatment of relapsed osteosarcoma. *Am J Clin Oncol* 1995;**18**:93–99.
205. Opanasopit, P., Sakai, M., Nishikawa, M., Kawakami, S., Yamashita, F., Hashida, M. Inhibition of liver metastasis by targeting immunomodulators using mannosylated liposome carriers. *J Control Release* 2002;**80**:283–294.

206. Li, H., Song, J.H., Park, J.S., Han, K. Polyethylene glycol-coated liposomes for oral delivery of recombinant human epidermal growth factor. *Int J Pharm* 2003;**258**:11–19.
207. Selzman, C.H., Shames, B.D., Reznikov, L.L., Miller, S.A., Meng, X., Barton, H.A., et al. Liposomal delivery of purified inhibitory-kappaBalpha inhibits tumor necrosis factor-alpha-induced human vascular smooth muscle proliferation. *Circ Res* 1999;**84**:867–875.
208. Postma, N.S., Crommelin, D.J., Eling, W.M., Zuidema, J. Treatment with liposome-bound recombinant human tumor necrosis factor-alpha suppresses parasitemia and protects against *Plasmodium berghei* k173-induced experimental cerebral malaria in mice. *J Pharmacol* Exp. *Ther* 1999;**288**:114–120.
209. Egbaria, K., Ramachandran, C., Kittayanond, D., Weiner, N. Topical delivery of liposomally encapsulated interferon evaluated by in vitro diffusion studies. *Antimicrob Agents Chemother* 1990;**34**:107–110.
210. du Plessis, J., Egbaria, K., Ramachandran, C., Weiner, N. Topical delivery of liposomally encapsulated gamma-interferon. *Antiviral Res* 1992;**18**:259–265.
211. Fleisher, D., Niemiec, S.M., Oh, C.K., Hu, Z., Ramachandran, C., Weiner, N. Topical delivery of growth hormone releasing peptide using liposomal systems: an in vitro study using hairless mouse skin. *Life Sci* 1995;**57**:1293–1297.
212. Yerushalmi, N., Arad, A., Margalit, R. Molecular and cellular studies of hyaluronic acid-modified liposomes as bioadhesive carriers for topical drug delivery in wound healing. *Arch Biochem Biophys* 1994;**313**:267–273.
213. Vutla, N.B., Betageri, G.V., Banga, A.K. Transdermal iontophoretic delivery of enkephalin formulated in liposomes. *J Pharm Sci* 1996;**85**:5–8.
214. Niemiec, S.M., Ramachandran, C., Weiner, N. Influence of nonionic liposmal composition on topical delivery of peptide drugs into pilosebaceous units: an in vivo study using the hamster ear model. *Pharm Res* 1995;**12**:1184–1188.
215. Gruber, S.A., Venkataram, S., Canafax, D.M., Cipolle, R.J., Bowers, L., Elsberry, D., et al. Liposomal formulation eliminates acute toxicity and pump incompatibility of parenteral cyclosporine. *Pharm Res* **6**:601–607.
216. Egbaria, K., Ramachandran, C., Weiner, N. Topical delivery of cyclosporin: evaluation of various formulations using in vitro diffusion studies in hairless mouse skin. *Skin Pharmacol* 1990;**3**:21–28.
217. Osanai, T., Nagai, Y. Suppression of experimental allergic encephalomyelitis with liposome-encapsulated protease inhibitor: therapy through the blood-brain barrier. *Neurochem Res* 1984;**9**:1407–1416.
218. Megaw, J., Gardner, K., Lerman, S. Intraocular liposome drug delivery system. *Invest Ophthalmol Vis Sci* 1981;**20**(suppl.):66.
219. Megaw, J. Delivery of liposomally encapsulated drugs to the ocular lens. *Ophth Res* 1983;**1**:211–234.
220. Han, Y.Y., Huang, L., Jackson, E.K., Dubey, R.K., Gillepsie, D.G., Carcillo, J.A. Liposomal atp or NAD+ protects human endothelial cells from energy failure in a cell culture model of sepsis. *Res Commun Mol Pathol Pharmacol* 2001;**110**:107–116.
221. Laham, A., Claperon, N., Durussel, J.J., Fattal, J., Puisieux, F., Couvreur, P. Rossignol, P. Liposomally entrapped adenosine triphosphate. Improved efficiency against experimental brain ischaemia in the rat. *J Chromatogr* 1988;**440**:455–458.
222. Konno, H., Matin, A.F., Maruo, Y., Nakamura, S., Baba, S. Liposomal ATP protects the liver from injury during shock. *Eur Surg Res* 1996;**28**:140–145.
223. Neveux, N., De Bandt, J.P., Chaumeil, J.C., Cynober, L. Hepatic preservation, liposomally entrapped adenosine triphosphate and nitric oxide production: a study of energy state and protein metabolism in the cold-stored rat liver. *Scand J Gastroenterol* 2002;**37**:1057–1063.
224. Niibori, K., Wroblewski, K.P., Yokoyama, H., Crestanello, J.A., Whitman, G.J. Bioenergetic effect of liposomal coenzyme Q10 on myocardial ischemia reperfusion injury. *Biofactors* 1999;**9**:307–313.
225. Niibori, K., Yokoyama, H., Crestanello, J.A., Whitman, G.J. Acute administration of liposomal coenzyme Q10 increases myocardial tissue levels and improves tolerance to ischemia reperfusion injury. *J Surg Res* 1998;**79**:141–145.
226. Xu, G.X., Xie, X.H., Liu, F.Y., Zang, D.L., Zheng, D.S., Huang, D.J., Huang, M.X. Adenosine triphosphate liposomes: encapsulation and distribution studies. *Pharm Res* 1990;**7**:553–557.

227. Liang, W., Levchenko, T.S., Torchilin, V.P. Encapsulation of ATP into liposomes by different methods: optimization of the procedure. *J Microencapsul* 2004;**21**:251–261.
228. Verma, D.D., Levchenko, T.S., Bernstein, E., Torchilin, V.P. ATP-Loaded liposomes effectively protect mechanical functions of the myocardium from global ischemia in an isolated rat heart model. *J Control Release* 2005;**108**(2–3):460–471.
229. Verma, D.D., Hartner, W.C., Levchenko, T.S., Bernstein, E.A., Torchilin, V.P. ATP-loaded liposomes effectively protect the myocardium in rabbits with an acute experimental myocardial infarction. *Pharm Res* 2005, **22**:2115–2120.
230. Liang, W., Levchenko, T., Khaw, B-A, Torchilin, V.P. ATP-containing immunoliposomes specific for cardiac myosin. *Curr Drug Delivery* 2004;**1**:1–7.
231. Gregoriadis, G. Liposomes as immunological adjuvants for peptide and protein antigens. In: Gregoriadis, G., Florence, A.T., Patel, H.M. (eds), *Liposomes in Drug Delivery*. Chur, Switzerland: Harwood Academic Publishers, 1993: p 77.
232. Friede, M. Liposomes as carriers of antigens. In: Philippot, J.R., Schuber, F. (eds), *Liposomes as Tools in Basic Research and Industry*. Boca Raton, F.L.: CRC Press, 1995: p 189.
233. Allison, A.C., Gregoriadis, G. Liposomes as immunological adjuvants. *Nature* 1974;**252**:252.
234. Manesis, E.K., Cameron, C.H., Gregoriadis, G. Hepatitis B surface antigen-containing liposomes enhance humoral and cell-mediated immunity to the antigen. *FEBS Lett* 1979;**102**:107–111.
235. Sanchez, Y., Ionescu-Matiu, I., Dreesman, G.R., Kramp, W., Six, H.R., Hollinger, F.B. Melnick, J.L. Humoral and cellular immunity to hepatitis B virus-derived antigens: Comparative activity of Freund's complete adjuvant, alum and liposomes. *Infect Immun* 1980;**30**:728–733.
236. El Guink, N., Kris, R.M., Goodman-Snitkoff, G., Small, P.A. Jr., Mannino, R.J. Intranasal immunization with proteoliposomes protects against influenza. *Vaccine* 1989;**7**:147–151.
237. Trudel, M., Nadon, F. Virosome preparation: differences between influenza and rubella haemagglutinin adsorption. *Can J Microbiol* 1981;**27**:958–962.
238. Steele, G. Jr., Ravikumar, T., Ross, D., King, V., Wilson, R.E., Dodson, T. Specific active immunotherapy with butanol-extracted tumor-associated antigens incorporated into liposomes. *Surgery* 1984;**96**:352–359.
239. Heath, T.D., Shek, P., Papahadjopoulos, D., investigators; Production of immunogens by antigen conjugation to liposomes. US patent 4565696, 1986.
240. Zhou, F., Kraehenbuhl, J.P., Neutra, M.R. Mucosal IgA response to rectally administered antigen formulations in IgA-coated liposomes. *Vaccine* 1995;**13**:637–644.
241. Sinha, R.K., Khuller, G.K. The protective efficacy of a liposmal encapsulated 30 kDa secretory protein of *Mycobacterium tuberculosis* H37Ra against tuberculosis in mice. *Immunol Cell Biol* 1997;**75**:461–466.
242. Guan, H.H., Budzynski, W., Koganty, R.R., Krantz, M.J., Reddish, M.A., Rogers, J.A., et al. Liposomal formulations of synthetic MUC1 peptides: effects of encapsulation versus surface display of peptides on immune responses. *Bioconjug Chem* 1998;**9**:451–458.
243. Griffiths, G.D., Phillips, G.J., Bailey, S.C. Comparison of the quality of protection elicited by toxoid and peptide liposomal vaccine formulations against ricin as assessed by markers of inflammation. *Vaccine* 1999;**17**:2562–2568.
244. Chikh, G.G., Kong, S., Bally, M.B., Meunier, J.C., Schutze-Redelmeier, M.P. Efficient delivery of Antennapedia homeodomain fused to CTL epitope with liposomes into dendritic cells results in the activation of CD8+ T cells. *J Immunol* 2001;**167**:6462–6470.
245. Lachman, L.B., Ozpolat, B., Rao, X.M. Cytokine-containing liposomes as vaccine adjuvants. *Eur Cytokine Netw* 1996;**7**:693–698.
246. Masuda, K., Horie, K., Suzuki, R., Yoshikawa, T., Hirano, K. Oral delivery of antigens in liposomes with some lipid compositions modulates oral tolerance to the antigens. *Microbiol Immunol* 2002;**46**:55–58.
247. Harokopakis, E., Hajishengallis, G., Michalek, S.M. Effectiveness of liposomes possessing surface-linked recombinant B subunit of cholera toxin as an oral antigen delivery system. *Infect Immun* 1998;**66**:4299–4304.
248. Ludewig, B., Barchiesi, F., Pericin, M., Zinkernagel, R.M., Hengartner, H., Schwendener, R.A. In vivo antigen loading and activation of dendritic cells via a liposomal peptide vaccine mediates protective antiviral and anti-tumour immunity. *Vaccine* 2000;**19**:23–32.

249. Chikh, G., Schutze-Redelmeier, M.P. Liposomal delivery of CTL epitopes to dendritic cells. *Biosci Rep* 2002;**22**:339–353.
250. Copland, M.J., Baird, M.A., Rades, T., McKenzie, J.L., Becker, B., Reck, F., et al. Liposomal delivery of antigens to human dendritic cells. *Vaccine* 2003;**21**:883–890.
251. Rao, M., Alving, C.R. Delivery of lipids and liposomal proteins to the cytoplasm and Golgi of antigen-presenting cells, *Adv Drug Deliv Rev* 2000;**41**:171–188.
252. Lipinski, C.A., Lombardo, F., Dominy, B.W., Feeney, P.J. Experimental and computational approaches to estimate solubility and permeability in drug discovery and development settings. *Adv Drug Deliv Rev* 2001;**46**:3–26.
253. Teicher, B.A. (ed.). *Anticancer Drug Delivery Guide*. Philadelphia: Humana Press, 1997.
254. Fernandez, A.M., Van Derpoorten, K., Dasnois, L., Lebtahi, K., Dubois, V., Lobl, T.J., et al. *N*-Succinyl-(beta-alanyl-L-leucyl-L-alanyl-L-leucyl)doxorubicin: an extracellularly tumor-activated prodrug devoid of intravenous acute toxicity. *J Med Chem* 2001;**44**:3750–3753
255. Yalkowsky, S.H. (ed.). *Techniques of Solubilization of Drugs*. New York: Marcel Dekker, 1981.
256. Shabner, B.A., Collings, G.M. (eds). *Cancer Chemotherapy: Principles and practice*. Philadelphia: JB Lippincott Co., 1990.
257. Yokogawa, K., Nakashima, E., Ishizaki, J., Maeda, H., Nagano, T., Ichimura, F. Relationships in the structure-tissue distribution of basic drugs in the rabbit. *Pharm Res* 1990;**7**:691–696.
258. Hageluken, A., Grunbaum, L., Nurnberg, B., Harhammer, R., Schunack, W., Seifert, R. Lipophilic beta-adrenoceptor antagonists and local anesthetics are effective direct activators of G-proteins. *Biochem Pharmacol* 1994;**47**:1789–1795.
259. Lipinski, C.A. Drug-like properties and the causes of poor solubility and poor permeability. *J Pharmacol Toxicol Methods* 2000;**44**:235–249.
260. Lasic, D.D., Papahadjopoulos, D. *Medical Applications of Liposomes*. New York: Elsevier, 1998: p 779.
261. Thompson, D., Chaubal, M.V. Cyclodextrins (CDS) – excipients by definition, drug delivery systems by function (part I: injectable applications). *Drug Deliv Technol* 2000;**2**:34–38.
262. Ray, R., Kibbe, A.H., Rowe, R., Shleskey, P., Weller, P. *Handbook of Pharmaceutical Excipients*. Washington, DC: APhA Publications, 2003.
263. Lasic, D.D. Mixed micelles in drug delivery. *Nature* 1992;**355**:279–280.
264. Elworthy, P.H., Florence, A.T., Macfarlane, C.B. (eds). *Solubilization by Surface Active Agents*. London: Chapman & Hall, 1968.
265. Attwood, D., Florence, A.T. (eds). *Surfactant Systems*. London: Chapman & Hall, 1983.
266. Kwon, G.D. Diblock copolymer nanoparticles for drug delivery. *Crit Rev Ther Drug Carrier Syst* 1998;**15**:481–512.
267. Kabanov, A.V., Batrakova, E.V., Alakhov, V.Yu. Pluronic block copolymers as novel polymer therapeutics for drug and gene delivery. *J Control Release* 2002;**82**:189–212.
268. Jones, M., Leroux, J. Polymeric micelles – a new generation of colloidal drug carriers. *Eur J Pharm Biopharm* 1999;**48**:101–111.
269. Torchilin, V.P. Structure and design of polymeric surfactant-based drug delivery systems. *J Control Release* 2001;**73**:137–172.
270. Zhang, L., Eisenberg, A. Multiple morphologies of 'crew-cut' aggregates of polystyrene-b-poly(acrylic acid) block copolymers. *Science* 1995;**268**:1728–1731.
271. Gref, R., Minamitake, Y., Peracchia, M.T., Trubetskoy, V.S., Torchilin, V.P., Langer, R. Biodegradable long-circulating polymeric nanospheres. *Science* 1994;**263**:1600–1603.
272. Gref, R., Domb, A., Quellec, P., Blunk, T., Muller, R.H., Verbavatz, J.M., Langer, R. The controlled intravenous delivery of drugs using PEG-coated sterically stabilized nanospheres. *Adv Drug Delivery Rev* 1995;**16**:215–234.
273. Kabanov, A.V., Chekhonin, V.P., Alakhov, V.Yu., Batrakova, E.V., Lebedev, A.S., Melik-Nubarov, N.S., et al. The neuroleptic activity of haloperidol increases after its solubilization in surfactant micelles. *FEBS Lett* 1989;**258**:343–345.
274. Hagan, S.A., Coombes, A.G.A., Garnett, M.C., Dunn, S.E., Davies, M.C., Illum, L., Davis, S.S. Polylactide-poly(ethylene glycol) copolymers as drug delivery systems, 1. Characterization of water dispersible micelle-forming systems. *Langmuir* 1996;**12**:2153–2161.

275. Inoue, T., Chen, G., Nakamae, K., Hoffman, A.S. An AB block copolymers of oligo(methyl methacrylate) and poly(acrylic acid) for micellar delivery of hydrophobic drugs. *J Control Release* 1998;**51**:221–229.
276. Müller, R.H. *Colloidal Carriers for Controlled Drug Delivery and Targeting: Modification, characterization, and in vivo distribution.* Boca Raton, FL: CRC Press, 1991.
277. Hunter, R.J. *Foundations of Colloid Science*, vol.1, New York: Oxford University Press, 1991.
278. Kuntz, R.M., Saltzman, W.M. Polymeric controlled delivery for immunization. *Trends Biotechnol* 1997;**15**:364–369.
279. Cammas, S., Suzuki, K., Sone, C., Sakurai, Y., Kataoka, K., Okano, T. Thermoresponsive polymer nanoparticles with a core-shell micelle structure as site specific drug carriers. *J Control Release* 1997;**48**:157–164.
280. Torchilin, V.P., Shtilman, M.I., Trubetskoy, V.S., Whiteman, K.R., Milstein, A.M. Amphiphilic vinyl polymers effectively prolong liposome circulation time in vivo. *Biochim Biophys Acta* 1994;**1195**:181–184.
281. Torchilin, V.P., Trubetskoy, V.S., Whiteman, K.R., Caliceti, P., Ferruti, P., Veronese, F.M. New synthetic amphiphilic polymers for steric protection of liposomes in vivo. *J Pharm Sci* 1995; **84**:1049–1053.
282. Miller, D.W., Batrakova, E.V., Waltner, T.O., Alakhov, V.Yu., Kabanov, A.V. Interactions of pluronic block copolymers with brain microvessel endothelial cells: evidence of two potential pathways for drug absorption. *Bioconj Chem* 1997;**8**:649–657.
283. Katayose, S., Kataoka, K. Remarkable increase in nuclease resistance of plasmid DNA through supramolecular assembly with poly(ethylene glycol)-poly(L-lysine) block copolymer. *J Pharm Sci* 1998;**87**:160–163.
284. Trubetskoy, V.S., Gazelle, G.S., Wolf, G.L., Torchilin, V.P. Block copolymer of polyethylene glycol and polylysine as a carrier of organic iodine: design of a long circulating particulate contrast medium for X-ray computed tomography. *J Drug Target* 1997;**4**:381–388.
285. Yokoyama, M., Miyauchi, M., Yamada, N., Okano, T., Sakurai, K., Kataoka, K., Ionue, S. Characterization and anticancer activity of the micelle-forming polymeric anticancer drug adriamycin-conjugated poly(ethylene glycol)-poly(aspartic acid) block copolymer. *Cancer Res* 1990;**50**: 1693–1700.
286. Harada, A., Kataoka, K. Novel polyion complex micelles entrapping enzyme molecules in the core. Preparation of narrowly-distributed micelles from lysozyme and poly(ethylene glycol)-poly(aspartic acid) block copolymer in aqueous medium. *Macromolecules* 1998;**31**:288–294.
287. La, S.B., Okano, T., Kataoka, K. Preparation and characterization of the micelle-forming polymeric drug indomethacin-incorporated poly(ethylene oxide)-poly(-benzyl L-aspartate) block copolymer micelles. *J Pharm Sci* 1996;**85**:85–90.
288. Kwon, G.S., Naito, M., Yokoyama, M., Okano, T., Sakurai, Y., Kataoka, Y. Block copolymer micelles for drug delivery: loading and release of doxorubicin. *J Control Release* 1997;**48**:195–201.
289. Jeong, Y.I., Cheon, J.B., Kim, S.H., Nah, J.W., Lee, Y.M., Sung, Y.K., Akaike, T., Cho, C.S. Clonazepam release from core-shell type nanoparticles in vitro. *J Control Release* 1998;**51**:169–178.
290. Kim, S.Y., Shin, I.G., Lee, Y.M., Cho, C.G., Sung, Y.K. Methoxy poly(ethylene glycol) and ε-caprolactone amphiphilic block copolymeric micelle containing indomethacin, II. Micelle formation and drug release behaviors. *J Control Release* 1998;**51**:13–22.
291. Allen, C., Yu, Y., Maysinger, D., Eisenberg, A. Polycaprolactone-b-poly(ethylene oxide) block copolymer micelles as a novel drug delivery vehicle for neurotrophic agents FK506 and L-685,818. *Bioconj Chem* 1998;**9**:564–572.
292. Ramaswamy, M., Zhang, X., Burt, H.M., Wasan, K.M. Human plasma distribution of free paclitaxel and paclitaxel associated with diblock copolymers. *J Pharm Sci* 1997;**86**:460–464.
293. Trubetskoy, V.S., Torchilin, V.P. Use of polyoxyethylene-lipid conjugates as long-circulating carriers for delivery of therapeutic and diagnostic agents, *Adv Drug Deliv Rev* 1995;**16**:311–320.
294. Lasic, D.D., Woodle, M.C., Martin, F.J., Valentincic, T. Phase behavior of 'stealth-lipid' decithin mixtures. *Period Biol* 1991;**93**:287–290.
295. Bedu-Addo, F.K., Tang, P., Xu, Y., Huang, L. Interaction of polyethyleneglycol-phospholipid conjugates with cholesterol-phosphatidylcholine mixtures: sterically stabilized liposome formulations. *Pharm Res* 1996;**13**:718–724.

296. Edwards, K., Johnsson, M., Karlsson, G., Silvander, M. Effect of polyethyleneglycol-phospholipids on aggregate structure in preparations of small unilamellar liposomes. *Biophys J* 1997;**73**:258–266.
297. Trubetskoy, V.S., Torchilin, V.P. Polyethyleneglycol based micelles as carriers of therapeutic and diagnostic agents. *STP Pharm Sci* 1996;**6**:79–86.
298. Lukyanov, A.N., Gao, Z., Mazzola, L., Torchilin, V.P.: Polyethylene glycol-diacyllipid micelles demonstrate increased accumulation in subcutaneous tumors in mice. *Pharm Res* 2002;**19**:1424–1429.
299. Lukyanov, A.N., Hartner, W.C., Torchilin, V.P.: Increased accumulation of PEG-PE micelles in the area of experimental myocardial infarction in rabbits. *J Control Release* 2004;**94**:187–193.
300. Wang, J., Mongayt, D.A., Lukyanov, A.N., Levchenko, T.S., Torchilin, V.P.: Preparation and in vitro synergistic anticancer effect of Vitamin K# and 1,8-diazabicyclo[5,4,0]undec-7-ene in poly(ethylene glycol)-diacyllipid micelles. *Int J Pharm* 2004;**272**:129–135.
301. Torchilin, V.P., Lukyanov, A.N., Gao, Z., Papahadjopoulos-Sternberg B: Immunomicelles: Targeted pharmaceutical carriers for poorly soluble drugs. *Proc Natl Acad Sci USA* 2003;**100**:6039–6044.
302. Krishnadas, A., Rubinstein, I., Onyuksel, H.: Sterically stabilized phospholipid mixed micelles: in vitro evaluation as a novel carrier for water-insoluble drugs. *Pharm Res* 2003;**20**:297–302.
303. Gao, Z., Lukyanov, A.N., Singhal, A., Torchilin, V.P.: Diacyl-polymer micelles as nanocarriers for poorly soluble anticancer drugs. *Nano Lett* 2002;**2**:979–982.
304. Gao, Z., Lukyanov, A.N., Chakilam, A.R., Torchilin, V.P. PEG-PE/Phosphatidylcholine mixed Immunomicelles specifically deliver encapsulated taxol to tumor cells of different origin and promote their efficient killing. *J Drug Target* 2003;**11**:87–92.
305. Alkan-Onyuksel, H., Ramakrishnan, S., Chai, H.B., Pezzuto, J.M. A mixed micellar formulation suitable for the parenteral administration of taxol, *Pharm Res* 1994;**11**:206–212.
306. Maeda, H., Wu, J., Sawa, T., Matsumura, Y., Hori, K. Tumor vascular permeability and the EPR effect in macromolecular therapeutics: a review. *J Control Release* 2000;**65**:271–284.
307. Weissig, V., Whiteman, K.R., Torchilin, V.P. Accumulation of protein-loaded long-circulating micelles and liposomes in subcutaneous Lewis lung carcinoma in mice. *Pharm Res* 1998;**15**:1552–1556.
308. Gabizon, A.A. Pegylated liposomal doxorubicin: metamorphosis of an old drug into a new form of chemotherapy. *Cancer Invest* 2001;**19**:424–436.
309. Papahadjopoulos, D., Allen, T.M., Gabizon, A., Mayhew, E., Matthay, K., Huang, S.K., et al. Sterically stabilized liposomes: improvements in pharmacokinetics and antitumor therapeutic efficacy. *Proc Natl Acad Sci USA* 1991;**88**:11460–11464.
310. Gabizon, A.A. Selective tumor localization and improved therapeutic index of anthracyclines encapsulated in long-circulating liposomes. *Cancer Res* 1992;**52**:891–896.
311. Hobbs, S.K., Monsky, W.L., Yuan, F., Roberts, W.G., Griffith, L., Torchilin, V.P., Jain, R.K. Regulation of transport pathways in tumor vessels: role of tumor type and microenvironment. *Proc Natl Acad Sci USA* 1998;**95**:4607–4612.
312. Helmlinger, G., Yuan, F., Dellian, M., Jain, R.K. Interstitial pH and pO_2 gradients in solid tumors in vivo: high-resolution measurements reveal a lack of correlation. *Nature Med* 1997;**3**:177–182.
313. Tannock, I.F., Rotin, D. Acid pH in tumors and its potential for therapeutic exploitation. *Cancer Res* 1989;**49**:4373–4384.
314. Kwon, G.S., Okano, T. Soluble self-assembled block copolymers for drug delivery. *Pharm Res* 1999;**16**:597–600.
315. Chung, J.E., Yokoyama, M., Aoyagi, T., Sakurai, Y., Okano, T. Effect of molecular architecture of hydrophobically modified poly(*N*-isopropylacrylamide) on the formation of thermoresponsive core-shell micellar drug carriers *J Control Release* 1998;**53**:119–131.
316. Kohori, F., Sakai, K., Aoyagi, T., Yokoyama, M., Sakurai, Y., Okano, T. Preparation an characterization of thermally responsive block copolymer micelles comprising poly(N-isopropylacrylamide-b-DL-lactide). *J Control Release* 1998;**55**:87–98.
317. Meyer, O., Papahadjopoulos, D., Leroux, J.C. Copolymers of *N*-isopropylacrylamide can trigger pH sensitivity to stable liposomes. *FEBS Lett* 1998;**41**:61–64.
318. Le Garrec, D., Taillefer, J., Van Lier, J.E., Lenaerts, V., Leroux, J.C. Optimizing pH-responsive polymeric micelles for drug delivery in a cancer photodynamic therapy model. *J Drug Target* 2002;**10**:429–437.

319. Yoo, H.S., Lee, E.A., Park, T.G. Doxorubicin-conjugated biodegradable polymeric micelles having acid-cleavable linkages. *J Control Release* 2002;**82**:17–27.
320. Chung, J.E., Yokoyama, M., Yamato, M., Aoyagi, T., Sakurai, Y., Okano, T. Thermo-responsive drug delivery from polymeric micelles constructed using block copolymers of poly(N-isopropylacrylamide) and poly(butylmethacrylate). *J Control Release* 1999;**62**:115–127.
321. Nagasaki, Y., Yasugi, K., Yamamoto, Y., Harada, A., Kataoka, K. Sugar-installed block copolymer micelles: their preparation and specific interaction with lectin molecules. *Biomacromolecules* 2001;**2**:1067–1070.
322. Vinogradov, S., Batrakova, E., Li, S., Kabanov, A. Polyion complex micelles with protein-modified corona for receptor-mediated delivery of oligonucleotides into cells. *Bioconj Chem* 1999;**10**: 851–860.
323. Leamon, C.P., Weigl, D., Hendren, R.W. Folate copolymer-mediated transfection of cultured cells. *Bioconj Chem* 1999;**10**:947–957.
324. Jule, E., Nagasaki, Y., Kataoka, K. Lactose-installed poly(ethylene glycol)-poly(D,L-lactide) block copolymer micelles exhibit fast-rate binding and high affinity toward a protein bed simulating a cell surface. A surface plasmon resonance study, *Bioconjug Chem* 2003;**14**:177–186.
325. Ogris, M., Brunner, S., Schuller, S., Kircheis, R., Wagner, E. PEGylated DNA/transferrin-PEI complexes: reduced interaction with blood components, extended circulation in blood and potential for systemic gene delivery. *Gene Ther* 1999;**6**:595–605.
326. Dash, P.R., Read, M.L., Fisher, K.D., Howard, K.A., Wolfert, M., Oupicky, D., et al. Decreased binding to proteins and cells of polymeric gene delivery vectors surface modified with a multivalent hydrophilic polymer and retargeting through attachment of transferrin. *J Biol Chem* 2000;**275**: 3793–3802.
327. Leamon, C.P., Low, P.S. Folate-mediated targeting: from diagnostics to drug and gene delivery. *Drug Discov Today* 2001;**6**:44–51.
328. Lee, E.S., Na, K., Bae, Y.H. Polymeric micelle for tumor pH and folate-mediated targeting. *J Control Release* 2003;**91**:103–113.
329. Chekhonin, V.P., Kabanov, A.V., Zhirkov, Y.A., Morozov, G.V. Fatty acid acylated Fab-fragments of antibodies to neurospecific proteins as carriers for neuroleptic targeted delivery in brain. *FEBS Lett* 1991;**287**:149–152.
330. Iakoubov, L.Z., Torchilin, V.P. A novel class of antitumor antibodies: nucleosome-restricted antinuclear autoantibodies (ANA) from healthy aged nonautoimmune mice. *Oncol Res* 1997;**9**: 439–446.
331. Torchilin, V.P., Iakoubov, L.Z., Estrov, Z. Therapeutic potential of antinuclear autoantibodies in cancer. *Cancer Ther* 2003;**1**:179–190.
332. Felgner, J.H., Kumar, R., Sridhar, C.N., Wheeler, C.J., Tsai, Y.J., Border, R., et al. Enhanced gene delivery and mechanism studies with a novel series of cationic lipid formulations. *J Biol Chem* 1994;**269**:2550–2561.
333. Ota, T., Maeda, M., Tatsuka, M., Cationic liposomes with plasmid DNA influence cancer metastatic capability. *Anticancer Res* 2002;22;4049–4052.
334. Kaiser, S., Toborek, M. High-efficiency transfection of human endothelial cells mediated by cationic lipids. *J Vasc Res* 2003;**38**:133–143.
335. Almofti, M.R., Harashima, H., Shinohara, Y., Almofti, A., Baba, Y., Kiwada, H. Cationic liposome-mediated gene delivery: biophysical study and mechanism of internalization. *Arch Biochem Biophys* 2003;**410**:246–253.
336. Hafez, I.M., Maurer, N., Cullis, P.R. On the mechanism whereby cationic lipids promote intracellular delivery of polynucleic acids. *Gene Ther* 2001;**8**:1188–1196.
337. Wang, J., Mongayt, D., Torchilin, V.P. Polymeric micelles for delivery of poorly soluble drugs: preparation and anticancer activity in vitro of paclitaxel incorporated into mixed micelles based on poly(ethylene glycol)-lipid conjugate and positively charged lipids. *J Drug Target* 2005;**13**:73–80.

CHAPTER 4

PRINCIPLES IN THE DEVELOPMENT OF INTRAVENOUS LIPID EMULSIONS

Joanna Rossi and Jean-Christophe Leroux

4.1 INTRODUCTION

4.2 EMULSION STABILITY

4.3 ELIMINATION MECHANISMS FOR LIPID EMULSIONS

4.4 BIODISTRIBUTION OF LIPID EMULSIONS

4.5 PREPARATION OF EMULSIONS FOR INTRAVENOUS ADMINISTRATION

4.6 LIPID EMULSIONS FOR THE DELIVERY OF NUCLEIC ACID-BASED DRUGS

4.7 CONCLUSIONS

4.1 INTRODUCTION

Emulsions can be defined as heterogeneous mixtures of two immiscible liquids, in which one phase is dispersed as fine droplets in the other. Small oil droplets dispersed in a continuous water phase is termed an 'oil-in-water' (o/w) emulsion. The opposite of this system is a 'water-in-oil' (w/o) emulsion, whereby the water phase is dispersed in an oily external medium. Among these types, only o/w emulsions can be used for intravenous administration [1]. Emulsions are thermodynamically unstable systems that will eventually destabilize into two separate phases. A third component, the surfactant or emulsifier, is added to stabilize the preparation by reducing the interfacial tension and increasing droplet–droplet repulsion through electrostatic and/or steric repulsive forces [2]. The addition of an emulsifying agent, however,

Role of Lipid Excipients in Modifying Oral and Parenteral Drug Delivery, Edited by Kishor M. Wasan

provides only kinetic stability. Even though emulsions are unstable systems, surface active agents may provide stability for several years, making the system useful for practical application [2].

Lipid emulsions have traditionally been used for parenteral nutrition to deliver essential fatty acids to patients unable to acquire them in food. As a result of the successful induction of lipid emulsions in parenteral nutrition, there has been increasing interest in developing emulsions as carriers for lipophilic drugs. Many intravenous lipid emulsion formulations are commercially available (Table 4.1) and a number of others are in clinical phase or in preclinical development (Table 4.2). Lipid emulsions are promising carriers for drug delivery as a result of their biocompatibility, reasonable stability, ability to solubilize high quantities of hydrophobic compounds, and relative ease of manufacture on an industrial scale [3, 4]. In addition, emulsions can protect the encapsulated drug against hydrolysis and enzymatic degradation in the blood compartment, reduce drug loss in infusion sets, lower the toxicity of cytotoxic compounds, and reduce the incidence of irritation and pain on injection [1, 4]. They can also provide a certain level of selectivity toward target tissues, increasing the therapeutic index of many drugs [5]. However, after intravenous injection, lipid emulsions can acquire apolipoproteins and be metabolized as natural fats or be recognized as foreign bodies and taken up by the cells of the mononuclear phagocyte system (MPS – also known as the reticuloendothelial system [RES]) [6]. Evading the MPS or natural fat metabolism is necessary when the encapsulated drug is to be delivered to non-MPS organs or liver parenchymal cells, respectively. The in vivo fate of lipid emulsions can be controlled to a certain extent by altering the physicochemical properties of the carrier, such as droplet size, composition and surface properties. This chapter discusses the main factors to consider when devleoping emulsions for intravenous injection.

TABLE 4.1 Several commercially available emulsions for intravenous injection

Product	Drug	Manufacturer	Indications
Diazemuls	Diazepam	Pfizer and Pharmacia	Anticonvulsive, sedative, muscle relaxant
Diazepam-Lipuro	Diazepam	B. Braun	General anesthesia
Diprivan[a]	Propofol	AstraZeneca	General anesthesia
Etomidate-Lipuro	Etomidate	B. Braun	General anesthesia
Limethason[a]	Dexamethasone palmitate	Mitsubishi Pharmaceutical	Rheumatoid arthritis
Liple (Lipo-PGE1)	Prostaglandin E1 (PGE1)	Mitsubishi Pharmaceutical	Vasodilator, platelet inhibitor
Propofol-Lipuro	Propofol	B. Braun	General anesthesia
Lipo-NSAID – Ropion[a]	Flurbiprofen axetil	Kaken Pharmaceuticals	Pain reliever
Vitalipid[a]	Vitamins A, D_2, E, K_1	Fresenius Kabi	Parenteral nutrition

[a]Based on the formulation of Intralipid (10 or 20% soybean oil; 1.2% egg lecithin; 2.5% glycerol).

TABLE 4.2 Some intravenously injectable emulsions in development and in clinical trials

Drug	Product name	Indications	Clinical[a]	Reference
Aclacinomycin A	–	Cancer chemotherapy	–	[110]
Amphotericin B	–	Treatment of fungal infections	–	[97, 98]
β-Elemene	SDP-111	Cancer chemotherapy	–	[111]
Cyclosporine A	–	Immunosuppressant	–	[112]
Docetaxel	SDP-014	Cancer chemotherapy	–	[111]
Perfluorooctyl bromide	Imavist	Ultrasound contrast agent	Phase III	[113]
Paclitaxel	TOCOSOL/S-8184	Cancer chemotherapy	Phase III	[72]
Paclitaxel	SDP-013	Cancer chemotherapy	–	[111]
Perfluorooctyl bromide	Oxygent	Artificial blood substitute	Phase III	[78]
Propofol	Ampofol	General anesthesia Sedation	Phase III	[114]
Propofol	IDD-D propofol	General anesthesia Sedation	Phase III	[115]
α-Tocopherol succinate	SDP-112	Cancer chemotherapy	–	[111]
Vincristine	–	Cancer chemotherapy	–	[116]

[a]Status as of March 2006.

4.2 EMULSION STABILITY

Emulsions are thermodynamically unstable systems and will inevitably break apart into separate oil and water phases. Emulsion instability is caused by the increase in surface free energy (ΔG) as small droplets are formed as a result of the enhanced surface area (ΔA). Adding a surfactant to the mixture reduces the interfacial tension (γ_{ow}) at the oil–water interface facilitating globule rupture during emulsification and stabilizes the preparation (Eqn 1).

$$\Delta G = \gamma_{ow} \, \Delta A \tag{1}$$

It is important to state that surfactants provide the emulsions only with kinetic stability, which delays the destabilization process. Nevertheless, surface-active agents can provide stability for several years, which is long enough for the system to be useful for practical purposes [2]. Emulsions that are thermodynamically stable are known as microemulsions. They are clear or translucent systems and do not require much energy input during emulsification. In contrast, emulsions are cloudy and require a greater amount of energy for emulsification [7]. The theory behind the formation of microemulsions is beyond the scope of this chapter.

4.2.1 Destabilization Processes

Emulsion destabilization can be characterized by three separate processes: flocculation, coalescence, and Ostwald ripening. Coalescence and Ostwald ripening are irreversible processes that lead to an increase in droplet size, requiring a large energy input to re-disperse the droplets. Flocculation, on the other hand, is reversible and occurs when droplets aggregate to form a clump of many individual droplets. The aggregated droplets move together as a cluster but each droplet still retains its separate identity. The interactions holding the droplets together are weak and can be broken by mild agitation. Even though floccules can be easily re-dispersed, they may eventually fuse together to form single, larger globules. The fusion of droplets is irreversible and is termed 'coalescence'. Ostwald ripening, which also increases droplet size, occurs in polydisperse formulations, wherein the smaller droplets are more soluble in the continuous phase than the larger ones. In this process, the oil from the smaller droplets dissolves in the aqueous phase and diffuses towards the larger droplets. This transfer of oil causes the big droplets to grow, while the smaller ones decrease in size. As the small droplets continue to shrink, the Ostwald ripening effect is enhanced. The progressive increase in droplet size over time will eventually lead to complete phase separation. Adding too much surfactant may promote Ostwald ripening because the excess surfactant will form micelles that enhance the solubility of the oil in the aqueous phase. Ostwald ripening can be reduced by increasing the viscosity of the continuous phase, decreasing polydispersity, or adding a third component that has a lower solubility in the continuous phase than the oil [8, 9].

Depending on the density differences between the dispersed and continuous phases, individual droplets or floccules can cream or sediment. If the dispersed phase is lower in density than the continuous phase, the droplets or floccules will rise to the surface, producing a highly concentrated layer of dispersed phase, which is known as a cream. In the case where the dispersed phase is higher in density than the continuous phase, a sediment will form at the bottom of the formulation. For o/w emulsions, creaming usually occurs because the oil phase is typically less dense than the aqueous phase. The rate of creaming or sedimentation can be linked to the size of the droplet by Stokes' equation (Eqn 2). According to this equation the limiting velocity of a falling sphere (υ) is:

$$\upsilon = \frac{2}{9}\frac{a^2 \Delta\rho}{\nu} g \qquad (2)$$

where α is the radius of the droplet, $\Delta\rho$ is the density difference between the dispersed and continuous phases, ν the viscosity of the continuous phase, and g the acceleration caused by gravity. Stokes' equation implies that droplets will rise or settle faster if the droplet size or the density difference between the dispersed and continuous phases is large, whereas an increase in continuous phase viscosity will slow down the separation process. As a result, creaming or sedimentation can be delayed by reducing droplet size, decreasing the density differences between the two phases and increasing the viscosity of the continuous phase. Not much emphasis is, however, being placed on density adjustments to produce stable emulsions because there are a limited number of oils approved for intravenous administration and these oils have similar densities.

Submicrometer emulsions have colloidal properties and as a result are less susceptible than coarse emulsions to the gravitational forces in Stokes' equation [10]. Nano-sized droplets are subjected to random brownian motion and consequently are less inclined to cream or sediment. Brownian motion does not, however, provide complete protection against instability because droplets may aggregate or coalesce on random collisions. Stability against these collisions depends on the attractive and repulsive forces acting on the droplets. Typically, emulsions are stabilized by either electrostatic or steric repulsive forces (or a combination of the two).

4.2.2 Electrostatic Stabilization

The balance between attractive van der Waals' forces and electrostatic repulsive forces is described in the theory of colloidal stability, termed 'DLVO' after its developers Derjaguin, Landau, Vervey, and Overbeek. If the net force is attractive, the droplets will either flocculate or coalesce. In contrast, if the net force is repulsive, the particles will repel each other and the system is stable. The attractive interaction between particles arises from van der Waals' forces and is experienced by all particles. Van der Waals' forces dominate at short separation distances and the strength of this attractive force can be determined from the magnitude of the Hamaker constant (A). Emulsions can overcome the attractive van der Waals' forces through electrostatic repulsion with charged emulsifying agents. Electrostatic repulsion is provided by the electric double-layer surrounding the droplet. The electric double-layer is characterized by an adsorbed layer of fixed counter-ions and a diffuse layer of ions that move freely with the fluid. Two approaching particles will experience a repulsive force as the electric double-layers overlap. The total potential of interaction between two droplets is the sum of the attractive van der Waals' forces and the electrostatic repulsive forces (Eqn 3);

$$V_T = V_A + V_R \tag{3}$$

where V_T is the total interaction potential, V_A represents the attractive van der Waals' forces, and V_R signifies the electrostatic repulsive forces. The potential energy of interaction between two droplets as a function of separation distance is illustrated in Figure 4.1. The repulsive barrier generated by the electric double-layer corresponds to the maximum in the curve. The height of the energy barrier determines the stability of the emulsion and depends on the ionization of the surfactants.

For the system to be stable, the energy barrier must be high enough such that the droplets do not have enough kinetic energy to surpass it and reach the primary minimum. At the primary minimum (maximum attractive potential), droplet coalescence readily occurs. Flocculation takes place at the secondary minimum and, contrary to coalescence, is reversible by providing a small amount of kinetic energy to overcome the weak attractive forces holding the droplets together. Flocculated

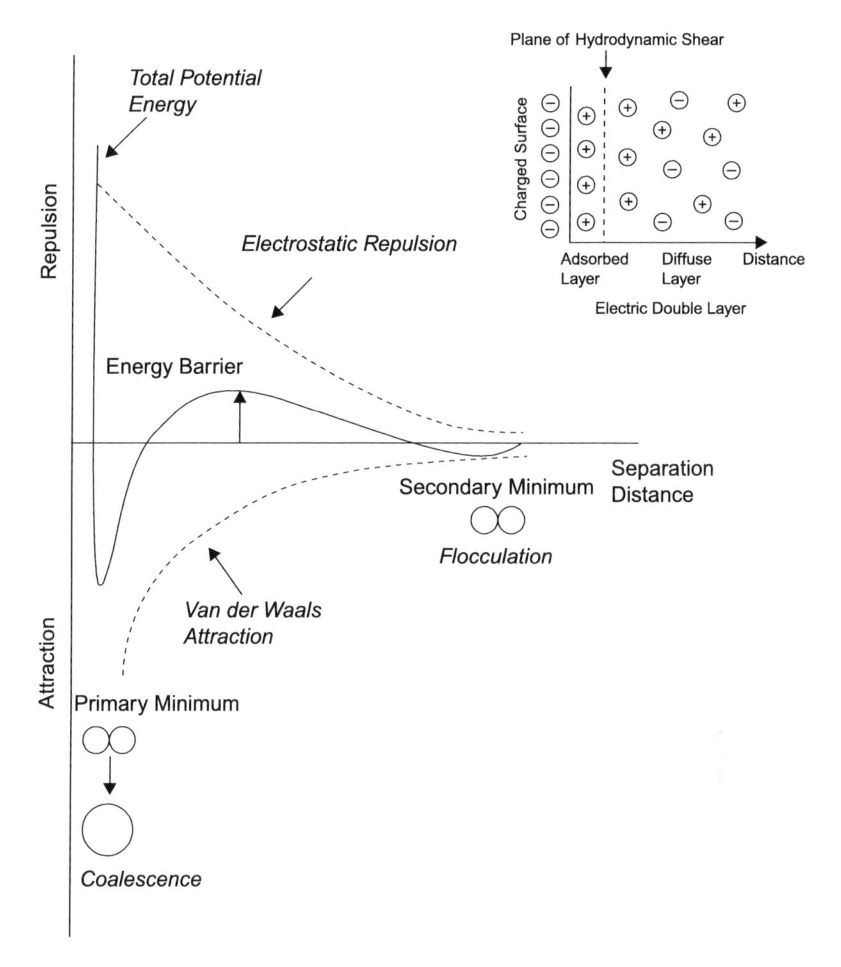

Figure 4.1 The total potential energy of interaction between two droplets as a function of separation distance (electric double-layer repulsion and van der Waals' attraction).

droplets are prevented from coalescing as a result of this repulsive energy barrier. If the flocculated droplets have enough energy to surpass the energy barrier, they will easily reach the primary minimum and coalesce. The strength of the electrostatic forces can be quantified by measuring the ζ (zeta) potential, which is the potential at the plane of hydrodynamic shear. Generally, emulsions are stabilized by electrostatic repulsive forces if the ζ potential is greater than $\pm 30\,\text{mV}$ [2, 8, 11]. An emulsion stabilized by electric double-layer repulsion can be destabilized if the concentration of electrolytes is increased above a critical value. Adding electrolytes to an emulsion decreases the electric double-layer repulsion potential, whereas the van der Waals' attractive potential remains unchanged. As electrolyte concentration increases, the repulsive forces stabilizing the colloid become weaker until the net force is attractive and stability is lost.

4.2.3 Steric Stabilization

Emulsions can also be stabilized by steric repulsion through the grafting of long-chain polymers at the emulsion interface. Steric repulsion is a non-DLVO interaction that occurs as a result of the unfavorable overlap of the polymer chains as two particles approach each other [8, 12]. Steric stabilization occurs at short interdroplet separation distances and can provide a strong barrier against coalescence [8]. Optimal steric repulsion can be achieved at high polymer surface density as desorption and chain rearrangement are minimized [8].

4.3 ELIMINATION MECHANISMS FOR LIPID EMULSIONS

After intravenous injection, lipid emulsions may be metabolized in a manner similar to chylomicrons, or might be recognized as foreign bodies and removed by the cells of the MPS. The mechanism of elimination from the body depends on the physicochemical properties of the emulsion. Both mechanisms of elimination can occur for a given lipid emulsion; however, one process may be favored over another. This section describes the two primary pathways of lipid emulsion elimination from the body.

4.3.1 Lipid Emulsions Metabolized as Endogenous Chylomicrons

Depending on the composition and surface properties, lipid emulsions may be recognized as chylomicrons and eliminated via the fat metabolism pathway. Chylomicrons are endogenous emulsions produced by the enterocytes of the small intestine after dietary lipids have been ingested. They are rich in triglycerides and possess apolipoproteins A-I, A-IV, and B-48 before entering the blood circulation (Figure 4.2) [13]. Chylomicrons are secreted into the lymph and enter the systemic circulation through the thoracic duct. After entering the blood, chylomicrons obtain the apolipoproteins Apo-C-II and Apo-E from the high-density lipoproteins (HDLs) and release Apo-A-IV. In the capillaries of adipose tissues and muscle, lipoprotein lipase (LPL) located on endothelial cells adsorbs on to the mature chylomicron and hydrolyzes the triglycerides to fatty acids [14]. The fatty acids are then absorbed mainly by adipose tissues and muscle. During lipolysis, a substantial amount of phospholipid, Apo-A and Apo-C is transferred to the HDLs and the size of the chylomicron is reduced considerably. The remnant chylomicrons, composed of mainly Apo-B-48, Apo-E, and cholesterol, are quickly removed from the blood by the liver. The uptake of remnant chylomicrons by the liver occurs via two Apo-E-specific recognition sites on parenchymal cells, which are the low-density lipoprotein receptor (LDLr) and the remnant receptor [15–17].

Injectable lipid emulsions differ from chylomicrons in that they do not have apolipoproteins on the surface before entering the bloodstream, although they may acquire them after systemic injection. Emulsions rich in triglycerides are known to acquire apolipoproteins (Apo-C-I, -C-II, -C-III, -E and possibly -A-IV), mainly from

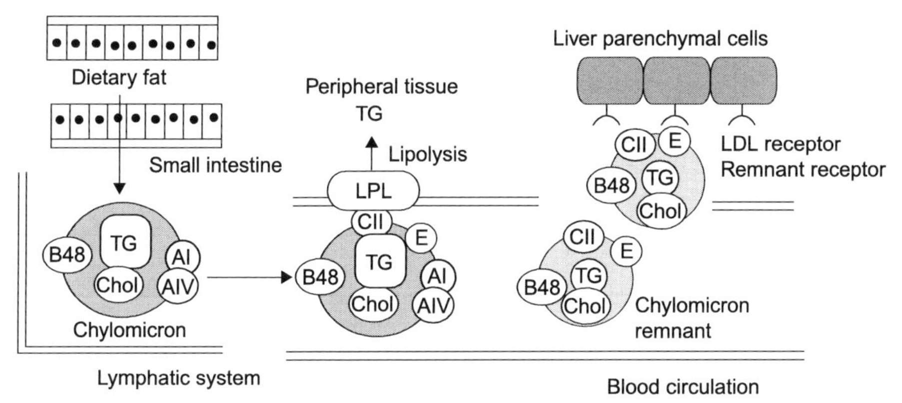

Figure 4.2 Absorption and metabolism of dietary fat. Dietary fats are metabolized and incorporated into chylomicrons in the small intestine. Then chylomicrons enter the blood circulation via the thoracic duct. During circulation, the triglycerides of chylomicrons are rapidly hydrolyzed via lipoprotein lipase (LPL) on the endothelial surfaces, and chylomicron remnants are produced. Finally, chylomicron remnants are cleared via the liver by the low-density lipoprotein (LDL) or remnant receptors. TG, triglyceride; Chol, cholesterol; AI, apolipoprotein AI; AIV, apolipoprotein AIV; B48, apolipoprotein B48; CII, apolipoprotein CII; E, apolipoprotein E. (Reprinted wih permission from Elsevier Ref. [13] Copyright 2000.)

HDLs, soon after injection into the systemic circulation, and are metabolized in a pathway comparable to that described for chylomicrons [15, 18, 19]. Among the apolipoproteins acquired, Apo-C-II and Apo-E are essential for LPL activation and uptake of remnant emulsions by the liver, respectively [19].

Elimination of the lipid emulsion via the pathway of natural fat metabolism may be desirable when the liver parenchymal cells are the target site. On the other hand, if the target site is not the liver, apolipoprotein adsorption on to the emulsion should be avoided. The metabolism of lipid emulsions as natural fats depends strongly on the type of emulsifier [20, 21], presence of cholesterol [22], and chain length of the triglyceride oil [23].

4.3.2 Elimination by the Mononuclear Phagocyte System

If the body recognizes the lipid emulsions as foreign, they will be captured by the cells of the MPS, mainly the Kupffer cells of the liver and the macrophages of the spleen, and removed from the systemic circulation. The MPS takes up the emulsions via endocytosis and localizes them in the lysosomal compartment, where they are degraded by enzymes [24]. The extent of emulsion clearance from the systemic circulation is enhanced by the adsorption of opsonins (proteins) on to the colloid surface. The bound proteins then interact with the receptors on monocytes and macrophages, facilitating endocytosis. Carriers that become bound to opsonins will be rapidly cleared from the blood and prevented from reaching the target site(s) [24]. Immunoglobins and complement components such as C1q and C3 fragments (C3b, iC3b) are well-known opsonins.

A major challenge in drug delivery using colloidal nano-carriers is to avoid clearance by the cells of the MPS when the target sites are non-MPS tissues. Overloading or saturating the MPS with large injection volumes has been shown to enhance the circulation time of lipid emulsions [25]. However, temporary impairment of the MPS may pose a health hazard to the patient [26]. Alternatively, the clearance rate of carriers from the blood can be altered by modifying the physicochemical properties of the emulsion, such as droplet size [27, 28] and surface characteristics [29]. This is discussed in detail in the next section.

4.4 BIODISTRIBUTION OF LIPID EMULSIONS

The biodistribution of an emulsion after systemic injection depends primarily on the droplet size, composition, and surface properties. A certain specificity toward the target site can be achieved by controlling the physicochemical properties of the emulsion. The principal factors that influence the biodistribution of emulsions has already been very thoroughly reviewed by Nishikawa [6]. This section provides a brief overview of these factors and has been updated with some recent work.

4.4.1 Effect of Lipid Emulsion Size

It is well known that droplet size greatly influences the uptake of the emulsions by the MPS [27, 28]. In general, larger particles are more susceptible to uptake by the MPS and are cleared more quickly from the systemic circulation. The influence of droplet size on the in vivo biodistribution of lipid emulsions was explored by Takino et al. [27]. The authors compared the biodistribution of large (250 nm) and small (100 nm) lipid emulsions composed of egg phosphatidylcholine (PC):soybean oil (1 : 1). [^{14}C]Cholesteryl oleate ([^{14}C]CO), a highly lipophilic compound (log P = 18.3) that does not undergo lipolysis by LPL and remains associated with the emulsion, was incorporated into each emulsion to track the elimination of the whole droplet [30]. The large PC emulsion was rapidly eliminated from the blood, with 60% of the injected emulsion being recovered in the liver within 10 min. The small PC emulsion, however, remained in the blood for longer and accumulated less in the liver.

Similarly, Lundberg et al. [28] reported that droplet size influenced emulsion clearance rate from plasma. They observed that the smallest emulsion (50 nm) survived the longest in plasma, whereas the larger emulsions (100 and 175 nm) were cleared more rapidly (Figure 4.3). The influence of emulsion-like lipid nanocapsule size (20, 50, and 100 nm) on the extent of complement activation and macrophage untake was evaluated by Vonarbourg et al. [31]. Similar to emulsions, lipid nanocapsules are core-shell structures with an oily internal phase that is stabilized by a monolayer of emulsifiers. They differ from lipid emulsions in the physicochemical properties of the hydrophilic/hydrophobic interface. In lipid nanocapsules, the emulsifiers form a semi-rigid shell, whereas the interface is more fluid in emulsions. The authors observed that larger lipid nanocapsules were stronger activators of the complement and taken up more by macrophages than the smaller ones.

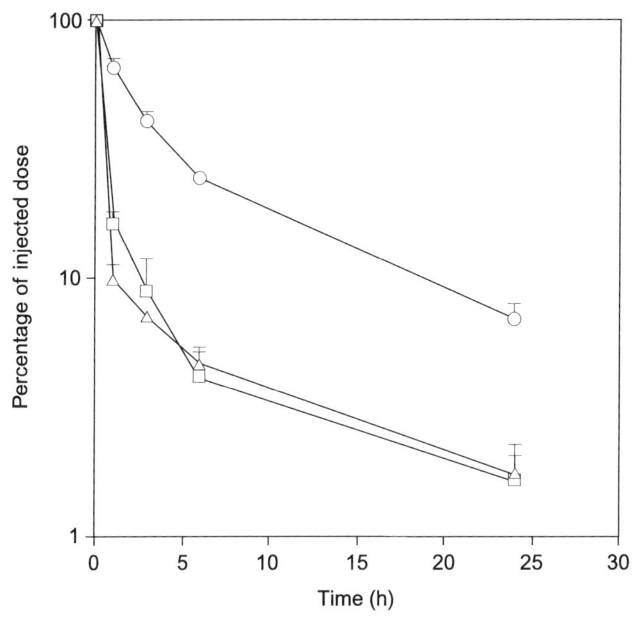

Figure 4.3 Effect of particle size on the clearance of cholesteryl oleate (CO) label from plasma as a function of time after intravenous administration into mice. The emulsions were composed of triolein (TO):1,2-dipalmitoyl-*sn*-glycero-3-phosphatidylcholine (DPPC):Polysorbate 80:polyethyleneglycol-modified 1,2-dipalmitoyl-*sn*-glycero-3-phosphatidylethanolamine (PEG_{2000}-DPPE) (2:1:0.4:0.1, w/w). The droplet sizes of the emulsions injected were 50 (○), 100 (□), and 175 nm (Δ). (Adapted with permission from Elsevier Ref. [28] Copyright 1996.)

The size of the lipid emulsion was also shown to influence lipolysis. Kurihara et al. [32] found that the rate of lipolysis was much faster for the small-sized emulsions (about 100 nm) in vitro compared with the larger ones (225–416 nm). However, after intravenous injection of these formulations in rats, they observed that the small-sized emulsions remained in the plasma longer than the larger ones, which is consistent with the studies of Takino [27] and Lundberg [28]. Consequently, even though small emulsions were better substrates for LPL, large emulsions were cleared from the blood faster, which suggests a greater uptake of by the MPS.

Droplet size also determines the ability of the emulsion to escape the systemic circulation through the blood capillaries and reach the extravascular space. Capillary walls are composed of a single layer of endothelial cells surrounded by a basement membrane. They are classified into three types, based on their wall structure: continuous (intact), fenestrated, or discontinuous (sinusoidal) [33]. Both fenestrated and discontinuous capillaries have pores in the endothelium, whereas continuous ones have tight junctions between adjacent endothelial cells [34]. Continuous capillaries have an intact subendothelial basement membrane and can be found in most regions of the body, such as the skin, connective tissue, skeletal and cardiac muscle, alveolar capillaries of the lung, and brain [33]. In fenestrated capillaries, the pores (fenestrae) are approximately 40–80 nm in diameter and they can be either open (unobstructed) or covered by a thin diaphragm [33]. These capillaries have a continuous subendothelial basement membrane and are situated in the intestinal mucosa, pancreas, glomerulus, peritubular capillaries, endocrine glands, the choroid plexus of the brain, and the ciliary body of the eye [33]. Discontinuous capillaries on the other hand, have large gaps between endothelial cells and are located in the liver, spleen, and bone marrow [33]. The basal membrane is either absent, which is the case for the liver or discontinuous (spleen and bone marrow) [34]. The largest pore size in the capillary endothelium is believed to be about 100 nm [35]. Nanoscopic drug carriers are generally too large to diffuse across the capillaries of continuous endothelium. Their best opportunity for escaping the systemic circulation is through the gaps between the endothelial cells of discontinuous capillaries. Consequently, colloidal drug carriers tend to accumulate in the liver, spleen, and bone marrow.

Control over carrier size can impart some selectivity for the extravascular space of tumoral sites, reducing anticancer drug toxicity toward healthy tissues. This selectivity can be achieved by taking advantage of the difference in capillary structure between tumors and normal tissues. Tumor vasculature is often characterized as porous or 'leaky', allowing enhanced permeation of colloidal particles across the endothelium and into the extravascular space. In addition, tumors have poor lymphatic drainage, allowing colloids to be retained in the tissue for longer periods of time [36]. This increased permeation and retention of colloids is called the enhanced permeation and retention (EPR) effect [37]. The optimum size range for colloidal particle accumulation in tumors is generally accepted to be about 50–200 nm [38]. Particles in this size range can be convected from the blood vessel into the extravascular space through the porous vasculature of the tumor. Depending on the porosity of the tumor capillaries, particles above 200 nm may not pass through the pores and will be eliminated more quickly by the MPS. On the other hand, particles less than 50 nm will easily extravasate through the discontinuous endothelium of the liver, spleen, and bone marrow.

As a rule of thumb, for successful accumulation of drug in the tumor by the EPR effect, the concentration of colloidal carriers in the plasma must remain high for more than 6 h [39]. The progressive extravasation of the carrier into the tumor tissue over several hours will result in increasing concentrations of anticancer drug in the vicinity of the cancer cells. Kurihara et al. [40] demonstrated that lipid emulsions under 230 nm in diameter could deliver more RS-1541, a highly lipophilic antitumor agent (13-*O*-palmitoyl-rhizoxin), to the tumor site (M5076 sarcoma cells) than larger droplets (Figure 4.4). The low concentrations of RS-1541 detected in the tumor for the larger emulsions is most probably a result of the impermeability of the leaky tumor capillaries to large particles and their faster removal rates from blood. It was also observed that emulsions larger than 220 nm reduced the toxicity of RS-1541 as shown by the higher maximum tolerated dose (MTD) with increasing size (Table 4.3). All emulsions regardless of size (70–380 nm) suppressed tumor growth and improved survival at the MTD. The medium-sized emulsions (220 nm), however, displayed the highest antitumor activity at the MTD as a result of the permeability of the tumor vasculature for the emulsions and reduced toxicity, permitting the injection of a higher dose. Hence, lipid emulsions can augment the delivery

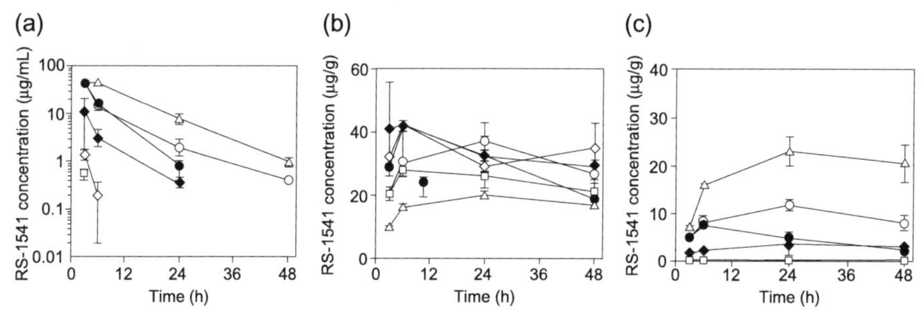

Figure 4.4 Concentrations of 13-*O*-palmitoyl-rhizoxin (RS-1541) in the (a) plasma, (b) liver, and (c) tumor after a single intravenous administration of various sizes of emulsion formulations and a surfactant solution of RS-1541 to mice bearing M5076 sarcoma at a dose of 5 mg/kg. The emulsion droplet sizes were 110 (Δ), 230 (○), 350 (◆), 410 (◇), 630 nm (□), and the surfactant solution (●). Each value represents the mean ± SE (standard error) of three mice. (Adapted with permission from Springer Science and Business Media Ref. [40] Copyright 1996.)

TABLE 4.3 Antitumor activity of RS-1541 emulsion formulations against M5076 sarcoma at the MTD

Mean diameter (nm)	Dose (MTD)[a] (mg/kg)	Tumor diameter[b] (%)	Tumor growth delay[c] (day)	ILS[d] (%)	Cure (on day 120)
Surfactant solution	6.0	213	17	62	0/6
70	4.5	166	24	66	0/6
100	4.5	113	29	69	0/6
220	15.0	13	>61	>224	4/6
380	40.0	18	56	>195	3/6

[a]RS-1541 was given in each formulation to M5076-bearing BDF1 mice via a single intravenous injection at maximum tolerated doses (MTDs) on day 13 after inoculation (six mice were used for each group).
[b]Tumor diameter on day 44 divided by that on treatment day.
[c]Days required for the tumors to reach again the diameter on treatment day following therapy.
[d]Increase in life span: ratio (%) of median survival days in a treatment group of mice to that in the control group (37 days).
Adapted with permission from Springer Science and Business Media Ref. [40] Copyright 1996.

of cytotoxic compounds to tumoral sites and reduce systemic toxicity by suitable selection of the droplet size.

4.4.2 Effect of Lipid Emulsion Composition and Emulsifiers

Composition of the Oil Phase

The composition of the internal phase has also been shown to alter the biodistribution of lipid emulsions. Lutz et al. [41] observed that lipid emulsions composed of medium-chain triglycerides (MCTs) were cleared from plasma more quickly than those prepared with long-chain triglycerides (LCTs). This is probably caused by the faster hydrolysis of MCTs by LPL and hepatic lipases compared with LCTs as a result of the greater solubility and mobility of shorter-chain triglycerides at the oil/water emulsion interface [42].

Adding free cholesterol has also been shown to alter the metabolism of triglyceride emulsions. Maranhao et al. [22] observed that emulsions with low free cholesterol content (<4% w/w) were metabolized in a manner similar to chylomicrons, as shown by the faster removal rate of triglycerides from the blood than CO as a result of LPL-mediated hydrolysis of the oil and greater uptake of CO than triglycerides by the liver. In contrast, emulsions with high free cholesterol (>6% w/w) displayed similar triglyceride and CO removal rates from blood, and equal uptake by the liver. The group also observed that emulsions containing high free cholesterol bound less Apo-A-I, Apo-A-IV, and Apo-C, and more Apo-E in vitro. Apo-C-II is essential for LPL binding and activation and hinders liver uptake, whereas Apo-E facilitates emulsion uptake by the liver. Hence, the presence of free cholesterol may modify the metabolism of the droplets by altering the binding of apolipoproteins onto the surface.

Phosphatidylcholine Composition

The biodistribution of emulsions can also be altered by the phospholipid emulsifier. Lenzo et al. [43] demonstrated that the nature of the PC affected the metabolism of the emulsion in rats. Five lipid emulsions with different phospholipid emulsifiers were prepared. The phospholipids selected were PC, 1,2-dioleoyl-*sn*-glycero-3-phosphatidylcholine (DOPC), 1,2-dimyristoyl-*sn*-glycero-3-phosphatidylcholine (DMPC), 1,2-dipalmitoyl-*sn*-glycero-3-phosphatidylcholine (DPPC), and 1-palmitoyl-2-oleoyl-*sn*-glycero-3-phosphatidylcholine (POPC). The average composition of each emulsion was similar and size was maintained at about 150 nm. The emulsions were radiolabeled with [^{14}C]triolein (TO) and [^{3}H]CO or dipalmitoylphosphatidyl-[*N*-methyl-^{3}H]choline to monitor the hydrolysis of the triglyceride oil by LPL, the clearance of the entire colloid particle, and the transfer of phospholipids to the HDLs, respectively.

The carriers emulsified with PC or POPC were metabolized in a manner similar to chylomicrons as shown by the rapid removal rate of [^{14}C]TO from plasma, consistent with hydrolysis by LPL, and the efficient uptake of [^{3}H]CO (remnant emulsions) by the liver. DPPC-based emulsions remained in plasma the longest and the triglycerides associated with this emulsion disappeared very slowly, suggesting that the emulsion was less susceptible to hydrolysis by LPL. Moreover, the phospholipid radiolabel did not transfer to HDLs. A possible explanation for the above observations is the difference in chain unsaturation among the five phospholipid emulsifiers. The authors hypothesized that rapid hydrolysis of the triglyceride oil by LPL and efficient transfer of phospholipids to HDLs requires a chain unsaturation at the glycerol 2-position.

4.4.3 Effect of Surface Charge

Lipid emulsions obtain their surface charge through the use of neutral, anionic, or cationic emulsifiers. Most emulsions used in drug delivery are either neutral or negatively charged because cationic carriers are more prone to aggregate in the presence of plasma proteins. This susceptibility for aggregation in the bloodstream is a result of the electrostatic interactions with negatively charged plasma proteins. It is generally accepted that surface charge has an effect on the rate of particle uptake by the MPS, although the connection is far from straightforward. Other surface properties, such as the nature of the emulsifier, may take precedence over the effects generated by surface charge. Davis et al. [44] found no clear correlation between ζ potential and the rate of emulsion uptake by mouse peritoneal macrophages, although emulsions with the weakest charge, prepared with the non-ionic surfactant poloxamer 338, had the slowest rate of uptake. Stossel et al. [45] found that emulsions with higher surface charge (positive or negative) were phagocytosed at a faster rate compared with neutral or weakly charged surfaces. Oku et al. [46] observed that uptake by the liver and spleen was greater for positively charged liposomes than for neutral or anionic ones. The higher accumulation of cationic liposomes in the MPS organs may be caused by both particle aggregation in the presence of serum and protein adsorption on to the colloid, which was observed to a lesser extent in neutral or anionic liposomes. Devine et al. [47] found that liposomes bearing a net positive or negative charge activated the complement in a dose-dependent manner, whereas no complement activation was observed for neutral liposomes. Interestingly, long-circulating cationic lipid emulsions have been reported in the literature by careful selection of the emulsifier [48].

Over the past few years, there has been increasing interest in developing cationic lipid–DNA complexes for the improved delivery of genetic material [49, 50]. An advantage of using cationic carriers is the enhanced cellular uptake via endocytosis over neutral or negatively charged carriers [51]. This is a result of the favorable electrostatic interactions of cationic particles with the negatively charged moieties on biological membranes. However, as a result of the tendency of cationic particles to aggregate in the presence of serum, the positive charge will need to be shielded, which will invariably reduce transfection efficiency in the absence of targeting ligands.

4.4.4 Long-circulating Lipid Emulsions

After intravenous administration, colloidal drug carriers are rapidly taken up by circulating monocytes and macrophages in the liver, spleen, and bone marrow. Avoiding the MPS is crucial when the emulsions are to be delivered to non-MPS cells. Prolonged circulation of the drug carrier is also necessary to achieve passive targeting of tumoral tissues via the EPR effect. Modification of the colloidal surface such that the carriers are invisible or 'stealth' to opsonins and macrophages is an approach investigated to increase the circulation time of submicrometer emulsions in blood.

Sphingomyelin

The presence of sphingomyelin (SM) at the oil/water interface has been shown to reduce the uptake of the emulsions by the MPS. Takino et al. [27] demonstrated that adding SM to a PC and soybean oil formulation increased the circulation time of the submicrometer emulsions in blood and decreased liver and spleen uptake. The emulsions were composed of PC:soybean oil (1:1) and PC:SM:soybean oil

(0.7 : 0.3 : 1) with [^{14}C]CO incorporated as a radiolabeled tracer. The AUC of the SM emulsion was 1.6 times larger than that of the one emulsified with PC only. Similarly, Redgrave et al. [21] observed that increasing the amount of SM enhanced the circulation time of the carrier in plasma (Figure 4.5) and reduced uptake by the liver

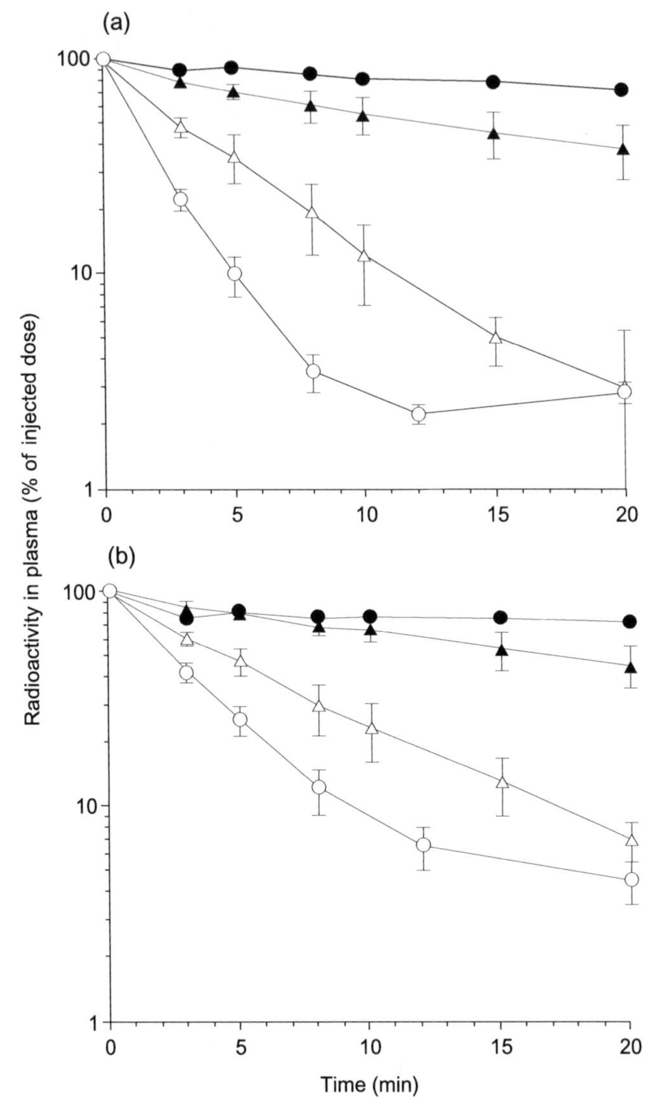

Figure 4.5 Radioactivity in plasma of triolein (TO) and cholesteryl oleate (CO) labels after injection of emulsions stabilized by mixtures of sphingomyelin (SM) with egg phosphatidylcholine (PC). TO-CO-cholesterol emulsions stabilized with mixtures of SM and PC were injected intravenously in conscious rats. Plotted are the data for labeled (a) TO and (b) CO incorporated in the emulsions remaining in the plasma at 3, 5, 8, 12, and 20 min after injection. Results are means ± SE of at least four experiments for each observation. SM 100% (●), SM/PC 50/50 (▲), SM/PC 25/75 (Δ), PC 100% (○). (Adapted with permission from Elsevier Ref. [21] Copyright 1992.)

(Figure 4.6). Even though SM and PC share a common phosphorylcholine polar head group, there are structural discrepancies between the two molecules that reflect their different physical properties in colloidal systems. SM has a high content of saturated acyl chains relative to naturally occurring PCs and has a stronger hydrogen bonding capacity, which may alter monolayer rigidity and interactions with blood components [52].

Polyethyleneglycol Lipids

A very widely used and effective method to avoid clearance by the MPS is to incorporate polyethyleneglycol (PEG) (also known as polyethylene oxide or PEO) at the colloid surface using a lipid derivative. PEG, a hydrophilic and flexible polymer, creates a zone of steric hindrance around the carrier which decreases the rate and extent of opsonin binding [53]. PEG is widely accepted for intravenous administration because it is a biocompatible, nontoxic, and nonimmunogenic polymer. Moreover, PEG-lipid derivatives are amphiphilic and as a result can be used as a co-emulsifier as well. Liu et al. [29] observed the influence of PEG molecular mass on the biodistribution of lipid emulsions composed of castor oil and PC. The PEG-lipid derivatives investigated were dioleoyl *N*-(monomethoxy-PEG-succinyl)phosphatidylethanolamine (PEG-DOPE) (M_r 1000, 2000, and 5000 Da) and PEO 20 sorbitan monooleate (Polysorbate 80). Emulsion droplet size was maintained at about 200 nm so that the circulation behavior was dependent only on the surface properties of the emulsions. It was observed that emulsion circulation time in blood

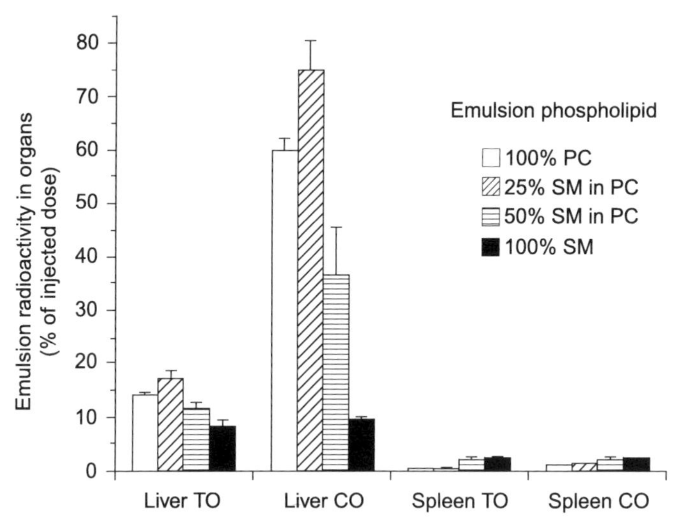

Figure 4.6 Radioactivity in the liver and spleen of triolein (TO) and cholesteryl oleate (CO) labels after injection of emulsions stabilized by mixtures of sphingomyelin (SM) with egg phosphatidylcholine (PC). TO-CO-cholesterol emulsions stabilized with mixtures of SM and PC were injected intravenously in conscious rats. Organ uptakes of radioactive TO and CO labels in the emulsions were measured 20 min after injection. Results are means ± SE of at least four experiments for each observation. By analysis of variance the differences between groups were statistically significant with $P < 0.01$ for liver TO, <0.001 for liver CO, <0.01 for spleen TO, and < 0.025 for the spleen CO. (Adapted with permission from Elsevier Ref. [21] Copyright 1992.)

depended on the length of the PEG chain. PEG_{2000}-DOPE and PEG_{5000}-DOPE kept the emulsions in the blood the longest. Approximately 60–70% of the injected dose remained in the blood after 30 min. PEG_{1000}-DOPE and Polysorbate 80 emulsions demonstrated comparable behavior in vivo with 47% of the injected dose remaining in the blood after 30 min. The high emulsion concentration observed in the blood for PEG_{2000}-DOPE and PEG_{5000}-DOPE translated into lower accumulation in the liver. Consequently, coating an emulsion surface with PEG of sufficient chain length can confer long circulating properties to submicrometer emulsions.

Hoarau et al. [38] evaluated two different processes to incorporate 1,2-distearoyl-*sn*-glycero-3-phosphatidylethanolamine-*N*-monomethoxy-[PEG] (PEG-DSPE) into lipid nanocapsules. The conventional method was the first investigated and involved the addition of PEG-DSPE with the other surfactants during the emulsification of the oil. The second method evaluated was after insertion, wherein an aqueous micelle solution of PEG-DSPE was added to the preformed lipid nanocapsules and then incubated for 90 min at 60°C. The authors observed that the post-insertion method enhanced the amount of PEG-DSPE that could be incorporated into the nanocapsule compared with the conventional process. For the conventional method, the amount of PEG_{2000}-DSPE and PEG_{5000}-DSPE could not exceed 3.4 and 1.5 mol% of the total surfactants, respectively, because physical instability would occur. In contrast, PEG-DSPE could be incorporated into the lipid nanocapsules at higher quantities (6–10 mol%) using the post-insertion method, regardless of the PEG chain length. Consequently, the pegylated lipid nanocapsules prepared by the post-insertion technique circulated longer in blood as a result of the greater PEG density at the surface (Figure 4.7), e.g. the AUC increased fivefold as the proportion of PEG_{2000}-DSPE increased from 1.7 to 10 mol%.

Other Methods to Enhance Circulation Time

Surfactants containing PEO chains such as, PEO-*b*-poly(propylene oxide)-*b*-PEO (PEO-*b*-PPO-*b*-PEO, poloxamers) and PEO-hydrogenated castor oil (cremophors) have also been investigated to enhance the hydrophilicity of emulsion surfaces to reduce opsonin binding and uptake by the MPS. Lee et al. [54] demonstrated that emulsions coated with poloxamer 338 reduced the amount of ibuprofen octyl ester delivered to the MPS organs. Ueda et al. [55] investigated the influence of ethylene oxide number in PEO-hydrogenated castor oil surfactants on menatetrenone clearance rate from plasma and distribution to MPS organs. They observed that a minimum of 20 ethylene oxide units (M_r = 880 Da) is required to prolong menatetrenone circulation time in plasma. The prolonged circulation of emulsions containing more than 20 ethylene oxide units translated into a lower accumulation of menatetrenone in the liver. Menatetrenone incorporated into emulsions with 10 ethylene oxide units were rapidly removed from plasma and taken up to a greater extent by the liver.

4.4.5 Active Targeting of Selected Cells

Drug delivery systems utilizing ligands that specifically recognize determinants on the surface of target cells have been extensively investigated in liposomes [56–58] and macromolecular prodrugs [59]. However, few studies have been done with emul-

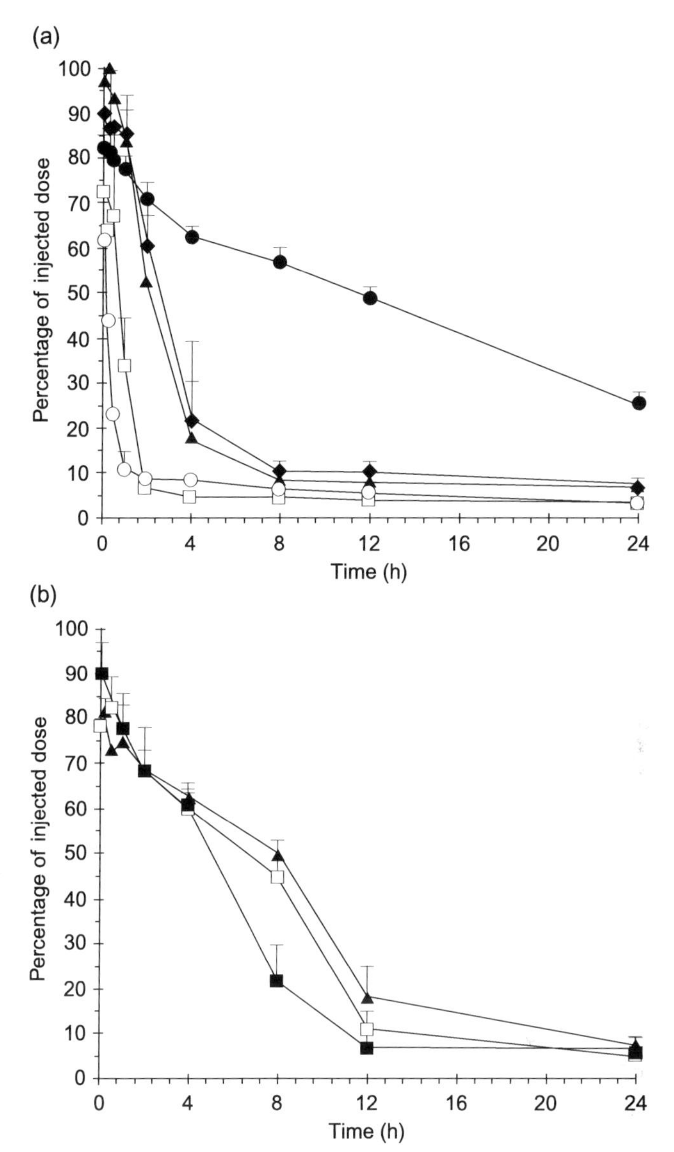

Figure 4.7 Blood concentration–time profile of stealth liposomes and different lipid nanocapsule formulations prepared by (a) the conventional or (b) the post-insertion method. Mean ± SD (standard deviation) ($n = 3 - 5$). (a) Pegylated liposomes (●), plain lipid nanocapsules (○), pegylated lipid nanocapsules with 1.7 mol% 1,2-distearoyl-*sn*-glycero-3-phosphatidylethanolamine-*N*-monomethoxy-[polyethyleneglycol] (PEG_{2000}-DSPE) (□), pegylated lipid nanocapsules with 1.4 mol% PEG_{5000}-DSPE (▲), and pegylated lipid nanocapsules with 3.4 mol% PEG_{2000}-DSPE (◆). (b) Pegylated lipid nanocapsules with 6 mol% PEG_{2000}-DSPE (■), pegylated lipid nanocapsules with 6 mol% PEG_{5000}-DSPE (□), and pegylated lipid nanocapsules with 10 mol% PEG_{2000}-DSPE (▲). Formulations were injected intravenously at a dose of 2 mg lipids/rat. (Adapted with permission from Springer Science and Business Media Ref. [38] Copyright 2004.)

sions. Incorporating ligands on to the emulsion interface is a promising method to enhance specificity towards the target site(s). For this method to be successful, lipid emulsions must have the appropriate ligand(s) anchored on to the surface, and must be able to reach the target cells, bind to the receptors, and either enter the cell or empty the contents in the vicinity of the cell.

Lipid Emulsions Associated with Apolipoprotein E

Apo-E has an affinity for both the remnant and LDLrs on hepatocytes and is an important mediator in the uptake of emulsions and lipoproteins by the liver. Incorporating Apo-E on lipid emulsions provides an opportunity to target hepatocytes. Rensen et al. [60] investigated the possibility of using lipid emulsions associated with Apo-E as drug carriers for a model antiviral prodrug, iododeoxyuridine-oleoyl (IDU-Ol_2), to target hepatocytes selectively for improved therapy of hepatitis B viral infection. The emulsions were prepared using natural lipids (PC, lysophosphatidylcholine, TO, and CO) and had a mean size of approximately 80 nm to mimic natural chylomicrons. The lipid emulsions were radiolabeled with [^{14}C]CO and [^{3}H]IDU-Ol_2 to track the in vivo distribution of the whole droplet and prodrug, respectively. After intravenous injection, lipid emulsions pre-loaded with Apo-E were removed faster from serum and were taken up more by the liver than the control emulsions (lipid emulsions not pre-loaded with Apo-E) (Figure 4.8). The uptake of the carrier by the liver reached about 70% of the injected dose for the Apo-E pre-loaded emulsion compared with only 30% for the control. The prodrug exhibited similar removal rates from serum and uptake by the liver as the carrier. The authors also showed that the carrier and prodrug were mainly taken up by parenchymal cells with little accumulation in the endothelial or Kupffer cells. Introducing lactoferrin, a glycoprotein that blocks Apo-E-mediated uptake of lipoproteins by parenchymal cells, before injecting the lipid emulsions, resulted in a considerable reduction in emulsion uptake by the liver.

Sugar-coated Emulsions to Target Hepatocytes

Incorporating Apo-E on to the lipid emulsion is a complex process that may cause reproducibility and stability issues [61]. Another method to enhance selectivity for hepatocytes is to incorporate sugars such as galactose on the surface of the lipid emulsion, to target the carbohydrate receptors on hepatocytes. Ishida et al. [61] investigated the biodistribution of galactosylated (Gal) and nongalactosylated emulsions after intravenous injection in mice. The results demonstrated that the Gal emulsion was more quickly removed from the blood compared with the bare emulsion, whereby the AUC values for the Gal emulsion and bare emulsions were 1.9 and 3.7 (percentage of dose $\times$ h/mL), respectively. In addition, the uptake of the Gal emulsion by the liver was 3.2 times greater than that of the bare emulsion. Moreover, Gal emulsions were taken up 7.4-times more by parenchymal cells than by non-parenchymal cells, compared with only 4.3 times for the bare emulsions. These findings suggest that introducing galactose on the surface of lipid emulsions is a promising method for delivering drugs to hepatocytes.

Antibody–peptide Conjugation onto Long-circulating Lipid Emulsions

Cancer cells often overexpress certain antigens or receptors, which provides another possible method to enhance the selectivity of anticancer drugs toward tumor tissues

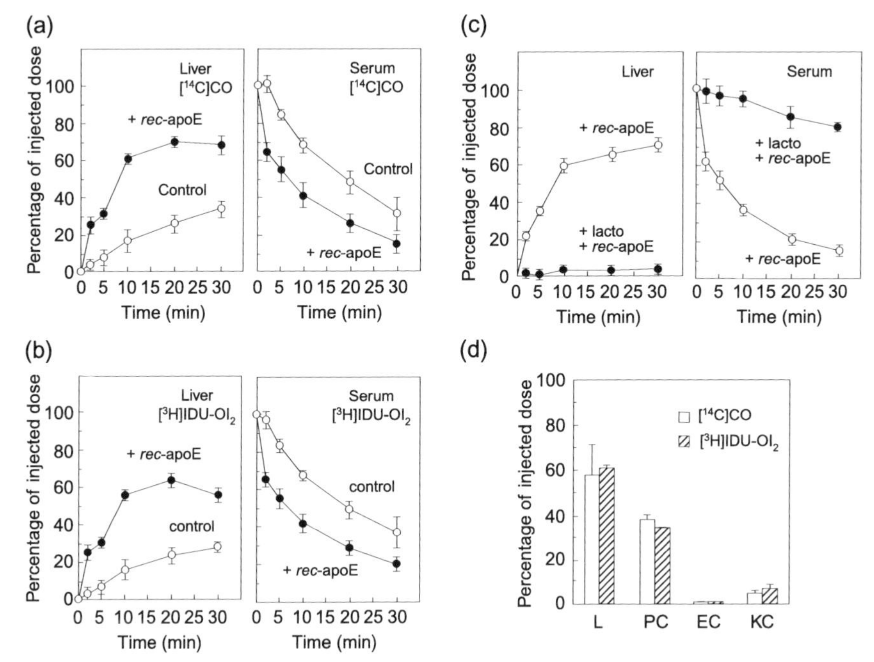

Figure 4.8 Liver uptake and serum decay of the control and human recombinant (*rec*) Apo-E-enriched emulsion iododeoxyuridine-oleoyl ([^{3}H]IDU-Ol$_2$) in rats, in the absence or presence of lactoferrin. Control and *rec* Apo-E-enriched emulsions, double-labeled with [1–^{14}C]cholesteryl oleate ([^{14}C]CO) and [^{3}H]IDU-Ol$_2$ were injected into fasted anaesthetized rats. (a, b, c) At the indicated times, the liver uptake and serum decay of (a) [^{14}C]CO and (b) [^{3}H]IDU-Ol$_2$ were determined. (c) The liver uptake and serum decay of *rec* Apo-E-enriched emulsions were also determined after preinjection of lactoferrin. (d) At 30 min after injection of emulsion *rec* Apo-E-IDU-Ol$_2$, the liver was perfused. Total liver (L) association was determined and parenchymal (PC), endothelial (EC), and Kupffer (KC) cells were subsequently isolated. Values are means ± SD of three experiments. (Adapted with permission from Macmillan Publishers Ltd Ref. [60] Copyright 1995.)

[58]. Antibodies, antibody fragments, or synthetic peptides can be incorporated onto the carrier surface specifically to recognize the antigen/receptors on cancer cells and offers a possible solution to the nonspecific and slow uptake of colloidal carriers by cancer cells. Linking antibodies to liposomes has been widely studied; however, much more progress is required for its successful application [58]. In theory, the techniques applied to liposomes can be carried over to lipid emulsions. Ideally, the tumor-targeting ability of antibodies can be coupled with the long-circulating properties of pegylated lipids. To avoid the interference between the PEG chain of pegylated lipid emulsions and the antibodies incorporated into the emulsion surface, antibodies linked to PEG chains have been developed in recent years. Lundberg et al. [62] successfully conjugated an anti-B-cell lymphoma monoclonal antibody (LL2) on to a lipid emulsion by coupling LL2 to PEG-DSPE. The immunoreactivity of the LL2-conjugated emulsion was tested by determining the binding affinity

to WN, the anti-idiotype antibody to LL2. The results showed that increasing the density of LL2 at the surface enhanced the binding of the emulsion to WN by up to 40 antibodies per droplet.

4.4.6 Drug Leakage from Emulsions

Drug retention within the droplet after intravenous administration is another important factor to consider when designing lipid emulsion, not only for dissolution purposes but also as carriers for lipophilic drugs. Controlling the biodistribution of the entire droplet will not enhance the therapeutic effect if the drug is released from the carrier before it reaches the target site. Takino et al. [27] suggested that the drug must have adequate lipophilicity (log $P > 9$) to remain sufficiently incorporated in the emulsion in the blood circulation. The influence of lipophilicity on drug leakage was observed by Kurihara et al. [63]. The group evaluated the pharmacokinetics of two lipophilic anticancer agents, rhizoxin and RS-1541, incorporated into lipid emulsions after intravenous injection in rats. The lipophilicities of rhizoxin and RS-1541 were very different with log P values of 1.9 and 13.8, respectively. After intravenous injection, rhizoxin was removed much more quickly from plasma than RS-1541 and distributed more to the liver, lung, and intestine. The different pharmacokinetic profiles can be attributed to the lower retention of rhizoxin within the lipid emulsion after injection. Similarly, Sakaeda et al. [64, 65] found that Sudan II, with a log P of 5.4, was rapidly released from the lipid emulsion in plasma. Consequently, the lipid emulsion did not alter the pharmacokinetics of Sudan II. The lipophilicity of a drug can be increased by chemically modifying it. A drawback of this approach is, however, that chemical modifications may reduce efficiency or completely inactivate the drug.

4.5 PREPARATION OF EMULSIONS FOR INTRAVENOUS ADMINISTRATION

To be used for intravenous administration, emulsions must be biocompatible, biodegradable, nontoxic, sterile, isotonic, physically and chemically stable, and nonimmunogenic [1]. Moreover, droplet size must be small enough to avoid forming pulmonary emboli. To achieve these requirements, the excipients, additives, and manufacturing conditions must be carefully selected. In addition, a complete physicochemical characterization of the emulsions is necessary, followed by a long-term stability testing schedule on all promising formulation candidates. This section describes the basic factors that need to be considered when developing emulsions for intravenous injection.

4.5.1 Excipient and Formulation Considerations

Internal Phase (Oils)

To comply with the essential requirement of biocompatibility, research on injectable emulsions has primarily focused on the use of vegetable oils (triglycerides) as the oil phase [66]. Triglyceride oils can be characterized as long, medium or short chain, depending on the number of carbon atoms per hydrocarbon chain. LCTs contain 14, 16, 18, 20, or 22 carbons in a fatty acid chain, which may or may not be unsaturated [24]. MCTs are derived from coconut oil and contain saturated fatty acids with chains of 6, 8, 10, or 12 carbons [1, 24, 67]. Lastly, short-chain triglycerides (SCTs), such as triacetin and tributyrin, have chain lengths of only two and four carbons, respectively. LCTs and MCTs should be considered in the initial stages of formulation because many of these oils are approved for injection and are found in a number of FDA (Food and Drug Administration) approved products [67] (Table 4.4). The choice of oil usually depends on the solubility and stability of the drug. MCTs have 100-fold greater water solubility than LCTs, and consequently are typically better solubilizers for drugs because most hydrophobic drugs have some polarity [24, 67]. Kan et al. [68] reported that triglycerides with short fatty acid chains were better solubilizers for paclitaxel, a lipophilic anticancer drug. They reported that paclitaxel solubility increased as the number of carbons per hydrocarbon chain decreased, e.g. tributyrin (C4) and tricaproin (C6) provided greater paclitaxel solubility than tricaprylin (C8) and other plants oils with a mixture of 6–22 carbons per hydrocarbon chain. Triacetin (C2) gave the highest paclitaxel solubility at a value of 75 mg/mL. Triacetin is not, however, approved for injection and might be difficult to emulsify as a result of its relatively high water solubility.

Vitamin E (DL-α-tocopherol) has been investigated as an alternative biocompatible oil to triglycerides to solubilize highly lipophilic drugs [69–71]. Constantinides et al. [70–72] have formulated a submicrometer emulsion of paclitaxel with high drug loading (8–10 mg/mL) using vitamin E as the internal phase and D-α-tocopheryl polyethyleneglycol 1000 succinate (TPGS) and poloxamer 407 as the emulsifiers (TOCOSOL-paclitaxel). This formulation is less toxic and has greater antitumor activity in mice bearing B16-melanoma tumors than the commercial formulation for paclitaxel (Taxol). At the MTD for Taxol (20 mg/kg), TOCOSOL-paclitaxel showed greater tumor regression than Taxol on a q3dx5 dosing schedule.

TABLE 4.4 List of oils used in commercial emulsions for parenteral nutrition

Oils	Commercial product name	Manufacturer
LCTs		
Cottonseed oil	Lipofundin	B. Braun
Safflower oil	Liposyn	Abbott Laboratories
Soybean oil	Intralipid	Kabi-Pharmacia
	Soyacal	Alpha Therapeutics
	Travamulsion	Travenol Laboratories
	Liposyn III	Abbott Laboratories
	Lipofundin S	B. Braun
	Trivé 1000	Egic
Safflower oil : soybean oil *LCTs + MCTs*	Liposyn II	Abbott Laboratories
Soybean oil : MCTs, (1 : 1)	Lipofundin MCT/LCT	B. Braun

LCT, long-chain triglyceride; MCT, medium-chain triglyceride.

Moreover, tumor growth was suppressed further at higher doses of this formulation (40 and 60 mg/kg). TOCOSOL-paclitaxel is currently in phase III clinical trials [72]. In addition to being a solubilizer for poorly soluble drugs, vitamin E may provide some therapeutic value. Bartels et al. [73] examined the influence of vitamin E, administered intravenously in an emulsion before surgery, on ischemia and reperfusion (I/R) injury in a double-blinded study on 68 patients. I/R injury is usually an outcome of liver surgery, which causes oxidative stress and cell damage. The results of the study indicated that administering vitamin E before surgery may reduce the impact of I/R injury in liver surgery. The antioxidant activity of vitamin E has also been shown to protect against doxorubicin-induced cardiotoxicity in animal studies [74, 75]. Moreover, vitamin E was found to enhance the anticancer activity of doxorubicin on human prostatic carcinoma cells in vitro [76].

Another possible internal phase for intravenous emulsions is perfluorocarbons. Emulsions containing perfluorochemicals have been investigated as contrast agents for diagnostic tissue imaging or as carriers for the transport of oxygen offering an alternative to blood transfusions [77, 78]. Perfluorochemicals are chemically inert, synthetic molecules containing carbon and fluorine atoms, and are capable of dissolving considerable amounts of oxygen [78]. They are hydrophobic and as a result require emulsification for dispersion in aqueous media. Several types of perfluorochemicals have been investigated such as, perfluorooctyl bromide ($C_8F_{17}Br$), perfluorodecyl bromide ($C_{10}F_{21}Br$) and perfluorodichlorooctane ($C_8F_{16}C_{12}$). Imavist (formally known as Imagent) and Oxygent are perfluorocarbon emulsions presently undergoing clinical trials as an ultrasound contrast agent and artificial blood substitute, respectively.

Emulsifiers

The purpose of surfactants is to emulsify the oil phase and provide physical stability against flocculation and coalescence during storage, which may be for extended periods of time. Surfactants provide physical stability by reducing the oil–water interfacial tension and promoting droplet–droplet repulsion. Injectable emulsions are frequently emulsified with natural lecithins obtained from either egg yolk or soybeans. These lipids are biocompatible and biodegradable, and have relatively good emulsifying properties [79]. Lecithins are differentiated by the nature of the head group, and the length and degree of saturation of the acyl chains. The head group can be phosphatidic acid (PA), ethanolamine (PE), serine (PS), or PC, and determines the surface charge of the emulsion. At pH 7, PE and PC head groups are uncharged, whereas PA and PS head groups are anionic. Surface charge can promote long-term emulsion stability by electrostatic repulsion and can influence its biodistribution in vivo.

The length and degree of saturation of the acyl chains greatly influences the gel–liquid phase transition temperature (T_c) and the surface properties of lipid bilayers (liposomes) and monolayers (emulsions). The T_c refers to the temperature at which the lipids shift from a highly ordered gel state to a less ordered fluid. Saturated lipids generally have phase transitions above room temperature (e.g. T_c of 1,2-distearoyl-*sn*-glycero-3-phosphatidylcholine (DSPC; C18:0) is 58°C) [12, 79]. Introducing unsaturations or reducing the length of the acyl chains decreases the T_c

substantially (e.g. T_c of DOPC (C18:1) is −22°C and of DMPC (C14:0) is 23°C) [12, 79]. Most natural phosphatides have chain lengths of 16–18 carbons; however, chains with as few as four carbons also exist. Nii et al. [79] observed that both PC acyl chain length and degree of chain unsaturation influenced the ability of the lipid to emulsify tricaprylin (glyceryl trioctanoate). PCs with shorter and saturated acyl hydrocarbon chains were more effective emulsifiers because they produced emulsions with smaller mean globule size with less change in appearance and droplet size over time.

Lipid emulsions are often co-emulsified with a biocompatible synthetic surfactant to enhance emulsification properties. An example of an oil requiring co-emulsification is vitamin E. Previous studies have shown that stable tocopherol emulsions cannot be prepared with lecithin as the sole emulsifier [72]. A possible explanation for this observation is the greater polarity of tocopherol compared with vegetable oils as a result of the presence of a hydroxyl group on the aromatic ring. The enhanced polarity of tocopherol may solubilize more lecithin in the emulsion core, resulting in less emulsifier at the interface to stabilize the system. Consequently, a more hydrophilic co-emulsifier is required to emulsify tocopherol. The synthetic surfactants approved for intravenous injection are few and include polysorbates, cremophors, and poloxamers.

To aid in the initial selection of emulsifiers, the hydrophile–lipophile balance (HLB) method is widely used. HLB is a system that classifies surfactants on an arbitrary scale, based on the relative proportions of the hydrophilic and hydrophobic parts on the molecule. Each surfactant is given a number, usually between 0 and 20. If the HLB value is high, the surfactant has a relatively large number of hydrophilic groups and is more soluble in water. In contrast, surfactants with a low HLB are more hydrophobic and will consequently be more easily dispersed in organic phases. In general, stable w/o emulsions are formed from surfactants with a low HLB, whereas those with a high HLB are typically used to make stable o/w emulsions. HLB values for several synthetic emulsifiers that are approved for intravenous injection are listed in Table 4.5. The HLB method also classifies the oil, but in terms of HLB 'required' ($HLB_{required}$). The $HLB_{required}$ specifies the HLB of the emulsifier that will produce the most stable emulsion. Oils are usually given two $HLB_{required}$ values, one to produce a stable o/w emulsion and the other for a stable w/o emulsion. This method allows the formulator to match the HLB of the emulsifiers with the $HLB_{required}$ of the oil to produce a stable emulsion. The $HLB_{required}$ values to produce a stable o/w emulsion for cottonseed oil, safflower oil, and soybean oil are 7.85, 7.72, and 7.66, respectively (Crodamol catalogue).

The HLB concept is advantageous in the initial screening stage of emulsion development because it reduces the number of emulsifiers to consider for a given type of oil. Although the formulator should be aware that the HLB method has serious limitations, arising from the fact that only the molecular structure of the individual surfactant is considered and the emulsion as a whole is ignored [8]. For instance, the HLB method does not take into account pertinent factors such as the conformation of the surfactant, salinity of the aqueous phase, or temperature [80]. Consequently, even if HLB and $HLB_{required}$ are correctly matched, the emulsion produced may not be stable.

TABLE 4.5 Several non-phospholipid surfactants approved for intravenous administration in at least one country or under clinical investigation

Chemical name	Common names	Properties	M_r	Reference	HLB	Reference
Poly(ethylene oxide) 35 castor oil	Cremophor EL	Nonionic	2515	[117]	13.5	[118]
Poly(ethylene oxide) 40 castor oil	Cremophor RH 40	Nonionic	N/A	–	14–16	[119]
Poly(ethylene oxide) 60 castor oil	Cremophor RH 60	Nonionic	N/A	–	15–17	[119]
Poly(ethylene oxide) 20 sorbitan monolaurate	Polysorbate 20	Nonionic	1225	[120]	16.9	[120]
Poly(ethylene oxide) 20 sorbitan monopalmitate	Polysorbate 40	Nonionic	1282	[120]	15.6	[120]
Poly(ethylene oxide) 20 sorbitan monooleate	Polysorbate 80	Nonionic	1310	[121]	15.0	[120]
PEO_{80}-*b*-PPO_{27}-*b*-PEO_{80}[a]	Poloxamer 188	Nonionic	8400	[122]	29	[118]
PEO_{101}-*b*-PPO_{56}-*b*-PEO_{101}[a,b]	Poloxamer 407	Nonionic	12600	[123]	22	[124]
D-α-Tocopheryl polyethyleneglycol 1000 succinate[b]	TPGS	Nonionic	1513	[121]	13	[72]
Poly(ethylene oxide)-15-hydroxystearate	Solutol HS-15	Non-ionic	958	–	14–16	–
Deoxycholic acid	–	Anionic	392	–	24	[72]
Glycocholic acid	–	Anionic	465	–	N/A	–

[a]Poly(ethylene oxide)-*b*-poly(propylene oxide)-*b*-poly(ethylene oxide).
[b]In phase III clinical trials.

Aqueous Phase

The isotonicity of an injectable emulsion is important in order to avoid disturbing the state of cells in contact with the formulation. The final osmolarity should be between 200 and 300 mosmol/kg and can be achieved by adding isotonizing agents such as glycerol, sorbitol, and xylitol to the aqueous phase. Glycerol is more commonly used and can be found in most parenteral emulsions including Intralipid, Lipofundin N, Liposyn, and Soyacal. Ionic agents such as sodium chloride can also adjust osmolarity; however, they should be avoided as the ions can destabilize emulsions (see earlier). The pH of the final emulsion may need to be adjusted and this can be done by adding small amounts of HCl or NaOH. The desired pH is usually between 7 and 8 to maintain physiological compatibility and minimize hydrolysis of the oil and phospholipids [81].

Antioxidants are often added to the emulsion to eliminate or reduce oxidation of the drug, oil, and emulsifier [8, 67]. Common antioxidants used for injectable formulations include α-tocopherol, deferoxamine mesylate, and ascorbic acid [81]. The formulation may also require the use of preservatives to resist microbial growth. Microorganisms may change the physicochemical properties of the emulsion, such as color, odor, pH, and physical stability, and may present a health hazard [8]. Common preservatives used in injectable preparations include phenol, cresol, and methyl, ethyl, or propyl esters of *p*-hydroxybenzoic acid [82].

4.5.2 Emulsion Preparation

The most common process for manufacturing emulsions is to incorporate the drug during the emulsification of the oil [66]. Another method of incorporating the active ingredient is to add a sterilized solution of the drug dissolved in a solvent to a preformed sterilized emulsion. This method is not often done because of stability issues that may be encountered, such as drug precipitation in the aqueous phase and emulsion cracking [1, 4].

The first step in emulsion preparation is usually to dissolve the water-soluble components (isotonizing agent and preservatives) in the aqueous phase and the lipophilic compounds (drug and perhaps the antioxidant) in the oil phase. The emulsifier can be dispersed in either phase. Both phases are typically heated and agitated to facilitate the dispersion of the various components [67]. The oil phase is then added to the aqueous phase. As the oil is added, the mixture is agitated with the aid of a medium-shear mixer and is usually heated. The rate of addition should be optimized because adding the oil phase too quickly can lead to incomplete dispersion of the oil in the aqueous phase [67]. The temperature and duration of heating in the premix stage depend largely on the thermosensitivity of the drug, oil, and emulsifier(s). This premix stage produces a coarse emulsion and can have a substantial impact on the final product [66]. A premix that is uniform with a droplet size under 20 μm generally produces a final emulsion that is more unimodal and physically stable [66].

After the premix stage, droplet size must be decreased to less than 5 μm in diameter and preferably below 1 μm, to avoid blocking the capillaries of the lungs. To produce emulsions with small droplet size, microfluidization or high-pressure homogenization is usually used. Microfluidization is a process whereby a liquid mixture is forced by high pressure through an interaction chamber, which splits the stream into two and then recombines them at ultrahigh velocities [83]. The product can be recycled to reduce droplet size further. The combination of high shear, turbulence, and cavitation generated by this apparatus can produce submicrometer emulsions with a narrow size distribution [84]. In high-pressure homogenization, fluid is forced at high pressure by means of a plunger pump through a very narrow channel. Depending on the type of homogenizer, the fluid may then collide head on with another high velocity stream or hit a hard-impact ring. Droplet size is reduced by cavitation, high shear forces, and high-speed collisions with other droplets [85]. Pressure, temperature, and number of passes are parameters that can be controlled and influence the efficiency of droplet size reduction.

After the desired droplet size is achieved, the formulation is filtered to remove large droplets or debris, and sterilized. Sterilization can be achieved by autoclaving or by filtration through a 0.22 μm cartridge filter. The heat generated by autoclaving can cause the oil and lecithin to hydrolyze, liberating free fatty acids, which will reduce the pH of the formulation. The conditions for sterilization by autoclaving will need to be selected carefully to minimize the degradation of heat-sensitive products. Filter sterilization, on the other hand, greatly reduces the heat burden on the emulsion, although this process does not provide the same guarantee for sterility as autoclaving [67]. Not all emulsions can be sterilized by filtration, because a mean droplet size of less than 200 nm is an essential requirement. Large droplets may clog the 0.22 μm cartridge filter and prevent sterile filtration [86]. The main manufacturing steps involved in the production of intravenous emulsions are outlined in Figure 4.9.

4.5.3 Emulsion Characterization

Injectable emulsions are often characterized for mean droplet diameter, size distribution, surface charge, and phase inversion temperature. The aforementioned properties are useful in predicting emulsion stability, biocompatibility and in vivo biodistribution. Control over droplet size and size distribution can impart some specificity toward target tissues and are also important predictors of biocompatibil-

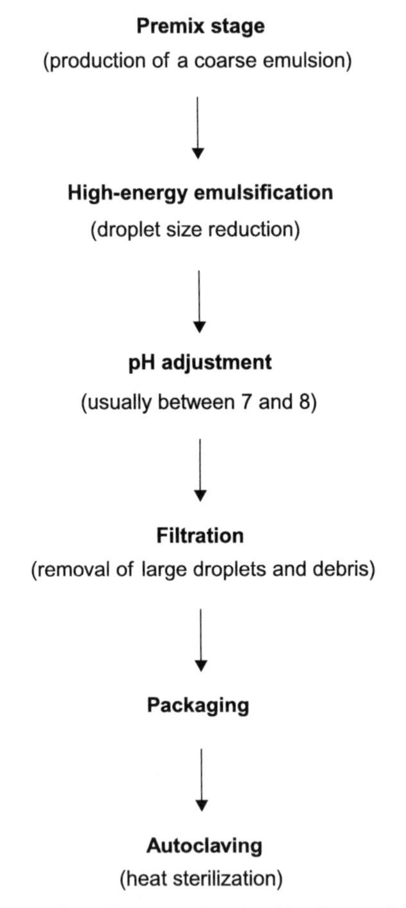

Figure 4.9 The main manufacturing steps involved in the production of intravenous emulsions.

ity because droplets larger than 5 μm can potentially form pulmonary emboli. The maximum allowable droplet size for intravenous administration is, however, unclear. Emulsions containing few droplets above 5 μm might not necessarily cause any adverse reaction because capillary blockage may be reversible by droplet degradation and large droplets may pass through small capillaries by deforming [87]. Burnham et al. [88] found that fat droplets larger than 7.5 μm in diameter could deform and pass through pulmonary vasculature without difficulty. There are a number of techniques available to measure mean droplet size and size distribution of emulsions. Dynamic light scattering (also known as photon correlation spectroscopy or quasielastic light scattering), atomic force microscopy, static light scattering (or intensity light scattering), and electron microscopy are frequently used to determine the size and size distribution for droplets below 1 μm [38, 89]. However, apart from atomic force microscopy and electron microscopy, the upper limit of detection on these instruments prevents the evaluation of the droplet size distribution above 5 μm. For detection of droplets larger than 5 μm, light obscuration, electrical-sensing zone (Coulter Counter), and optical microscopy are appropriate methods [90–92].

Surface charge measurements are also useful indicators of emulsion biocompatibility. Surfaces with a net positive charge are more likely to aggregate in the bloodstream in the presence of plasma proteins than negatively charged or neutral droplets. Charged surfaces can also impart physical stability to the emulsion by preventing/reducing coalescence on random collisions through electrostatic repulsion (see earlier). The surface charge of an emulsion droplet can be obtained through measurements of ζ potential using laser Doppler anemometry [93]. Lastly, phase inversion temperature, the temperature at which the emulsion changes from o/w to w/o or vice versa, can be a useful predictor of emulsion stability during temperature-altering processes such as heating, emulsification and sterilization [94].

4.5.4 Stability Measurement

The stability of an emulsion formulation is vital for its use in clinical applications. The formulation must display physical, chemical, and microbial stability for at least 1 year, if not more. The difficulty in emulsion formulation is that the system must be stable in aqueous solution, as opposed to polymeric micelles [95] or nanoparticles [96], which can be easily stored as a lyophilized powder whereby they have less opportunity to destabilize. Creaming and a visible layer of oil are classic signs of a physically unstable emulsion, whereas formulation discoloration is a typical indication of chemical instability. A long-term stability testing schedule should be performed on all promising formulation candidates, whereby each emulsion is stored at various temperatures ranging from 4 to 50°C [81]. The emulsions should be monitored for changes in size, pH, drug content, ζ potential, viscosity, electrical conductivity, and chemical composition [81].

Physical Stability

The long-term stability of an emulsion is difficult to estimate and only time can actually tell you whether the formulation is stable. Waiting for extended periods of time to find out whether a number of formulation efforts are stable is very impractical and there are methods available to accelerate stability testing. Most accelerated tests induce physical instability by increasing the number of collisions between globules. Accelerated tests based on sample heating are not, however, reliable because they do not reflect the environment of samples kept under storage conditions. Heating the sample not only enhances collisions but also diminishes the protective action of adsorbed surfactants, increases the solubility of all components, promotes the degradation of heat-sensitive products, alters the electric double-layer, and reduces surfactant adsorption at the emulsion interface, which can cause a potentially stable emulsion to destabilize, leading to erroneous results [2, 81].

Steam sterilization is considered an acceptable temperature-raising accelerated test because it approximates the environment that an emulsion would experience during autoclaving. Excessive shaking and freezing–thawing cycles are other commonly used accelerated stability testing processes because these techniques predict the conditions that a formulation will be subjected to during transportation and storage [81]. Another good method to estimate emulsion stability is to make size measurements frequently several weeks after the formulations are prepared. Emulsions that increase in size over time during the first few weeks in storage will invariably destabilize [2]. If there is no change in size, the formulator can have some hope that the emulsions will be physically stable long term [2].

Chemical Stability

Injectable emulsions can undergo chemical changes by oxidation and hydrolysis of the oil and/or emulsifier [67]. Chemical instability can be detected by formulation discoloration and changes in pH caused by the increase in free fatty acids. Chemical instability can be reduced by storing the emulsions refrigerated, protected from light, and in sealed containers with a layer of an inert gas, typically argon. These precautionary methods will reduce hydrolysis of the oil and emulsifier(s) [67]. Degradation of the encapsulated drug can also occur during storage. Consequently, the integrity of the encapsulated drug over time must also be determined [67].

Some groups have investigated storing lipid emulsions as a lyophilized powder to overcome some of the stabilized issues encountered when stored in solution [97, 98]. Freeze–drying emulsions is, however, difficult because the droplets may crack during the lyophilization process; it is rarely done. Bensouda et al. [99] evaluated the influence of a number of cryoprotective agents (glucose, mannitol, sorbitol, maltose, lactose, glycine, and dextran) on the success of the freeze–drying process. Glucose, maltose, sorbitol, and lactose provided protection against changes in particle size, whereas mannitol, dextran, and glycine offered no protective action.

4.6 LIPID EMULSIONS FOR THE DELIVERY OF NUCLEIC ACID-BASED DRUGS

Gene therapy is the science in which nonfunctional genes are substituted, altered, or supplemented for the treatment of genetic or acquired diseases. The difficulty in the successful application of gene therapy is the complexity of delivering functional genetic material such as plasmid DNA, antisense oligonucleotides (ODNs), or small interfering RNA into the cell. This is a result of their rapid degradation in plasma and their inability to cross cell membranes as a result of their hydrophilic and polyanionic nature, and relatively large size. Viral vectors have been extensively investigated for gene delivery [100]. However, concerns about host inflammatory and immune responses have created a demand for nonviral vectors [101–103]. As such, cationic liposomes are commonly investigated because they enhance transfection of DNA into the cell [104, 105]. Nevertheless, a major shortcoming of this technology is the formation of large aggregates with time and the reduced transfection of the liposome–DNA complex in the presence of serum [106].

Cationic lipid emulsions have been considered as alternative nonviral gene delivery vectors to liposomes. Complexation occurs through electrostatic interactions between the nucleic acids and the cationic lipid emulsifiers. The lipid emulsion–DNA complex can be prevented from forming large aggregates in the presence of serum by co-emulsification with an appropriate nonionic surfactant [107]. Yi et al. [108] prepared a cationic lipid emulsion–DNA complex that retained more than 60% transfection efficiency in the presence of 90% (v/v) serum. This formulation was composed of soybean oil, 1,2-dioleoyl-*sn*-glycero-3-trimethylammonium propane (DOTAP) as the cationic surfactant and co-emulsified with 1,2-dioleoyl-*sn*-glycero-3-phosphoethanolamine (DOPE) and 1-palmitoyl-2-oleoyl-*sn*-glycero-3-phosphoethanolamine-N-[PEG 2000] (PEG_{2000}-POPE).

The relatively high transfection efficiency of the lipid emulsion–plasmid DNA complex can be attributed to its ability to resist aggregation in the presence of serum, which may be caused by steric stabilization of the PEO chains of PEG_{2000}-POPE. Indeed, Kim et al. [109] observed that adding a co-emulsifier containing a PEO chain such as Polysorbate 80 or the Brij series (PEO 4-, 7-, 10-, or 23-lauryl ether) produced DNA-complexed emulsions that could resist changes in size in the presence of serum. In contrast, surfactants without a PEO group such as sorbitan monooleate (Span 80), mannide oleate (Montanide 80), and oleyl alcohol aggregated in the presence of serum and during DNA complexation. The authors also observed that the presence of PEO in both Polysorbate 80 and Brij interfered with the electrostatic interactions between DNA and the emulsion interface. DNA and cationic lipid interactions were reduced when the Polysorbate 80 content or PEO chain length in the Brij series increased. Despite the progress, much still needs to be done for the successful application of cationic emulsions to deliver genetic material in vivo. Indeed, cationic lipid emulsions will most likely face the same problems as other nonviral gene delivery systems.

4.7 CONCLUSION

Lipid emulsions are promising carriers for highly lipophilic drugs as a result of their biocompatibility, reasonable shelf-life, aptness for large-scale manufacture, and ability to solubilize large quantities of drug in their oily core. Lipid emulsions can also alter the biodistribution of incorporated drugs and enhance specificity toward target tissues by passive and active methods. Evading the MPS or natural fat metabolism is necessary when the encapsulated drug is to be delivered to non-MPS organs or liver parenchymal cells, respectively. Long-circulating lipid emulsions can be obtained by reducing droplet size and grafting long-chain hydrophilic polymers, such as PEO, to the emulsion interface. Moreover, active targeting using ligands that recognize specific determinants on cells can enhance specificity and uptake by target cells. It should be kept in mind that control over droplet biodistribution alone will not enhance the therapeutic effect because the drug may leak out of the carrier before reaching the target site. In general, drugs with higher lipophilicities (log $P > 9$) are retained better within the emulsion after intravenous administration. In addition to being carriers for lipophilic drugs, lipid emulsions have also been adapted to deliver genetic material and are an alternative nonviral vector to liposomes.

Several therapeutic lipid emulsions are commercially available for clinical use and other formulations are presently undergoing clinical trials. The use of intravenous lipid emulsions in the clinic as drug carrier vehicles will expand as new and less toxic formulations are discovered.

ACKNOWLEDGEMENTS

Financial support from the Natural Sciences and Engineering Research Council of Canada and the Canada Research Chair Program is acknowledged.

REFERENCES

1. Klang, S., Benita, S. Design and evaluation of submicron emulsions as colloidal drug carriers for intravenous administration. In: Benita, S. (ed.), *Submicron Emulsions in Drug Targeting and Delivery*. Amsterdam: Harwood Academic Publishers, 1998:119–152.
2. Friberg, S.E., Quencer, L.G., Hilton, M.L. Theory of emulsions. In: Lieberman, H.A., Rieger, M.M., Banker, G.S. (eds), *Pharmaceutical Dosage Forms*. New York: Marcel Dekker, Inc., 1996:53–90.
3. Buszello, K., Muller, B.W. Emulsions as drug delivery systems. In: Swarbrick, J. (ed.), *Pharmaceutical Emulsions and Suspensions*. New York: Marcel Dekker, Inc., 2000:191–228.
4. Prankerd, R.J., Stella, V.J. The use of oil-in-water emulsions as a vehicle for parenteral drug administration. *J Parenter Sci Technol* 1990;**44**:139–149.
5. Barratt, G. Colloidal drug carriers: achievements and perspectives. *Cell Mol Life Sci* 2003;**60**:21–37.
6. Nishikawa, M., Takakura, Y., Hashida, M. Biofate of fat emulsions. In: Benita, S. (ed), *Submicron Emulsions in Drug Targeting and Delivery*. Amsterdam: Harwood Academic, 1998:99–118.
7. Lawrence, M.J., Rees, G.D. Microemulsion-based media as novel drug delivery systems. *Adv Drug Deliv Rev* 2000;**45**:89–121.
8. Eccleston, G.M. Emulsions and microemulsions. In: Swarbrick, J., Boylan, J.C. (eds), *Encyclopedia of Pharmaceutical Technology*. New York: Marcel Dekker, Inc., 2002:1066–1085.
9. Welin-Berger, K., Bergenstahl, B. Inhibition of Ostwald ripening in local anesthetic emulsions by using hydrophobic excipients in the disperse phase. *Int J Pharm* 2000;**200**:249–260.
10. Sarker, D.K. Engineering of nanoemulsions for drug delivery. *Curr Drug Deliv* 2005;**2**:297–310.
11. Chansiri, G., Lyons, R.T., Patel, M.V., Hem, S.L. Effect of surface charge on the stability of oil/water emulsions during steam sterilization. *J Pharm Sci* 1999;**88**:454–458.
12. Washington, C. Stability of lipid emulsions for drug delivery. *Adv Drug Deliv Rev* 1996;**20**:131–145.
13. Kawakami, S., Yamashita, F., Hashida, M. Disposition characteristics of emulsions and incorporated drugs after systemic or local injection. *Adv Drug Deliv Rev* 2000;**45**:77–88.

14. Olivecrona, T., Bengtsson-Olivecrona, G., Ostergaard, P., Liu, G., Chevreuil, O., Hultin, M. New aspects on heparin and lipoprotein metabolism. *Haemostasis* 1993;**23**(suppl 1):150–160.
15. Carpentier, Y.A., Dupont, I.E. Advances in intravenous lipid emulsions. *World J Surg* 2000;**24**:1493–1497.
16. Rensen, P.C.N., de Vrueh, R.L.A., Kuiper, J., Bijsterbosch, M.K., Biessen, E.A.L., van Berkel, T.J.C. Recombinant lipoproteins: lipoprotein-like lipid particles for drug targeting. *Adv Drug Deliv Rev* 2001;**47**:251–276.
17. de Faria, E., Fong, L.G., Komaromy, M., Cooper, A.D. Relative roles of the LDL receptor, the LDL receptor-like protein, and hepatic lipase in chylomicron remnant removal by the liver. *J Lipid Res* 1996;**37**:197–209.
18. Ferezou, J., Bach, A.C. Structure and metabolic fate of triacylglycerol- and phospholipid-rich particles of commercial parenteral fat emulsions. *Nutrition* 1999;**15**:44–50.
19. Olivecrona, G., Olivecrona, T. Clearance of artificial triacylglycerol particles. *Curr Opin Clin Nutr Metab Care* 1998;**1**:143–151.
20. Lenzo, N.P., Martins, I., Mortimer, B.C., Redgrave, T.G. Effects of phospholipid composition on the metabolism of triacylglycerol, cholesteryl ester and phosphatidylcholine from lipid emulsions injected intravenously in rats. *Biochim Biophys Acta* 1988;**960**:111–118.
21. Redgrave, T.G., Rakic, V., Mortimer, B.-C., Mamo, J.C.L. Effects of sphingomyelin and phosphatidylcholine acyl chains on the clearance of triacylglycerol-rich lipoproteins from plasma. Studies with lipid emulsions in rats. *Biochim Biophys Acta* 1992;1126:65–72.
22. Maranhao, R.C., Tercyak, A.M., Redgrave, T.G. Effects of cholesterol content on the metabolism of protein-free emulsion models of lipoproteins. *Biochim Biophys Acta* 1986;**875**:247–255.
23. Hultin, M., Mullertz, A., Zundel, M.A., Olivecrona, G., Hansen, T.T., Deckelbaum, R.J., et al. Metabolism of emulsions containing medium- and long-chain triglycerides or interesterified triglycerides. *J Lipid Res* 1994;**35**:1850–1860.
24. Tamilvanan, S. Oil-in-water lipid emulsions: implications for parenteral and ocular delivering systems. *Prog Lipid Res* 2004;**2004**:489–533.
25. Ueda, K., Ishida, M., Inoue, T., Fujimoto, M., Kawahara, Y., Sakaeda, T., Iwakawa, S. Effect of injection volume on the pharmacokinetics of oil particles and incorporated menatetrenone after intravenous injection as O/W lipid emulsions in rats. *J Drug Target* 2001;**9**:353–360.
26. Waitzberg, D.L., Lotierzo, P.H., Logullo, A.F., Torrinhas, R.S., Pereira, C.C., Meier, R. Parenteral lipid emulsions and phagocytic systems. *Br J Nutr* 2002;**87**(suppl 1):S49–S57.
27. Takino, T., Konishi, K., Takakura, Y., Hashida, M. Long circulating emulsion carrier systems for highly lipophilic drugs. *Biol Pharm Bull* 1994;**17**:121–125.
28. Lundberg, B.B., Mortimer, B.C., Redgrave, T.G. Submicron lipid emulsions containing amphipathic polyethylene glycol for use as drug-carriers with prolonged circulation time. *Int J Pharm* 1996;**134**:119–127.
29. Liu, F., Liu, D. Long-circulating emulsions (oil-in-water) as carriers for lipophilic drugs. *Pharm Res* 1995;**12**:1060–1064.
30. Takino, T., Nakajima, C., Takakura, Y., Sezaki, H., Hashida, M. Controlled biodistribution of highly lipophilic drugs with various parenteral formulations. *J Drug Target* 1993;**1**:117–124.
31. Vonarbourg, A., Passirani, C., Saulnier, P.P.S., Leroux, J.C., Benoit, J.P. Evaluation of pegylated lipid nanocapsules versus complement system activation and macrophage uptake. *J Biomed Mater Res A* 2006; in press.
32. Kurihara, A., Shibayama, Y., Yasuno, A., Ikeda, M., Hisaoka, M. Lipid emulsions of palmitoylrhizoxin: effects of particle size on blood dispositions of emulsion lipid and incorporated compound in rats. *Biopharm Drug Dispos* 1996;**17**:343–353.
33. Jain, R.K. Determinants of tumor blood flow: a review. *Cancer Res* 1988;**48**:2641–2658.
34. Crommelin, D.J.A., Hennink, W.E., Storm, G. Drug targeting systems: fundamentals and applications to parenteral drug delivery. In: Hillery, A.M., Lloyd, A.W., Swarbrick, J. (eds), *Drug Delivery and Targeting*. New York: Taylor & Francis, 2001:117–144.
35. Wisse, E. An electron microscopic study of the fenestrated endothelial lining of rat liver sinusoids. *J Ultrastruct Res* 1970;**31**:125–150.
36. Maeda, H. SMANCS and polymer-conjugated macromolecular drugs: advantages in cancer chemotherapy. *Adv Drug Deliv Rev* 2001;**46**:169–185.

37. Maeda, H., Wu, J., Sawa, T., Matsumura, Y., Hori, K. Tumor vascular permeability and the EPR effect in macromolecular therapeutics: a review. *J Control Release* 2000;**65**:271–284.
38. Hoarau, D., Delmas, P., David, S., Roux, E., Leroux, J.C. Novel long-circulating lipid nanocapsules. *Pharm Res* 2004;**21**:1783–1789.
39. Greish, K., Fang, J., Inutsuka, T., Nagamitsu, A., Maeda, H. Macromolecular therapeutics: advantages and prospects with special emphasis on solid tumour targeting. *Clin Pharmacokinet* 2003;**42**:1089–1105.
40. Kurihara, A., Shibayama, Y., Mizota, A., Yasuno, A., Ikeda, M., Sasagawa, K., et al. Enhanced tumor delivery and antitumor activity of palmitoyl rhizoxin using stable lipid emulsions in mice. *Pharm Res* 1996;**13**:305–310.
41. Lutz, O., Lave, T., Frey, A., Meraihi, Z., Bach, A.C. Activities of lipoprotein lipase and hepatic lipase on long- and medium-chain triglyceride emulsions used in parenteral nutrition. *Metabolism* 1989;**38**:507–513.
42. Deckelbaum, R.J., Hamilton, J.A., Moser, A., Bengtsson-Olivecrona, G., Butbul, E., Carpentier, Y.A., et al. Medium-chain versus long-chain triacylglycerol emulsion hydrolysis by lipoprotein lipase and hepatic lipase: implications for the mechanisms of lipase action. *Biochemistry* 1990;**29**:1136–1142.
43. Lenzo, N.P., Martins, I., Mortimer, B.C., Redgrave, T.G. Effects of phospholipid composition on the metabolism of triacylglycerol, cholesterol ester and phosphatidylcholine from lipid emulsions injected intravenously in rats. *Biochim Biophys Acta* 1988;**960**:111–118.
44. Davis, S.S., Hansrani, P. The influence of emulsifying agents on the phagocytosis of lipid emulsions by macrophages. *Int J Pharm* 1985;**23**:69–77.
45. Stossel, T.P., Mason, R.J., Hartwig, J., Vaughan, M. Quantitative studies of phagocytosis by polymorphonuclear leukocytes: use of emulsions to measure the initial rate of phagocytosis. *J Clin Invest* 1972;**51**:615–624.
46. Oku, N., Tokudome, Y., Namba, Y., Saito, N., Endo, M., Hasegawa, Y., et al. Effect of serum protein binding on real-time trafficking of liposomes with different charges analyzed by positron emission tomography. *Biochim Biophys Acta* 1996;**1280**:149–154.
47. Devine, D.V., Wong, K., Serrano, K., Chonn, A., Cullis, P.R. Liposome-complement interactions in rat serum: implications for liposome survival studies. *Biochim Biophys Acta* 1994;**1191**:43–51.
48. Yang, S.C., Benita, S. Enhanced absorption and drug targeting by positively charged submicron emulsions. *Drug Dev Res* 2000;**50**:476–486.
49. Chesnoy, S., Huang, L. Structure and function of lipid-DNA complexes for gene delivery. *Annu Rev Biophys Biomol Struct* 2000;**29**:27–47.
50. Pedroso de Lima, M.C., Simoes, S., Pires, P., Faneca, H., Duzgunes, N. Cationic lipid-DNA complexes in gene delivery: from biophysics to biological applications. *Adv Drug Deliv Rev* 2001;**47**:277–294.
51. Blau, S., Jubeh, T.T., Haupt, S.M., Rubinstein, A. Drug targeting by surface cationization. *Crit Rev Ther Drug Carrier Syst* 2000;**17**:425–465.
52. Barenholz, Y., Thompson, T.E. Sphingomyelins in bilayers and biological membranes. *Biochim Biophys Acta* 1980;**604**:129–158.
53. Harris, J.M., Martin, N.E., Modi, M. Pegylation: a novel process for modifying pharmacokinetics. *Clin Pharmacokinet* 2001;**40**:539–551.
54. Lee, M.-J., Lee, M.-H., Shim, C.-K. Inverse targeting of drugs to reticuloendothelial system-rich organs by lipid microemulsion emulsified with poloxamer 338. *Int J Pharm* 1995;**113**:175–187.
55. Ueda, K., Yamazaki, Y., Noto, H., Teshima, Y., Yamashita, C., Sakaeda, T., Iwakawa, S. Effect of oxyethylene moieties in hydrogenated castor oil on the pharmacokinetics of menatetrenone incorporated in o/w lipid emulsions prepared with hydrogenated castor oil and soybean oil in rats. *J Drug Target* 2003;**11**:37–43.
56. Hilgenbrink, A.R., Low, P.S. Folate receptor-mediated drug targeting: from therapeutics to diagnostics. *J Pharm Sci* 2005;**94**:2135–2146.
57. Noble, C.O., Kirpotin, D.B., Hayes, M.E., Mamot, C., Hong, K., Park, J.W., et al. Development of ligand-targeted liposomes for cancer therapy. *Exp Opin Ther Targets* 2004;**8**:335–353.
58. Sapra, P., Allen, T.M. Ligand-targeted liposomal anticancer drugs. *Prog Lipid Res* 2003;**42**:439–462.

59. Ohya, Y., Oue, H., Nagatomi, K., Ouchi, T. Design of macromolecular prodrug of cisplatin using dextran with branched galactose units as targeting moieties to hepatoma cells. *Biomacromolecules* 2001;**2**:927–933.
60. Rensen, P.C.N., van Dijk, M.C.M., Havenaar, E.C., Bijsterbosch, M.K., Kruijt, J.K., van Berkel, T.J.C. Selective liver targeting of antivirals by recombinant chylomicrons – a new therapeutic approach to hepatitis B. *Nat Med* 1995;**1**:221–225.
61. Ishida, E., Managit, C., Kawakami, S., Nishikawa, M., Yamashita, F., Hashida, M. Biodistribution characteristics of galactosylated emulsions and incorporated probucol for hepatocyte-selective targeting of lipophilic drugs in mice. *Pharm Res* 2004;**21**:932–939.
62. Lundberg, B.B., Griffiths, G., Hansen, H.J. Conjugation of an anti-B-cell lymphoma monoclonal antibody, LL2, to long-circulating drug-carrier lipid emulsions. *J Pharm Pharmacol* 1999;**51**:1099–1105.
63. Kurihara, A., Shibayama, Y., Mizota, A., Yasuno, A., Ikeda, M., Hisaoka, M. Pharmacokinetics of highly lipophilic antitumor agent palmitoyl rhizoxin incorporated in lipid emulsions in rats. *Biol Pharm Bull* 1996;**19**:252–258.
64. Sakaeda, T., Takahashi, K., Nishihara, Y., Hirano, K. O/W lipid emulsions for parenteral drug delivery. I. Pharmacokinetics of the oil particles and incorporated Sudan II. *Biol Pharm Bull* 1994;**17**:1490–1495.
65. Sakaeda, T., Hirano, K. O/W lipid emulsions for parenteral drug delivery. II. Effect of composition on pharmacokinetics of incorporated drug. *J Drug Target* 1995;**3**:221–230.
66. Collins-Gold, L.C., Lyons, R.T., Bartholow, L.C. Parenteral emulsions for drug delivery. *Adv Drug Deliv Rev* 1990;**5**:189–208.
67. Floyd, A.G. Top ten considerations in the development of parenteral emulsions. *PSTT* 1999;**2**:134–143.
68. Kan, P., Chen, Z.B., Lee, C.J., Chu, I.M. Development of nonionic surfactant/phospholipid o/w emulsion as a paclitaxel delivery system. *J Control Release* 1999;**58**:271–278.
69. Han, J., Davies, S.S., Papandreou, C., Melia, C.D., Washington, C. Design and evaluation of an emulsion vehicle for paclitaxel. I. physiochemical properties and plasma stability. *Pharm Res* 2004;**21**:1573–1580.
70. Constantinides, P.P., Lambert, K.J., Tustian, A.K., Schneider, B., Lalji, S., Ma, W., et al. Formulation development and antitumor activity of a filter-sterilizable emulsion of paclitaxel. *Pharm Res* 2000;**17**:175–182.
71. Constantinides, P.P., Tustian, A., Kessler, D.R. Tocol emulsions for drug solubilization and parenteral delivery. *Adv Drug Deliv Rev* 2004;**56**:1243–1255.
72. Constantinides, P.P., Han, J., Davis, S.S. Advances in the use of tocols as drug delivery vehicles. *Pharm Res* 2006;**23**:243–255.
73. Bartels, M., Biesalski, H.K., Engelhart, K., Sendlhofer, G., Rehak, P., Nagel, E. Pilot study on the effect of parenteral vitamin E on ischemia and reperfusion induced liver injury: a double blind, randomized, placebo-controlled trial. *Clin Nutr* 2004;**23**:1360–1370.
74. Myers, C.E., McGuire, W., Young, R. Adriamycin: amelioration of toxicity by alpha-tocopherol. *Cancer Treat Rep* 1976;**60**:961–962.
75. Sonneveld, P. Effect of alpha-tocopherol on the cardiotoxicity of adriamycin in the rat. *Cancer Treat Rep* 1978;**62**:1033–1036.
76. Ripoll, E.A., Rama, B.N., Webber, M.M. Vitamin E enhances the chemotherapeutic effects of adriamycin on human prostatic carcinoma cells in vitro. *J Urol* 1986;**136**:529–531.
77. Bentley, P.K., Johnson, O.L., Washington, C., Lowe, K.C. Uptake of concentrated perfluorocarbon emulsions into rat lymphoid tissues. *J Pharm Pharmacol* 1993;**45**:182–185.
78. Spahn, D.R. Blood substitutes. Artificial oxygen carriers: perfluorocarbon emulsions. *Crit Care* 1999;**3**:R93–R97.
79. Nii, T., Ishii, F. Properties of various phosphatidylcholines as emulsifiers or dispersing agents in microparticle preparations for drug carriers. *Colloids Surf B Biointerfaces* 2004;**39**:57–63.
80. Salager, J-L. Formulation concepts for the emulsion maker. In: Nielloud, F., Marti-Mestres, G. (eds), *Pharmaceutical Emulsions and Suspensions*. New York: Marcel Dekker, Inc., 2000:19–72.
81. Benita, S., Levy, M.Y. Submicron emulsions as colloidal drug carriers for intravenous administration: comprehensive physicochemical characterization. *J Pharm Sci* 1993;**82**:1069–1079.

82. Dubois, E., Lambert, G. Émulsions Parentérales. In: Falson-Rieg, F., Faivre, V., Pirot, F. (eds), *Nouvelles Formes Médicamenteuses*. New York: Lavoisier, 2004:237–258.
83. Washington, C., Davis, S.S. The production of parenteral feeding emulsions by microfluidizer. *Int J Pharm* 1988;**44**:169–176.
84. Pinnamaneni, S., Das, N.G., Das, S.K. Comparison of oil-in-water emulsions manufactured by microfluidization and homogenization. *Pharmazie* 2003;**58**:554–558.
85. Patravale, V.B., Date, A.A., Kulkarni, R.M. Nanosuspensions: a promising drug delivery strategy. *J Pharm Pharmacol* 2004;**56**:827–840.
86. Lidgate, D.M., Trattner, T., Shultz, R.M., Maskiewicz, R. Sterile filtration of a parenteral emulsion. *Pharm Res* 1992;**9**:860–863.
87. Koster, V.S., Kuks, P.F.M., Lange, R., Talsma, H. Particle size in parenteral fat emulsions, what are the true limitations? *Int J Pharm* 1996;**134**:235–238.
88. Burnham, W.R., Hansrani, P.K., Knott, C.E., Cook, J.A., Davis, S.S. Stability of a fat emulsion based intravenous feeding mixture. *Int J Pharm* 1983;**13**:9–22.
89. Haskell, R.J. Characterization of submicron systems via optical methods. *J Pharm Sci* 1998;**87**:125–129.
90. Driscoll, D.F., Etzler, F., Barber, T.A., Nehne, J., Niemann, W., Bistrian, B.R. Physicochemical assessments of parenteral lipid emulsions: light obscuration versus laser diffraction. *Int J Pharm* 2001;**219**:21–37.
91. Han, J., Davis, S.S., Washington, C. Physical properties and stability of two emulsion formulations of propofol. *Int J Pharm* 2001;**215**:207–220.
92. Driscoll, D.F. Stability and compatibility assessment techniques for total parenteral nutrition admixtures: setting the bar according to pharmacopeial standards. *Curr Opin Clin Nutr Metab Care* 2005;**8**:297–303.
93. Rabinovich-Guilatt, L., Couvreur, P., Lambert, G., Goldstein, D., Benita, S., Dubernet, C. Extensive surface studies help to analyse zeta potential data: the case of cationic emulsions. *Chem Phys Lipids* 2004;**131**:1–13.
94. Fernandez, P., Andre, V., Rieger, J., Kuhnle, A. Nano-emulsion formation by emulsion phase inversion. *Colloids Surf A Physicochem Eng Aspects* 2004;**251**:53–58.
95. Fournier, E., Dufresne, M.H., Smith, D.C., Ranger, M., Leroux, J.C. A novel one-step drug-loading procedure for water-soluble amphiphilic nanocarriers. *Pharm Res* 2004;**21**:962–968.
96. De Jaeghere, F., Allemann, E., Leroux, J.C., Stevels, W., Feijen, J., Doelker, E., Gurny, R. Formulation and lyoprotection of poly(lactic acid-co-ethylene oxide) nanoparticles: influence on physical stability and in vitro cell uptake. *Pharm Res* 1999;**16**:859–866.
97. Brime, B., Frutos, P., Bringas, P., Nieto, A., Ballesteros, M.P., Frutos, G. Comparative pharmacokinetics and safety of a novel lyophilized amphotericin B lecithin-based oil-water microemulsion and amphotericin B deoxycholate in animal models. *J Antimicrob Chemother* 2003;**52**:103–109.
98. Brime, B., Molero, G., Frutos, P., Frutos, G. Comparative therapeutic efficacy of a novel lyophilized amphotericin B lecithin-based oil-water microemulsion and deoxycholate-amphotericin B in immunocompetent and neutropenic mice infected with *Candida albicans*. *Eur J Pharm Sci* 2004;**22**:451–458.
99. Bensouda, Y., Cave, G., Seiller, M., Puisieux, F. Freeze-drying of emulsions – influence of congealing on granulometry research of a cryoprotective agent. *Pharm Acta Helv* 1989;**64**:40–44.
100. Young, L.S., Searle, P.F., Onion, D., Mautner, V. Viral gene therapy strategies: from basic science to clinical application. *J Pathol* 2006;**208**:299–318.
101. Yang, Y., Li, Q., Ertl, H.C., Wilson, J.M. Cellular and humoral immune responses to viral antigens create barriers to lung-directed gene therapy with recombinant adenoviruses. *J Virol* 1995;**69**:2004–2015.
102. Zhou, H.S., Liu, D.P., Liang, C.C. Challenges and strategies: the immune responses in gene therapy. *Med Res Rev* 2004;**24**:748–761.
103. Liu, F., Huang, L. Development of non-viral vectors for systemic gene delivery. *J Control Release* 2002;**78**:259–266.
104. Felgner, J.H., Kumar, R., Sridhar, C.N., Wheeler, C.J., Tsai, Y.J., Border, R., et al. Enhanced gene delivery and mechanism studies with a novel series of cationic lipid formulations. *J Biol Chem* 1994;**269**:2550–2561.

105. Audouy, S.A., de Leij, L.F., Hoekstra, D., Molema, G. In vivo characteristics of cationic liposomes as delivery vectors for gene therapy. *Pharm Res* 2002;**19**:1599–1605.
106. Escriou, V., Ciolina, C., Lacroix, F., Byk, G., Scherman, D., Wils, P. Cationic lipid-mediated gene transfer: effect of serum on cellular uptake and intracellular fate of lipopolyamine/DNA complexes. *Biochim Biophys Acta* 1998;**1368**:276–288.
107. Hara, T., Liu, F., Liu, D., Huang, L. Emulsion formulations as a vector for gene delivery in vitro and in vivo. *Adv Drug Deliv Rev* 1997;**24**:265–271.
108. Yi, S.M., Yune, T.Y., Kim, T.W., Chung, H., Choi, Y.W., Kwon, I.C., et al. A cationic lipid emulsion/DNA complex as a physically stable and serum-resistant gene delivery system. *Pharm Res* 2000;**17**:314–320.
109. Kim, T.W., Kim, Y.J., Chung, H., Kwon, I.C., Sung, H.C., Jeong, S.Y. The role of non-ionic surfactants on cationic lipid mediated gene transfer. *J Control Release* 2002;**82**:455–465.
110. Wang, J., Maitani, Y., Takayama, K. Antitumor effects and pharmacokinetics of aclacinomycin A carried by injectable emulsions composed of vitamin E, cholesterol, and PEG-lipid. *J Pharm Sci* 2002;**91**:1128–1134.
111. Source: ADVENTRX Pharmaceuticals, Inc.
112. Suzuki, Y., Masumitsu, Y., Okudaira, K., Hayashi, M. The effects of emulsifying agents on disposition of lipid-soluble drugs included in fat emulsion. *Drug Metab Pharmacokinet* 2004;**19**:62–67.
113. Source: Alliance Pharmaceutical Corp.
114. Song, D., Hamza, M.A., White, P.F., Byerly, S.I., Jones, S.B., Macaluso, A.D. Comparison of a lower-lipid propofol emulsion with the standard emulsion for sedation during monitored anesthesia care. *Anesthesiology* 2004;**100**:1072–1075.
115. Source: SkyePharma, Inc.
116. Junping, W., Takayama, K., Nagai, T., Maitani, Y. Pharmacokinetics and antitumor effects of vincristine carried by microemulsions composed of PEG-lipid, oleic acid, vitamin E and cholesterol. *Int J Pharm* 2003;**251**:13–21.
117. Balakrishnan, A., Rege, B.D., Amidon, G.L., Polli, J.E. Surfactant-mediated dissolution: contributions of solubility enhancement and relatively low micelle diffusivity. *J Pharm Sci* 2004;**93**: 2064–2075.
118. Bogman, K., Erne-Brand, F., Alsenz, J., Drewe, J. The role of surfactants in the reversal of active transport mediated by multidrug resistance proteins. *J Pharm Sci* 2003;**92**:1250–1261.
119. Rowe, R.C., Sheskey, P.J., Weller, P.J. *Handbook of Pharmaceutical Excipients*. 4. London: Pharmaceutical Press and American Pharmaceutical Association, 2003:474–478.
120. Ueda, K., Fujimoto, M., Noto, H., Kawaguchi, Y., Sakaeda, T., Iwakawa, S. Effect of oxyethylene moiety in polyoxyethylene sorbitan esters on the pharmacokinetics of menatetrenone incorporated in O/W lipid emulsions prepared with polyoxyethylene sorbitan esters and soybean oil in rats. *J Pharm Pharmacol* 2002;**54**:1357–1363.
121. Strickley, R.G. Solubilizing excipients in oral and injectable formulations. *Pharm Res* 2004;**21**:201–230.
122. Maskarinec, S.A., Hannig, J., Lee, R.C., Lee, K.Y. Direct observation of poloxamer 188 insertion into lipid monolayers. *Biophys J* 2002;**82**:1453–1459.
123. Millar, J.S., Cromley, D.A., McCoy, M.G., Rader, D.J., Billheimer, J.T. Determining hepatic triglyceride production in mice: comparison of poloxamer 407 with Triton WR-1339. *J Lipid Res* 2005;**46**:2023–2028.
124. Handa, T., Eguchi, Y., Miyajima, K. Effects of cholesterol and cholesteryl oleate on lipolysis and liver uptake of triglyceride/phosphatidylcholine emulsions in rats. *Pharm Res* 1994;**11**:1283–1287.

CHAPTER 5

PROTEIN ADSORPTION PATTERNS ON PARENTERAL LIPID FORMULATIONS: KEY FACTOR DETERMINING THE IN VIVO FATE

Rainer H. Müller and Torsten M. Göppert

5.1 INTRODUCTION

5.2 CONCEPT OF DIFFERENTIAL PROTEIN ADSORPTION

5.3 ANALYTICAL PROCEDURE

5.4 PLASMA PROTEIN ADSORPTION PATTERNS ON O/W EMULSIONS

5.5 PLASMA PROTEIN ADSORPTION PATTERNS ON SOLID LIPID NANOPARTICLES

5.6 CONCLUSION

5.1 INTRODUCTION

The in vivo fate of a drug is no longer mainly determined by the properties of the drug, but to a great extent by the delivery system, which should allow a controlled and localized release of the drug to the site of action. One approach of controlled drug delivery is the incorporation of the drug into parenteral colloidal drug carriers [1], such as liposomes [2, 3], polymeric nanoparticles [4, 5], or lipid formulations [6–8]. The existence of different carrier systems raises the question as to which of them might be the most suitable for the desired purpose. Aspects to consider are, for example, toxicity, scaling up feasibility, costs, stability or drug loading capacity and finally the possibility of intravenous drug targeting.

Role of Lipid Excipients in Modifying Oral and Parenteral Drug Delivery, Edited by Kishor M. Wasan

Liposomes were described for the first time by Bangham et al. in the 1960s [9] and have become one of the most extensively investigated carrier systems [10–14]. Meanwhile, there are some trade products on the market, e.g. AmBisome, Doxil, or DaunoXome. The advantages of these products are the reduction in toxic side effects and the increase in efficacy of the treatment [15, 16]. However, there are still some chemical and physical stability problems of liposome dispersions, which can lead to aggregation or drug leakage during storage [15, 17]. Other obstacles to the development of liposome formulations are difficulties in upscaling or nonspecific clearance by the mononuclear phagocytic system (MPS) [18].

Advantages of polymeric nanoparticles made from either nonbiodegradable (e.g. polystyrene) [19] or biodegradable polymers (e.g. polyalkylcyanoacrylate) [20] are controlled release and protection of the incorporated drugs [21]. However, the cytotoxicity of the polymers, even those that are accepted for use as microparticles and implants (e.g. polylactic acid), after internalization into the cells is a problematic aspect [22]. Furthermore, there is a lack of large-scale production methods that are acceptable to the regulatory authorities and at the same time cost-effective. Therefore, polymeric nanoparticles have so far not been relevant for the pharmaceutical market.

To avoid the toxicity problem, nanoparticles can be produced by using lipids made from physiological compounds, either liquid or solid at room temperature. Since the 1960s, oil-in-water emulsions (o/w emulsions, fat emulsions) have found wide use as parenteral nutrition (e.g. Intralipid, Lipofundin, Lipovenoes) [23–25]. As a result of their good tolerability in vivo they have also become an interesting delivery system for lipophilic drugs, which can be incorporated easily into the droplets (e.g. Diprivan, Diazemuls) [26, 27]. The main advantages are cost-effective large-scale production and the reduction of side effects caused at the injection side [28] or at certain organs (e.g. nephrotoxicity of amphotericin B) [29]. Nevertheless, the number of market products is very limited, because most drugs of commercial interest exhibit too low a solubility in the available registered oils (e.g. soybean oil or medium-chain triglyceride or MCT) [30]. However, a novel approach was developed recently that allows the formulation of drugs with poor solubility in both water and oils (e.g. amphotericin B). The SolEmuls technology allows the localization of the drugs in the emulsifier layer at the oil–water interface of the o/w emulsions [30, 31]. A very important point is the fact that the existing industrial production lines can be used, because SolEmuls technology is a very simple process using high-pressure homogenization. Major drawbacks of drug-containing o/w emulsions are, however, critical physical instability, which can be caused by the incorporated drug, lacking protection of incorporated chemically labile drugs, and the missing opportunity of controlled drug release.

Therefore, in the middle of the 1990s, the attention of various research groups was focused on alternative lipid formulations made from solid lipids. The so-called solid lipid nanoparticles (SLNs) offer the advantages of fat emulsions while at the same time minimizing the associated problems [7, 32, 33]. Basically, lipids can be used that are well tolerated by the body (e.g. glycerides composed of fatty acids that are present in the emulsions for parenteral nutrition) [34, 35] and large-scale production can easily be performed using high-pressure homogenization [36, 37].

Moreover, drug mobility in a solid lipid is considerably lower compared with a liquid oil. Therefore, SLNs show a controlled drug release [7, 38] and increased drug stability [39, 40].

Independent of the carrier system used, to achieve drug targeting via the intravenous route it is necessary to avoid recognition by the MPS. The second prerequisite is the direction of the carriers, which are successfully avoiding the MPS, to the desired site of action. The key factors for both challenges are the plasma proteins adsorbing onto the surface of such carriers immediately after intravenous injection. This chapter describes the technology for determining plasma protein adsorption patterns on fat emulsions and SLNs. The composition of the adsorption patterns (opsonins, dysopsonins, targeting moiety) and the relevance for the organ distribution is discussed, e.g. to create blood circulating carriers or target them to the brain.

5.2 CONCEPT OF DIFFERENTIAL PROTEIN ADSORPTION

Since the 1950s attempts have been made to correlate the organ distribution of intravenous injected drug carriers with their physicochemical properties, such as particle size [41], hydrophobicity [1], or electrical charge [42]. However, at the end of the 1980s, it was realized – and meanwhile has generally been accepted – that the blood proteins, adsorbing on the particle surface after their injection, are the crucial factors for the in vivo organ distribution [8, 43–45]. Depending on the particle surface properties, certain proteins will be preferentially adsorbed, leading to the adherence of the particles to cells with the appropriate receptor on the surface ('concept of differential adsorption' [43, 46]). Therefore, there is an intercorrelation of the physicochemical surface characteristics of the drug carriers, the plasma protein adsorption patterns, and the resulting organ distribution (Figure 5.1). The knowledge of this basic correlation can be exploited to develop site-specific intravenous drug carriers in a controlled way.

Proteins with opsonic function (e.g. the immunoglobulin IgG) [47] or complement factor C4γ [48]) lead to a fast uptake of the carriers by the MPS. In con-

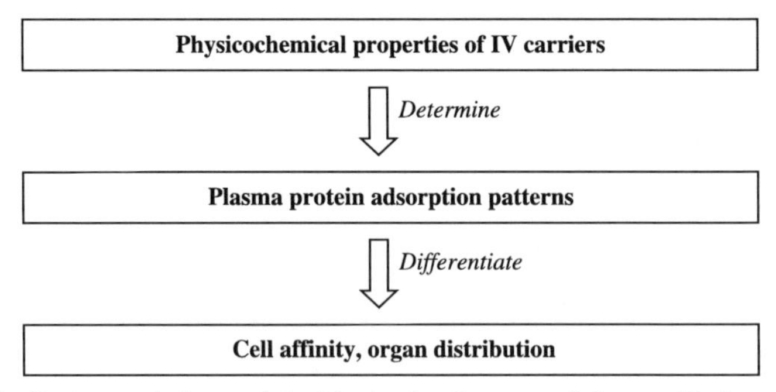

Figure 5.1 Basic correlation exploited in the development of site-specific intravenous drug carriers.

trast, if these opsonins are missing on the particle surface, or even more if dysopsonins are adsorbed (e.g. albumin [49] or IgA [50]), the carriers can circulate in the bloodstream and be used as circulating depots for the controlled release of drugs. This was previously shown for different particles that had been surface modified with hydrophilic polyethylene oxide chains [11, 51–53] or Poloxamine 908 [54–56]. Moreover, to achieve an active drug targeting with these nonrecognizing carriers, it is important to have, in addition, enrichment of a protein-mediating uptake into the target cells, e.g. apolipoprotein E (Apo-E) for brain targeting [57–59]. Apo-E plays an important role in the transport of lipoproteins into the brain via the low-density lipoprotein (LDL) receptor on the blood–brain barrier (BBB) [60, 61]. Müller et al. observed an enrichment of Apo-E on Polysorbate 80-modified polybutylcyanoacrylate (PBCA) nanoparticles [57], which were able to deliver various drugs (e.g. dalargin [62, 63], loperamide [64, 65], tubocurarine [66], doxorubicin [67, 68]) to the brain. Thus, it is possible that Apo-E-adsorbing drug carriers mimic lipoprotein particles, leading to their brain uptake by endocytic processes. To prove this thesis, the negative control of the experiments was taken (particles without Polysorbate 80, dalargin on the surface), before intravenous injection Apo-E was adsorbed on the surface, and after this the particles were able to deliver dalargin to the brain [69].

As a result of their importance for organ distribution, it was decided to develop a reliable method that is able to determine the plasma protein adsorption patterns on colloidal drug carriers. A powerful tool for detecting all proteins together is the high-resolution two-dimensional polyacrylamide gel electrophoresis (2-DE). The basic principles of 2-DE with colloidal drug carriers is described next.

5.3 ANALYTICAL PROCEDURE

5.3.1 2-DE Analysis

Two-dimensional polyacrylamide gel electrophoresis has a unique capacity for the resolution of complex mixtures of proteins, permitting the simultaneous analysis of hundreds or even thousands of gene products [70, 71]. It is divided into two different steps: in the first dimension proteins are separated within IPG strips according to their charge (isoelectric point, p*I*) by isoelectric focusing (IEF), and in the second dimension within polyacrylamide gradient slab gels according to their molecular mass (M_r) by SDS-PAGE (sodium dodecylsulfate–poylacrylamide gel electrophoresis). For the analysis of plasma protein adsorption patterns on colloidal drug carriers, 2-DE was performed essentially as described by Hochstrasser et al. for standard plasma samples [72]. The protocol was modified and established by Blunk et al. [46, 73] and is divided into following steps:

1. Sample preparation:
   - **(a)** incubation of particle suspension (containing constant surface areas) in plasma or serum for 5 min at 37°C
   - **(b)** separation of the particles from bulk medium by centrifugation.

2. Desorption of the adsorbed proteins from particle surface (according to Cook and Retzinger [74]).
3. Electrophoretic separation and silver staining of the proteins (according to Hochstrasser et al. [72]).
4. Identification of the spots by matching with master maps of human plasma Anderson and Anderson [75].
5. Data processing by MELANIE III software (according to Appel et al. [76]).

The protein adsorption patterns of colloidal drug carriers clearly differ from plasma bulk solution (Figure 5.2), this means that the proportion of certain proteins is reduced (e.g. the IgG γ chain) or increased (e.g. Apo-C-III).

5.3.2 Sample Preparation: The Crucial Factor for 2-DE Analysis of Colloidal Drug Carriers

Separation Problem

The prerequisite for 2-DE analysis of protein adsorption patterns is the availability of a valid separation method for the particles from the plasma. In the case of particles with higher densities than plasma (e.g. polymeric particles [46]), the particles were centrifuged for 90 min at 15 000*g* and the supernatant was easily discarded. Afterwards, the particle pellet was redispersed in the washing medium and centrifuged again. This washing procedure was repeated twice.

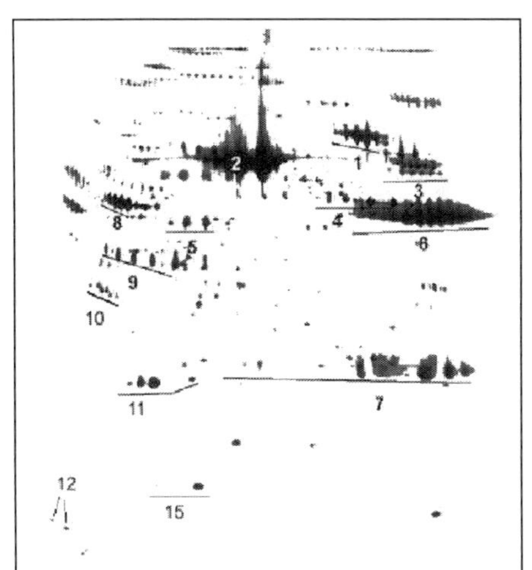

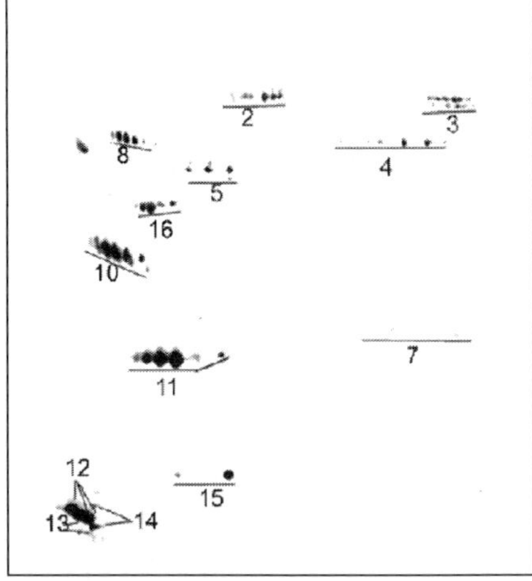

Figure 5.2 Two-dimensional polyacrylamide gel electrophoresis (2-DE) gels of human plasma (left) and solid lipid nanoparticles (SLN) (incubated in human plasma for 5 min), respectively (right). The entire gels are shown: p*I* 4.0–9.0 (from left to right, nonlinear), M_r 250 to 6 kDa (top to bottom, nonlinear). Indicated proteins: (1) transferrin, (2) albumin, (3) fibrinogen α chain, (4) fibrinogen β chain, (5) fibrinogen γ chain, (6) IgG γ chain, (7) Ig light chain, (8) α_1-antitrypsin, (9) haptoglobin β chain, (10) Apo-J, (11) Apo-A-I, (12) Apo-C-III, (13) Apo-C-II, (14) Apo-A-II, (15) transthyretin, (16) Apo-A-IV.

A specific problem of lipid formulations is the possible coalescence during the centrifugation steps as a result of excessively large centrifugal forces. A decrease in surface area can diminish the amount of protein available for the analysis. Moreover, the curvature of the surface might impair the adsorption patterns. However, Harnisch and Müller observed, for o/w emulsions only a slight decrease in surface area (95.2% of the surface remained), which did not affect protein analysis, when centrifugation was performed under optimized conditions (60 min at 15000 g, three washing steps) [77]. The emulsion droplets formed a coherent and stable top layer, whereas plasma formed the bottom layer. Plasma was removed by making a hole in the lipid top layer using a needle and carefully removing the plasma using a syringe.

Nevertheless, the standard separation method centrifugation cannot generally be applied to SLNs. In most cases, SLNs possess a density value relatively close to the density value of water, for example with lipids commonly used for SLNs – such as triglycerides (e.g. Dynasan), partial glycerides (e.g. Compritol), or fatty acids (e.g. stearic acid). Therefore, in such cases an alternative separation method has to be used. Other separation methods are the use of a magnetic field in the case of magnetite particles, established by Thode et al. [78], or gel filtration (size exclusion chromatography) for liposomes, established by Diederichs [79]. The latter protocol was modified and transferred successfully to SLNs by Göppert and Müller [80]. Briefly, the elution of material from a Sepharose 2B column was monitored by ultraviolet (UV) absorbance of each collected 1 mL fraction. Via the specific adsorption coefficients of plasma and particles, and a two-equation system, the calculation of the concentrations of each component in each fraction was possible. The equations used were:

$$A_{279\text{nm}} = [\text{SLN}] \times AC_{\text{SLN, 279nm}} + [\text{CP}] \times AC_{\text{CP, 279nm}} \quad (1)$$

$$A_{350\text{nm}} = [\text{SLN}] \times AC_{\text{SLN, 350nm}} + [\text{CP}] \times AC_{\text{CP, 350nm}} \quad (2)$$

where A is the absorption and AC is the specific absorption coefficient. The information in brackets is the concentration of SLNs and plasma, respectively. From this, elution fractions 3 and 4 were identified as being practically free of plasma and were subjected to 2-DE.

The SLNs used for establishing the gel filtration were prepared in a way that they had a sufficiently low density to be additionally separated by the standard centrifugation method. This was realized by using the lipid cetyl palmitate. The bulk density of cetyl palmitate is 0.816–0.819 g/cm^3, so it was sufficiently low for centrifugation, which was performed as described for o/w emulsions by Harnisch and Müller [77]. The adsorption patterns obtained after separation with both methods were qualitatively (Figure 5.3) and quantitatively (Figure 5.4) identical, showing the suitability of the gel filtration method for SLNs, which are too close in density to the density of water.

Although gel filtration has been allocated as an accurate method to separate SLNs from plasma, centrifugation was regarded as the most suitable separation method (if applicable) because of its speed (all samples could be separated at the same time, whereas gel filtration works only with a single assay) and ease of use (gel filtration demands regeneration of the column after each run).

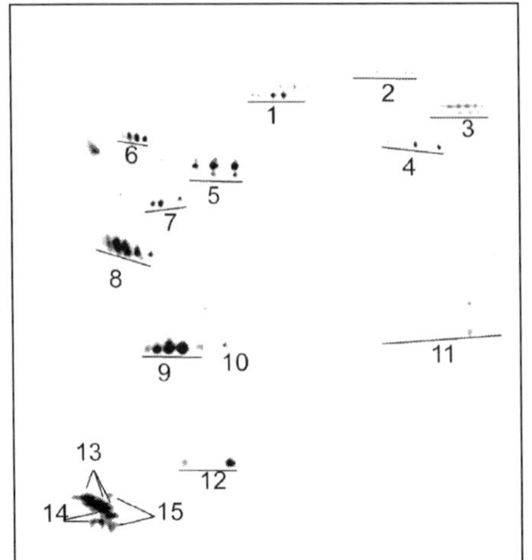

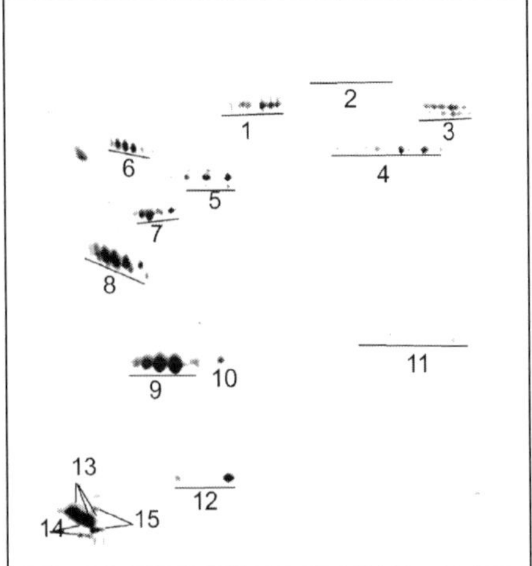

Figure 5.3 Plasma protein adsorption patterns on cetyl palmitate-solid lipid nanoparticles (SLNs), stabilized with Poloxamer 188 (P188-SLN). Influence of separation method: (a) gel filtration, (b) centrifugation; p*I* 4.0–9.0 (from left to right, nonlinear), M_r 250 to 6 kDa (top to bottom, nonlinear). Indicated proteins: (1) albumin, (2) IgM μ chain, (3) fibrinogen α chain, (4) fibrinogen β chain, (5) fibrinogen γ chain, (6) α_1-antitrypsin, (7) Apo-A-IV, (8) Apo-J, (9) Apo-A-I, (10) pro-Apo-A-I, (11) Ig light chain, (12) transthyretin, (13) Apo-C-III, (14) Apo-C-II, (15) Apo-A-II. (Modified after Göppert and Müller [80].)

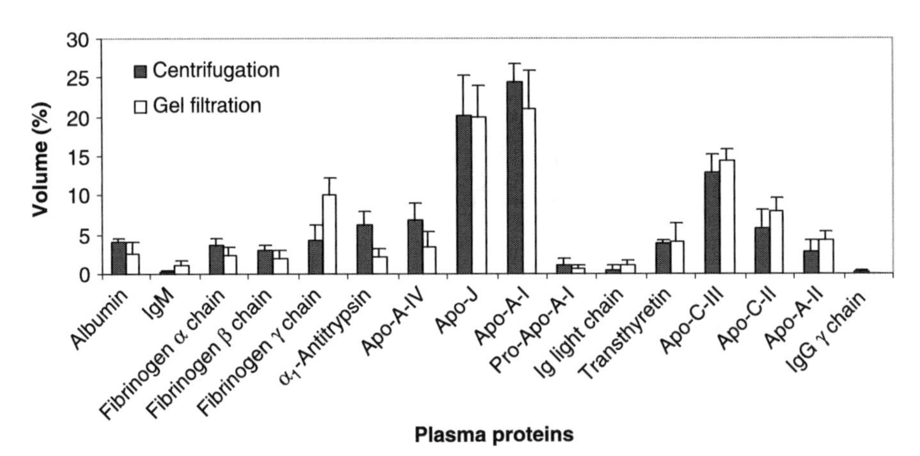

Figure 5.4 Semiquantitative plasma protein composition on P-solid lipid nanoparticles (SLNs) analyzed after applying two different separation methods of the particles from plasma, expressed as percentage of the overall detected protein pattern; error bars represent the standard deviation (SD; $n = 3$). (Reproduced with permission from Göppert and Müller [80].)

Washing Problem

The composition of the washing medium has an effect on the resulting adsorption pattern as has been shown for polymeric nanoparticles [73] and iron oxide particles for magnetic resonance imaging (MRI) [81]. The most frequently used washing media – distilled water and phosphate buffer, pH 7.4 – were therefore investigated for o/w emulsions [77] as well as for cetyl palmitate SLNs [82]. Using distilled water, IgG was observed to be the predominant protein (up to 50% of the total amount of detected proteins on emulsions [77] and up to 30% on SLNs [82]). When changing the washing medium to 20 mmol/L phosphate buffer, pH 7.4, the IgG adsorption was practically nil (Figure 5.5). A similar effect had been described for polymeric nanoparticles and identified as an artificial adsorption caused by a lower solubility of IgG in distilled water [73]. It was, therefore, concluded that phosphate buffer is the most suitable washing medium.

Adsorption Kinetics

Protein adsorption on to various solid surfaces (e.g. glass, metal) has been reported to be time dependent [83–88]. Routinely, the particles were incubated for 5 min in undiluted citrated plasma. An incubation time of 5 min was chosen because in general the first 5 min after an intravenous injection are decisive for the fate of the particles. In case recognition by the MPS occurs, up to 90% of the injected dose is taken up by the macrophages within the first 5 min [89]. In cases where particles survived these first 5 min, prolonged blood circulation was found [55].

The adsorption patterns have, however, to be regarded as a product of a sequence of adsorption of more plentiful proteins with lower affinity and their displacement by less plentiful proteins with higher affinity to the investigated surface. This displacement phenomenon often occurs within the first seconds or even within

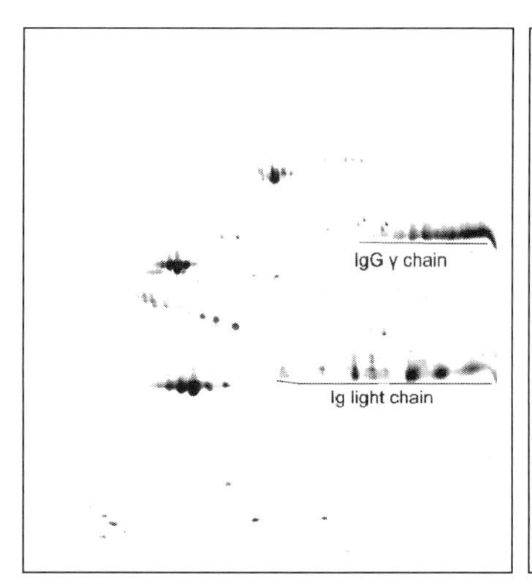

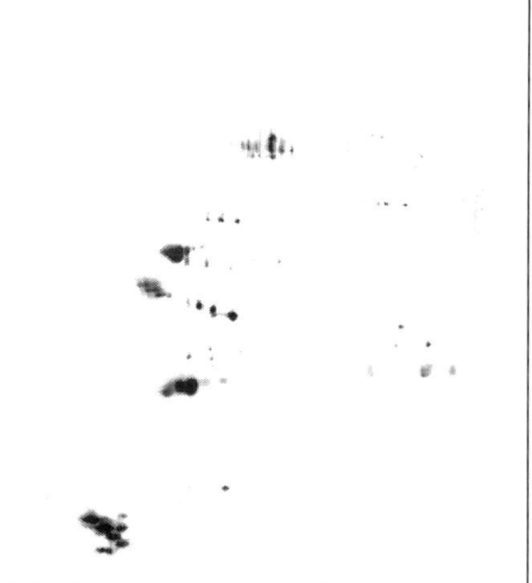

Figure 5.5 Plasma protein adsorption patterns on solid lipid nanoparticles (SLNs) after washing with distilled water (left, artificial IgG spot) and after washing with 20 mmol/L phosphate buffer pH 7.4 (right).

a fraction of a second and is called 'Vroman effect' [83, 84]. Therefore, if particle separation from excess plasma is time-consuming, the initially adsorbed proteins are not detectable. Nevertheless, by diluting the plasma sufficiently, the concentration of the displacing proteins with higher affinity would be decreased to such an extent that the resistance times of the plentiful proteins that adsorb first would be prolonged, and therefore will become detectable [85, 87].

It was shown that the extent of the 'Vroman effect' depends on the surface material on to which the proteins are adsorbed [87, 90]: Blunk et al. [91] demonstrated a competitive displacement of proteins on the surface of polystyrene model particles. Albumin, the most abundant protein in human plasma (3500–5000 mg/100 mL), was displaced by fibrinogen (200–450 mg/100 mL), which in turn was displaced by apolipoproteins, particularly Apo-C-III (12–14 mg/100 mL) and Apo-J (3.5–10.5 mg/100 mL) [91].

Harnisch and Müller observed no 'Vroman effect' on emulsion droplets [92]. With increasing plasma concentration, the total amount of major proteins increased steadily whereas their percentage, related to the overall amount of adsorbed proteins, remained almost unchanged (Figures 5.6 and 5.7). The authors concluded that the differences in the adsorption kinetics are the result of the different surface properties of the systems, caused by either their different physical (condition solid versus liquid) or their chemical (polystyrene versus lipid) condition.

In the case of SLNs, which have a solid matrix similar to the polystyrene model particles but also a lipid matrix similar to the o/w emulsions, Göppert and Müller observed a transient adsorption of fibrinogen [93]. With increasing plasma concentrations the amounts of fibrinogen steadily decreased (from 60% fibrinogen of the overall protein amount with 1.2% plasma to 6.8% with 75% plasma), whereas the amount of the apolipoproteins steadily increased (from 25% with 1.2% plasma to 68% with 75% plasma) (Figures 5.8 and 5.9). This was in agreement with the findings of Vroman et al. [83] and Blunk et al. [91] with solid surfaces.

Furthermore, the adsorption kinetics of plasma proteins on surface-modified SLNs over a period of time (0.5 min to 4 h) were investigated, which should be more relevant for the organ distribution of long-circulating carriers than the 'Vroman effect' after a split second. Previously, it has been shown that polymeric model particles surface-modified with Poloxamer 908 (P908) circulated in the bloodstream [55, 56], whereas particles modified with Poloxamer 407 (P407) accumulated in the bone marrow [94–96]. Moreover, Blunk et al. showed distinct quantitative changes in the adsorbed amounts of various apolipoproteins on the surface of polystyrene model particles, which were surface modified with these surfactants [91]: with increasing incubation time, the amount of Apo-C-III decreased drastically from 18% after 0.5 min to below 1% after 240 min. The amount of Apo-E and Apo-A-I (0.1% and 2%, respectively, after 0.5 min) increased to 9% and 8.5%, respectively. Obviously, there was a displacement among the apolipoproteins and it is unambiguous that such changes could easily lead to variations in the organ distribution of long-circulating carriers.

However, in contrast to the surface-modified polystyrene particles, the composition of the plasma proteins adsorbed on chemically similar surface-modified SLNs (P407-SLN and P908-SLN, respectively) was much more stable [93]. It has

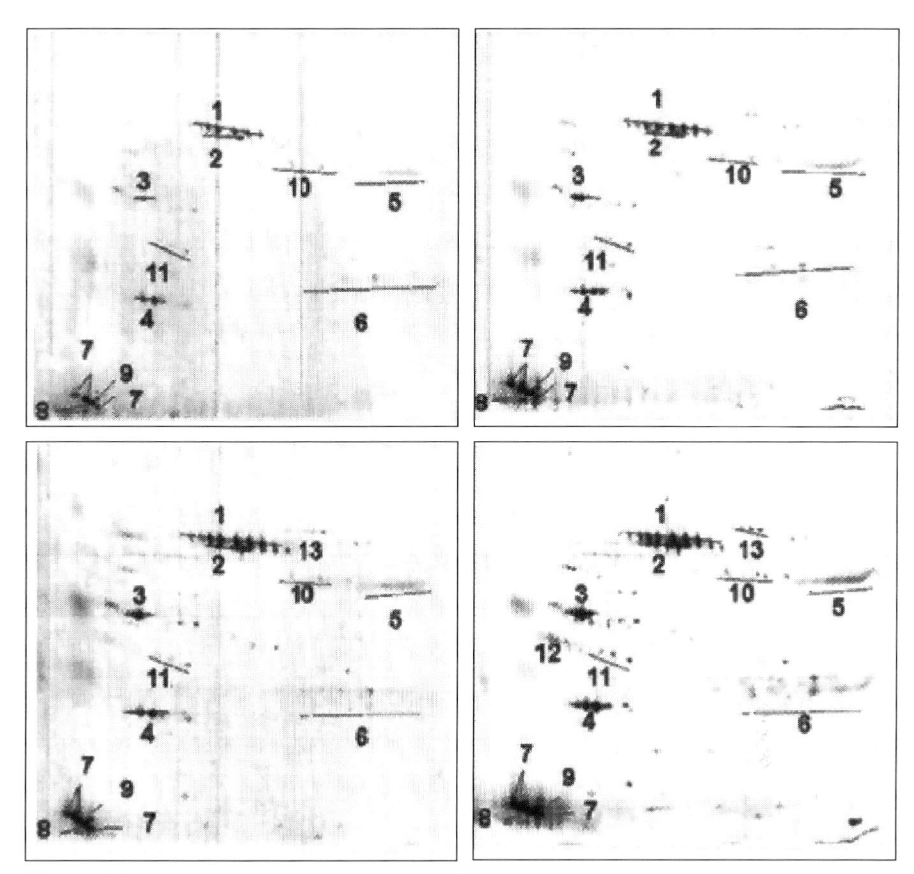

Figure 5.6 Protein adsorption patterns on Lipofundin MCT 20% oil-in water (o/w) emulsion, which resulted from an incubation time of 5 min each. Plasma concentration in the incubation media: top left, 11%; top right, 33%; bottom left, 55%; bottom right, 75%. Indicated spots: (1) IgD δ chain, (2) albumin, (3) Apo-A-IV, (4) Apo-A-I, (5) IgG γ chain, (6) Ig light chain, (7) Apo-C-III, (8) Apo-C-II, (9) Apo-A-II, (10) Apo-H, (11) Apo-E, (12) Apo-J, (13) complement factor C3β. (Reproduced with permission from Harnisch and Müller [92].)

been shown that no protein desorption from P407-SLN and P908-SLN occurred (Figure 5.10). Contact with plasma over 4 h led entirely to an increase of the total amount of the proteins adsorbed (Figure 5.11). However, size measurements of SLNs revealed that there was only a slightly decrease in particle size [93]. Thus, the increase of the total amount of the proteins was not caused by a surface degradation of the SLNs in plasma. In fact, it was concluded that, with increasing incubation time, further proteins will be adsorbed on to the first layer of proteins adsorbed, whereas their percentages, related to the overall amount of proteins, remained almost unchanged (Figure 5.12) [93].

As a result, the less pronounced time dependence of protein adsorption on surface-modified lipid formulations might be absolutely beneficial for site-specific

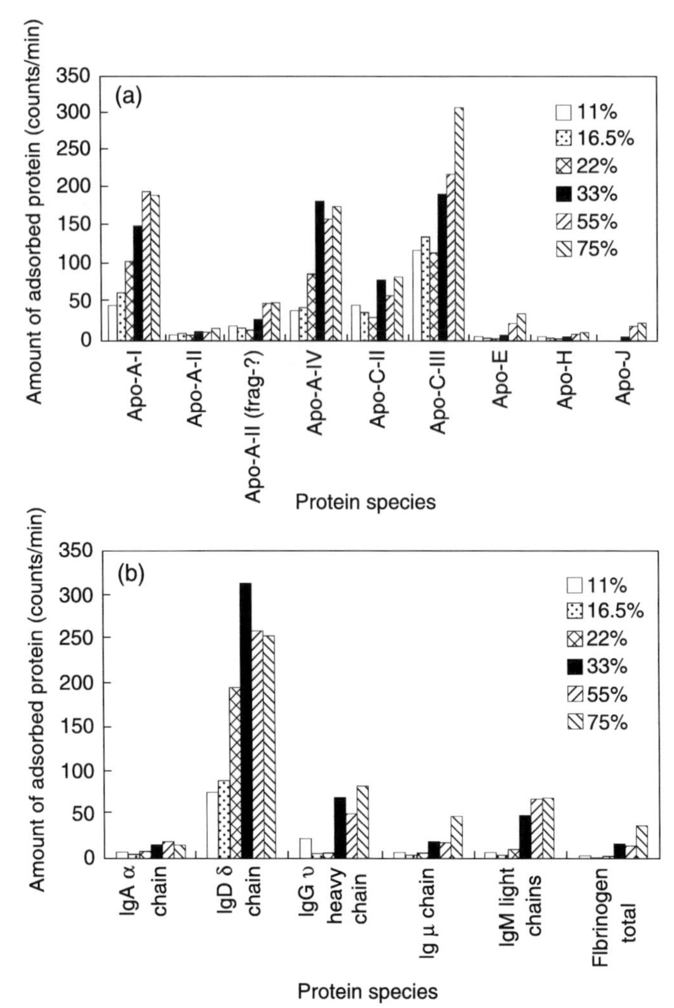

Figure 5.7 Amount of major proteins on the two-dimensional gels of Lipofundin MCT 20% o/w (oil-in-water) emulsion. (a) The increasing adsorption of apolipoproteins with increasing plasma concentration in the incubation medium; (b) the increasing adsorption of certain immunoglobulins and fibrinogen. (Reproduced with permission from Harnisch and Müller [92].)

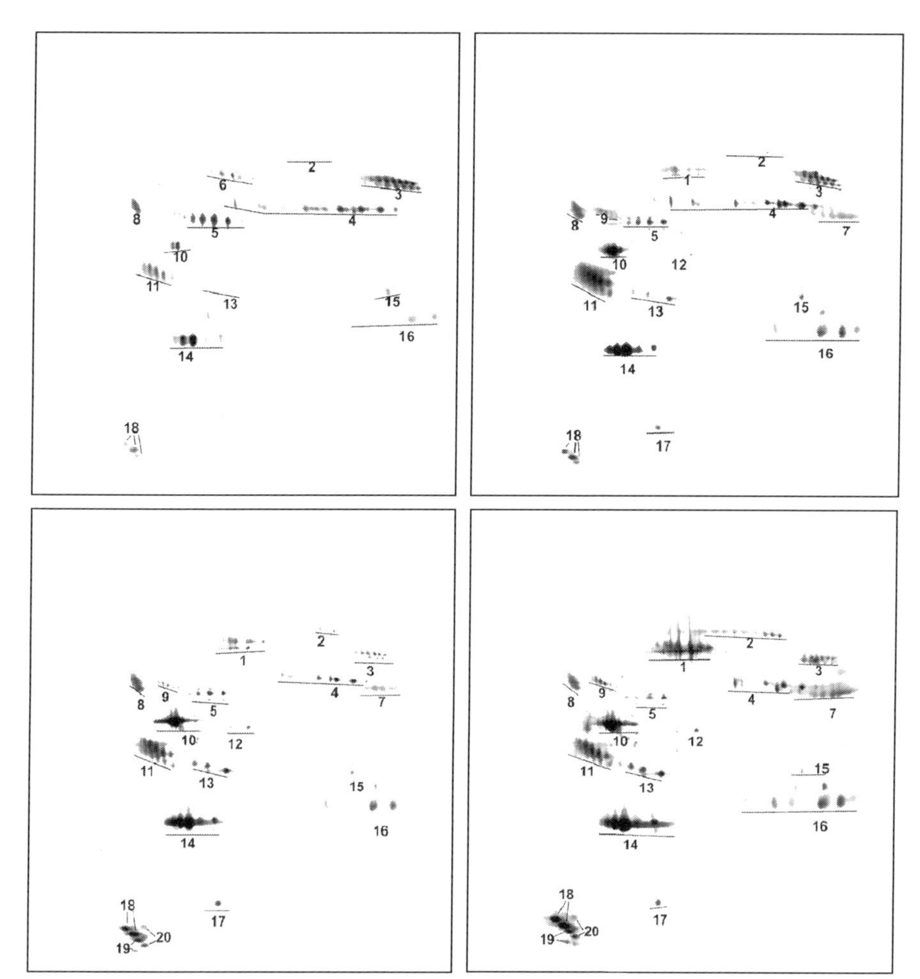

Figure 5.8 Protein adsorption patterns on solid lipid nanoparticles (SLNs), which resulted from an incubation time of 5 min each. Plasma concentration in the incubation media: top left, 1.2%; top right, 12%; bottom left, 55%; bottom right, 75%. Indicated spots: (1) albumin, (2) IgM μ chain (3) fibrinogen α chain, (4) fibrinogen β chain, (5) fibrinogen γ chain, (6) IgD δ chain, (7) IgG γ chain, (8) α2-HS-glycoprotein, (9) α_1-antitrypsin, (10) Apo-A-IV, (11) Apo-J, (12) haptoglobin β chain, (13) Apo-E, (14) Apo-A-I, (15) C4γ, (16) Ig light chain, (17) transthyretin, (18) Apo-C-III, (19) Apo-C-II, (20) Apo-A-II. (Modified after Göppert and Müller [93].)

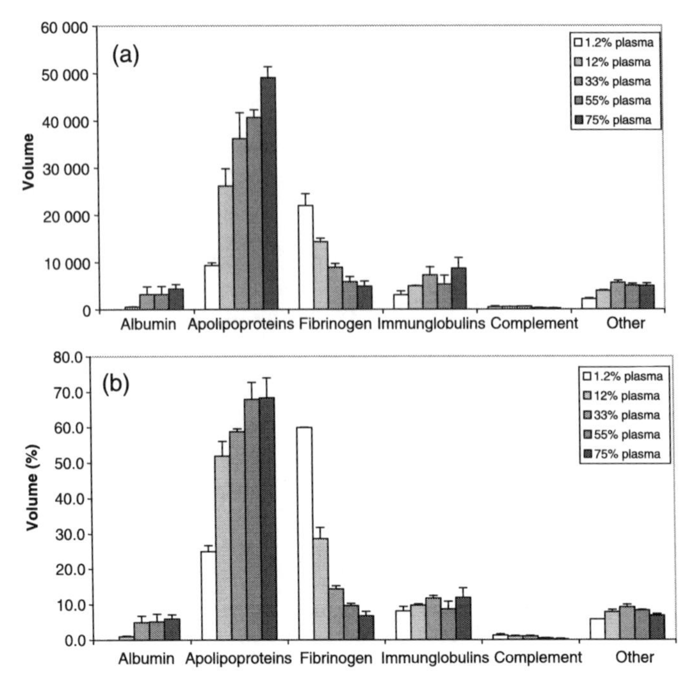

Figure 5.9 (a) Total amounts and (b) percentages of the major proteins adsorbed from the different plasma dilutions on solid lipid nanoparticles (SLNs) (error bars represent the SD ($n = 2$)). (Reproduced with permission from Göppert and Müller [93].)

drug delivery, because the adsorption pattern may be better exploited for this purpose than an adsorption pattern that is very dependent on contact time with the blood. Indeed, it was shown that there is also a 'Vroman effect' on SLNs, although this displacement of fibrinogen was restricted to the very early stages of adsorption.

Serum versus Plasma

Routinely, the particles were incubated in citrated plasma, because it contains all the proteins including the clotting cascade, and the adsorption of a single protein strongly depends on the composition of the mixture from which the adsorption appears (see above) [45, 97]. Moreover, in various studies with different types of particles the adsorption of fibrinogen was demonstrated [46, 98–101]. Many authors therefore described the use of plasma as an incubation medium to be a reasonable compromise for the in vitro evaluation of colloidal drug carriers [46, 102–104].

It is important, however, to bear in mind that the addition of an anticoagulant such as sodium citrate – which chelates bivalent cations such as Ca^{2+} – leads to inactivation of the amplifying systems such as complement or coagulation [105, 106]. Leroux et al. demonstrated the importance of the incubation medium with polylactic acid (PLA) nanoparticles, which were analyzed in vitro with regard to their uptake by cells of the MPS after incubation in plasma and serum [107]. Drastic dif-

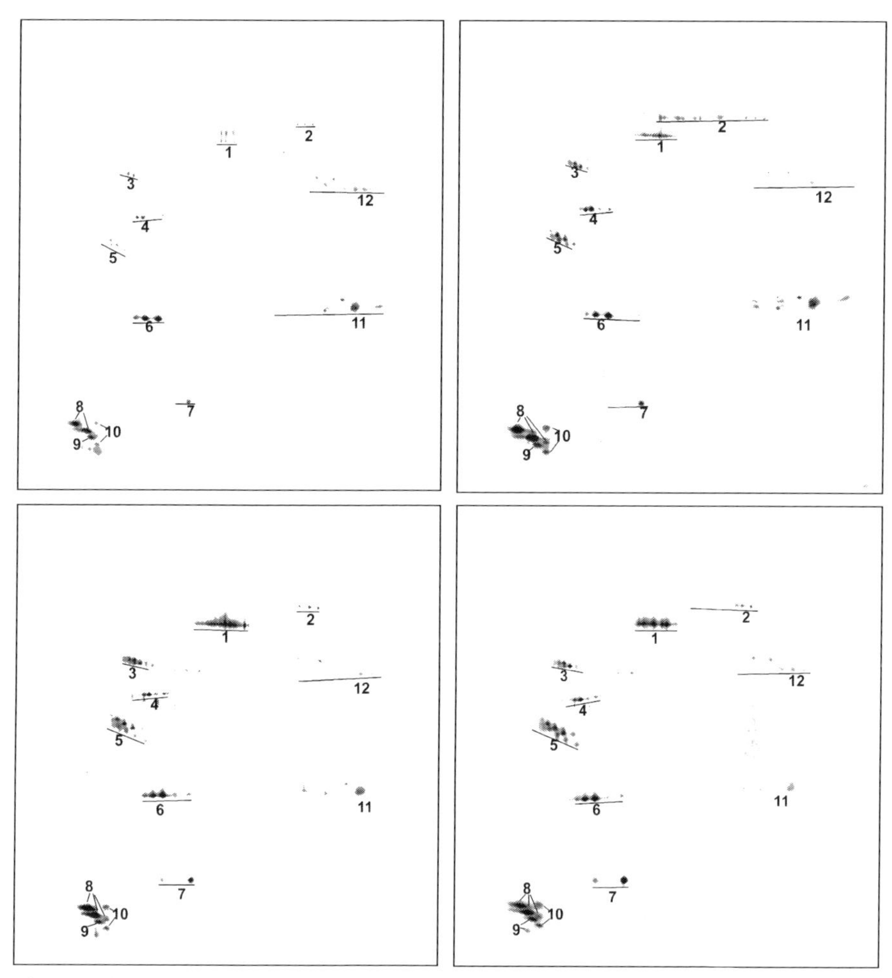

Figure 5.10 Plasma protein adsorption patterns on P407-SLN (Poloxamer 407–solid lipid naonoparticle) after different incubation times (sample preparation: gel filtration). Top left, 0.5 min; top right, 5 min; bottom left, 30 min; bottom right, 240 min. The entire gels are shown; p*I* 4.0–9.0 (from left to right, nonlinear), M_r 250 to 6 kDa (top to bottom, nonlinear). (1) Albumin, (2) IgM μ chain (3) α_1-antitrypsin, (4) Apo-A-IV, (5) Apo-J, (6) Apo-A-I, (7) transthyretin, (8) Apo-C-III, (9) Apo-C-II, (10) Apo-A-II, (11) Ig light chain, (12) IgG γ chain. (Reproduced with permission from Göppert and Müller [93].)

ferences in the uptake by monocytes were detected, which was attributed to an involvement of complement proteins [107] active only in serum. Lück et al. showed, with the polystyrene model particles, that 2-DE analysis provides the possibility of distinguishing between physical adsorption of C3 and covalent binding of C3b (a prerequisite for complement activation) to particulate surfaces. Therefore, when a correlation between the adsorption patterns on colloidal drug carriers detected in vitro and their in vivo behavior is approached, it is important to consider both plasma

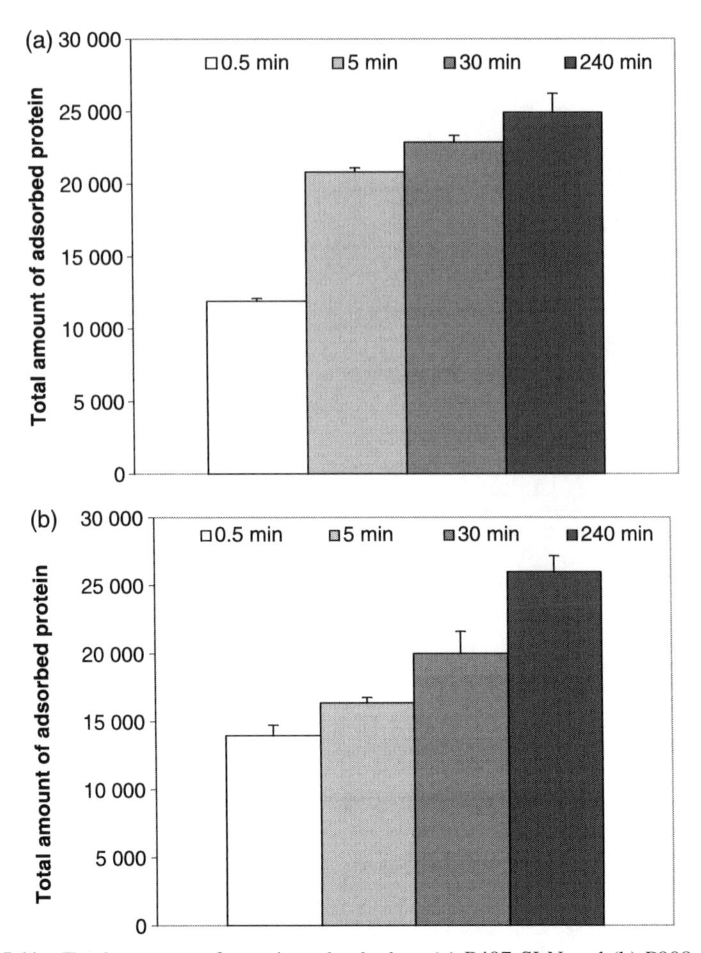

Figure 5.11 Total amounts of proteins adsorbed on (a) P407-SLN and (b) P908-SLN after different incubation time (error bars represent SD ($n = 2$)). (Reproduced with permission from Göppert and Müller [93].)

and serum. As 2-DE analysis is very time-consuming and cost-intensive, this could be realized only with selected lipid formulations, e.g. Poloxamer 235 surface-modified cetyl palmitate-SLN (P235-SLN). As fibrinogen is not present in serum, the adsorption pattern detected on P235-SLN after incubation in serum was consequently found to be free of fibrinogen chains (Figures 5.13 and 5.14). However, the overall pattern was not found to be distinctly different from the plasma protein adsorption pattern (Figure 5.13 and 5.14). Both adsorption patterns were dominated by apolipoproteins (about 66% of the overall detected protein amount). The most striking observations were the absence of complement factors and the comparatively strong enrichment of the IgG γ chain (Figure 5.14). Obviously, this protein was able to bind to sites, which would be occupied by fibrinogen after incubation in plasma.

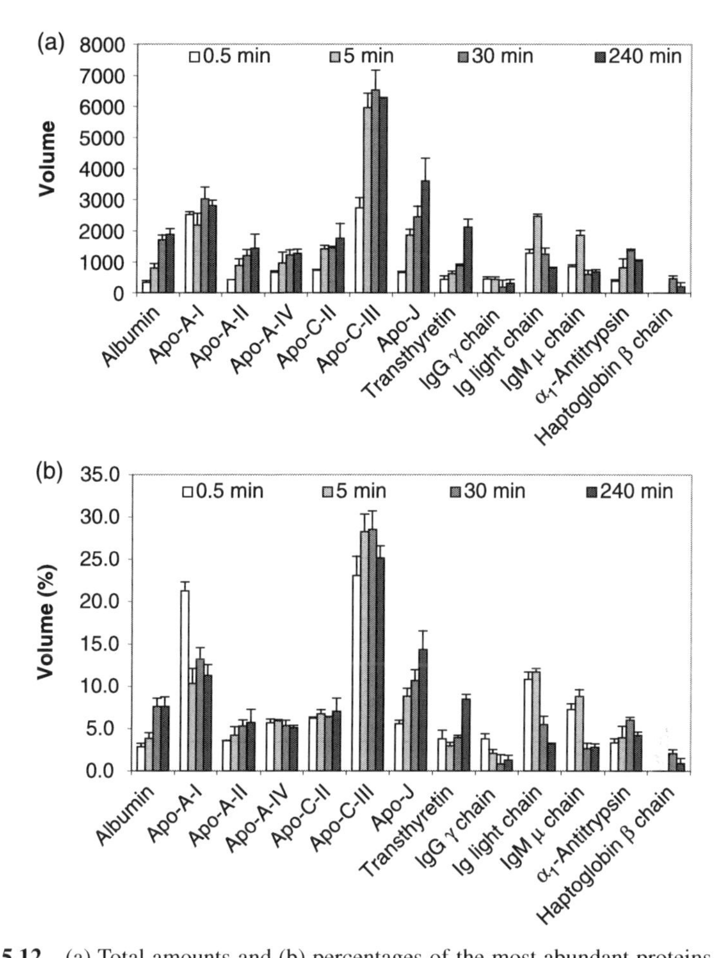

Figure 5.12 (a) Total amounts and (b) percentages of the most abundant proteins adsorbed on P407-SLN after different incubation time (error bars represent SD ($n = 2$)). (Reproduced with permission from Göppert and Müller [93].)

5.4 PLASMA PROTEIN ADSORPTION PATTERNS ON O/W EMULSIONS

5.4.1 Influence of Lipid Composition of o/w Emulsions on Protein Adsorption

An important aspect of the use of parenteral fat emulsions is that they show little uptake by the macrophages of the MPS, because MPS impairment by pronounced phagocytosis exposes patients to the danger of an increased incidence of infectious diseases. It has been shown that MCT-based emulsions (e.g. Lipofundin MCT) cause lower MPS impairment compared with other commercial emulsions, e.g. Lipofundin N (soybean oil based) [108]. Therefore, plasma protein adsorption patterns of both

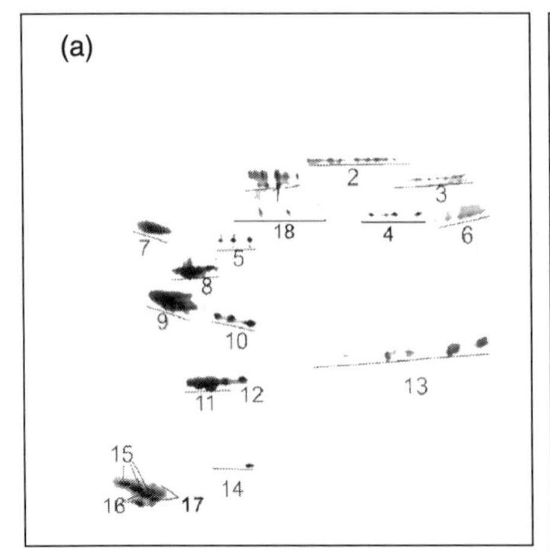

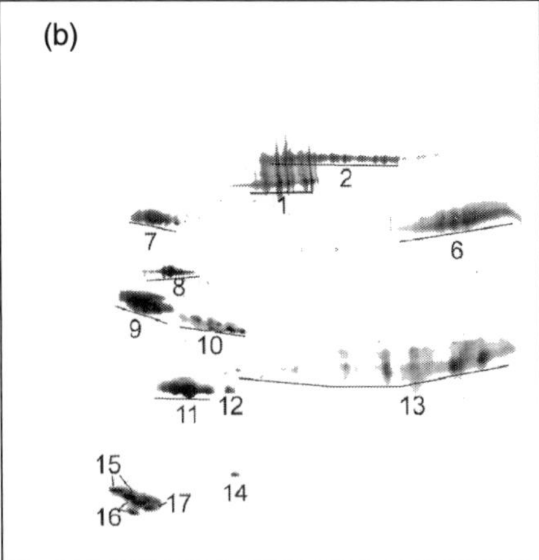

Figure 5.13 Plasma protein adsorption patterns on P235-SLN incubated in plasma (left), and serum (right), respectively. (1) Albumin, (2) IgM μ chain, (3) fibrinogen α chain, (4) fibrinogen β chain, (5) fibrinogen γ chain, (6) IgG γ chain, (7) α_1-antitrypsin, (8) Apo-A-IV, (9) Apo-J, (10) Apo-E, (11) Apo-A-I, (12) pro-Apo-A-I, (13) Ig light chain, (14) transthyretin, (15) Apo-C-III, (16) Apo-C-II, (17) Apo-A-II, (18) Apo-H.

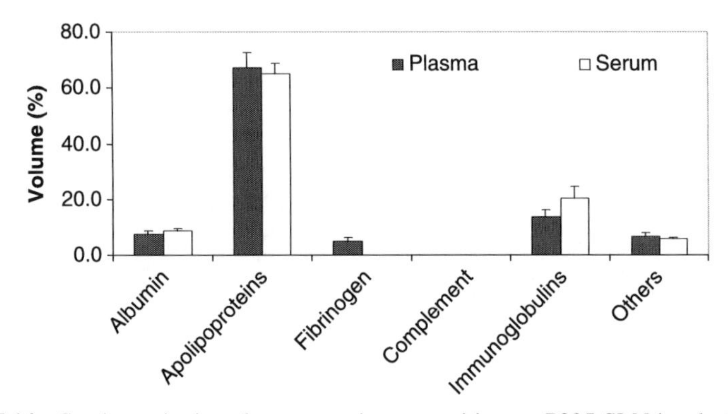

Figure 5.14 Semiquantitative plasma protein composition on P235-SLN incubated in plasma and serum, respectively, are expressed in percentage of the overall detected protein amount; error bars represent SD ($n = 2$).

formulations were determined [109]. Figure 5.15 shows close-ups of the 2-DE gels of Lipofundin MCT 10% and Lipofundin N 10% emulsions. The patterns were found to be qualitatively identical. Moreover, they were dominated by apolipoproteins (as shown earlier). The great affinity of apolipoproteins to lipid formulations is explained by their relatively flexible molecular structure (α-helical segments), which can change its conformation when adsorbing on to surfaces ('soft proteins') [110]. The lipid-bound state is thermodynamically favored and they are inherently adsorbed on to the surface of hydrophobic lipids such as triglycerides, cholesterol, and cholesterol esters, forming lipoproteins in the blood [111].

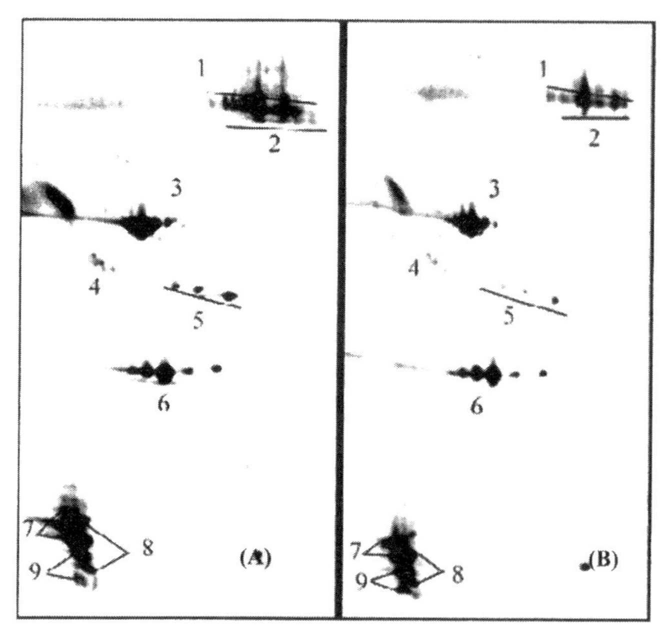

Figure 5.15 Close-ups of two-dimensional polyacrylamide electrophoresis gels showing the lower left part, pI 4.4–5.7 (from left to right, nonlinear) and M_W 200 to 6 kDa (top to bottom, nonlinear). (A) Lipofundin MCT 10%, (B) Lipofundin N 10%. (1) IgD, (2) albumin, (3) Apo-A-IV, (4) Apo-J, (5) Apo-E, (6) Apo-A-I, (7) Apo-C-III, (8) Apo-A-II, (9) Apo-C-II. (Adapted from Schmidt [109].)

In Figure 5.16 the percentages of the major proteins on the 2-DE gels are compared. The patterns were also found to be quantitatively rather similar. However, the higher percentage of Apo-E on Lipofundin MCT in comparison to Lipofundin N (3.1% vs 1.6%) was very interesting, considering its lower MPS impairment. Endogenous lipoproteins such as chylomicrons or very-low-density lipoprotein (VLDL) remnants can be removed from blood circulation directly through interaction with the LDL receptor, which is mainly expressed on hepatocytes [112, 113] (but also at the BBB, see earlier). Thus, the higher enrichment of Apo-E on Lipofundin MCT could explain its lower MPS impairment as a result of a faster clearance from the bloodstream via the LDL receptor [109]. These results could be confirmed by investigating the adsorption patterns of Lipofundin MCT 20% and Lipofundin N 20% [109]. Thus, it was concluded that different lipid compositions of o/w emulsions might lead to different protein adsorption patterns and consequently to a different behavior in vivo [109].

5.4.2 Influence of Surface Modification of o/w Emulsions on Protein Adsorption

Efficient surface modifiers that can be used to reduce the MPS uptake are polyethylene oxide (PEO)-containing block co-polymers, i.e. poloxamers and poloxamines

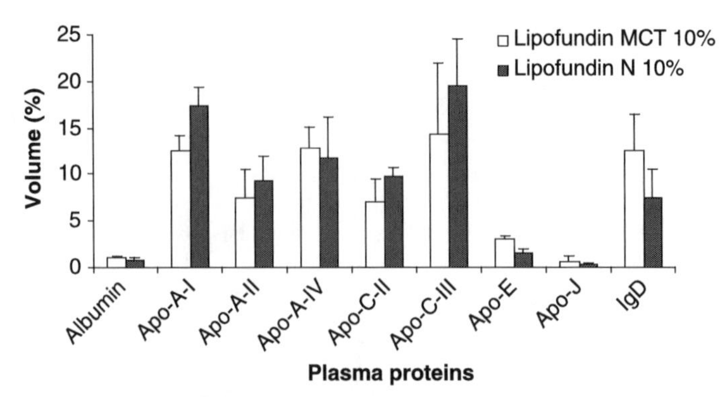

Figure 5.16 Major proteins adsorbed on Lipofundin MCT 10% and Lipofundin N 10%, respectively. Error bars represent SD ($n = 2$). (Adapted from Schmidt [109].)

TABLE 5.1 Characterization data of the different polymers

	P184	P235	P237	P188	P238	P407	P338	908
M_r	2900	4600	7700	8350	10800	11500	14000	25000
n	13	27	62	75	97	98	128	121
m	30	39	35	30	39	67	54	16
Apo (%)	86.2	90.6	91.8	83.8	86.3	90.5	86.3	84.0
Apo-E (%)	4.3	3.5	2.6	0.2	0.2	0.3	0.2	0.1

P, poloxamer; 908, Poloxamine 908; n, polyethylene oxide units; m, polypropylene oxide units.
Percentages of the overall detected protein amount: Apo (%), percentage of all apolipoproteins; Apo-E (%), percentage of Apo-E.
Data from Hochstrasser et al. [72].

(see earlier). Therefore, o/w emulsions (20%, w/w, soy bean oil) were surface modified with a range of these polymers (2.5%, w/w) (Table 5.1) to assess the resulting plasma protein adsorption patterns. It was shown that the percentages of apolipoproteins on these emulsions was even higher (84.0–91.8%) (Table 5.1) than on Lipofundin emulsions (about 60%, see earlier) [73]. This could be explained by a pronounced stealth effect of the block co-polymers, which prevents larger proteins (e.g. albumin or immunoglobulins) from being adsorbed on to the surface. Thus, percentages of these proteins decreased whereas percentages of smaller proteins (e.g. apolipoproteins or transthyretin) increased. Looking at the complete adsorption patterns, even when using poloxamers of low M_r, opsonins such as immunoglobulins, fibrinogen or complement factors were not detected or in a very low amount. Figure 5.17 is an exemplary case of the apolipoprotein adsorption on four different types of such emulsions (E-908, E-238, E-407, and E-235, respectively). It is striking that Apo-E was preferentially adsorbed when the emulsions were stabilized with poloxamers with a low number of PEO units (i.e. E-184, E-235, E-237) (Figure 5.18). On o/w emulsions stabilized with polymers with longer PEO chains ($n > 70$) Apo-E was detected in only very low amounts (about 0.2%). However, the adsorption of the

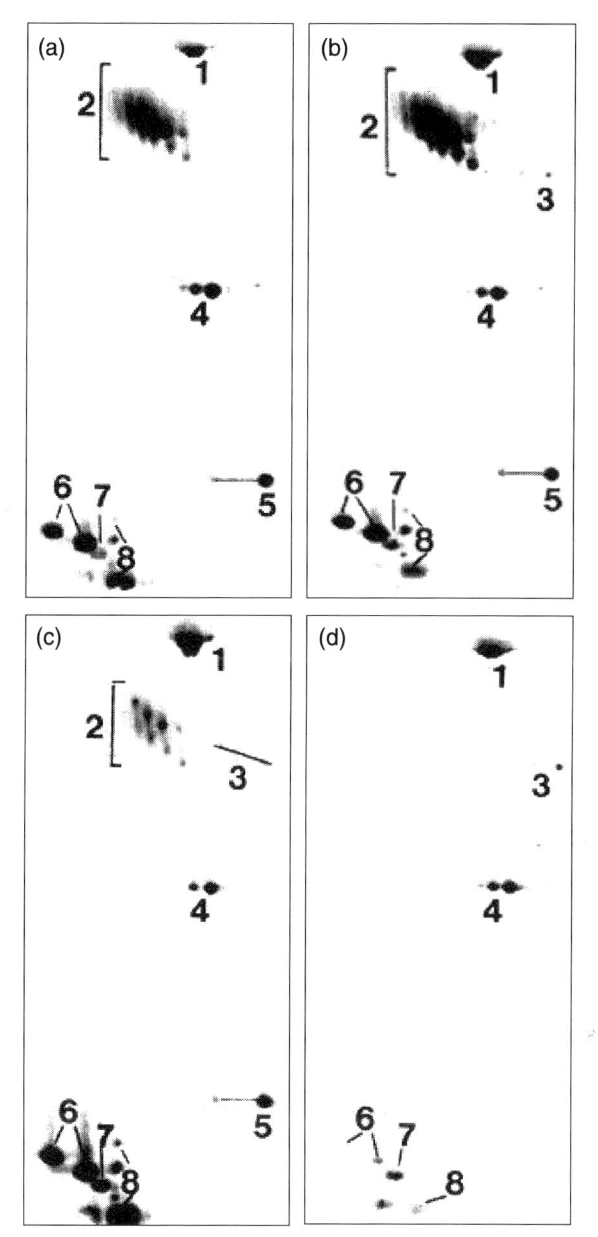

Figure 5.17 Close-ups of two-dimensional polyacrylamide electrophoresis gels showing the lower left part, pI 4.4–6.0 (from left to right, non-linear) and M_r 46 to 6 kDa (top to bottom, nonlinear). Oil-in-water (o/w) emulsions (20%, w/w, soybean oil) surface modified with (a) Poloxamine 908 (E-908), (b) Poloxamer 238 (E-238), (b) Poloxamer 407 (E-407), (d) Poloxamer 235 (E-235). (1) Apo-A-IV, (2) Apo-J, (3) Apo-E, (4) Apo-A-I, (5) transthyretin, (6) Apo-C-III, (7) Apo-C-II, (8) Apo-A-II. (Adapted from Blunk [73].)

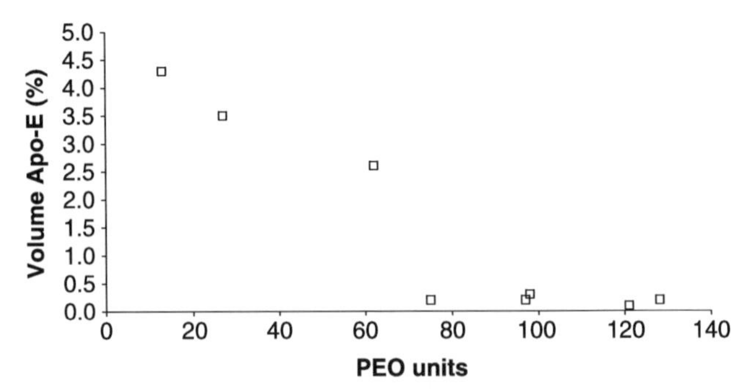

Figure 5.18 Percentages adsorbed Apo-E on soybean oil emulsions (20%, w/w) in dependence on polyethylene oxide (PEO) chain length of the used surfactants. (Adapted from Blunk [73].)

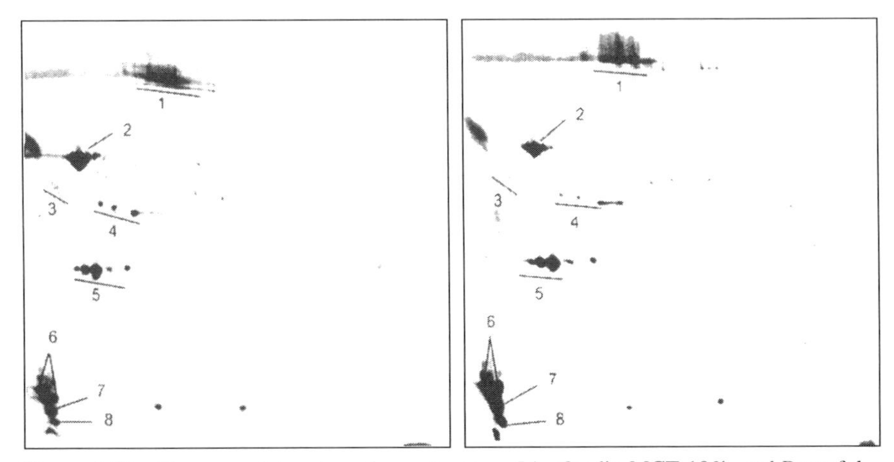

Figure 5.19 Plasma protein adsorption patterns on Lipofundin MCT 10%, and Propofol-Lipuro 1% (right), respectively. (1) Albumin, (2) Apo-A-IV, (3) Apo-J, (4) Apo-E, (5) Apo-A-I, (6) Apo-C-III, (7) Apo-A-II, (8) Apo-C-II. (Adapted from Schmidt [109].)

other apolipoproteins (i.e. Apo-A-I, Apo-C-III, Apo-C-II, and Apo-A-II) did not show any obvious relationship to the chain length of the polymers, which could be explained by their smaller required space as a result of their lower M_r [114].

5.4.3 Influence of Drug Loading of o/w Emulsions on Protein Adsorption

Recently, the importance of the localization of the drug has been shown [109]. In small and lipophilic molecules (e.g. propofol), which are located inside the oil droplets, it was reported that the plasma protein patterns were very similar to the patterns obtained with emulsions without drug (Figure 5.19). Both 2-DE gels were

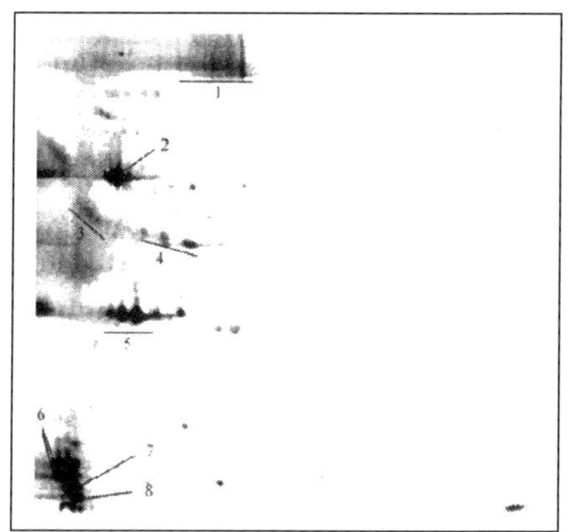

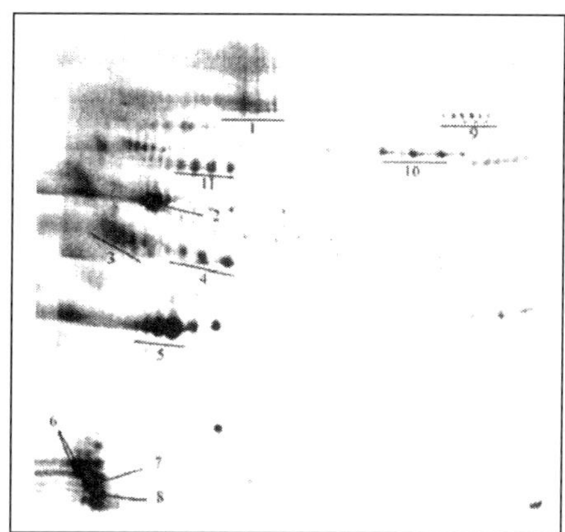

Figure 5.20 Plasma protein adsorption patterns on Lipofundin MCT 20%, and Lipofundin MCT 20% + 0.1% amphotericin B (right), respectively. (1) Albumin, (2) Apo-A-IV, (3) Apo-J, (4) Apo-E, (5) Apo-A-I, (6) Apo-C-III, (7) Apo-A-II, (8) Apo-C-II, (9) fibrinogen α chain, (10) fibrinogen β chain, (11) fibrinogen γ chain. (Adapted from Schmidt [109].)

qualitatively similar and, considering the standard deviation (SD) typically obtained with this method, no significant quantitative difference could be found. Identical results were obtained with other lipophilic drugs with low M_r, e.g. diazepam and etomidate [115]. Therefore, the authors concluded that incorporated drugs will have no influence on the plasma protein adsorption pattern and consequently no impact on the resulting organ distribution [109].

In amphiphilic drugs, which are located in the lecithin layer at the oil–water interface of the o/w emulsions (e.g. amphotericin B, SolEmuls technology [30]), a different result was found [109]. In fact, the 2-DE gels did not differ greatly with regard to qualitative aspects. However, addition of 0.1% amphotericin B to Lipofundin MCT 20% emulsions resulted in additional adsorption of fibrinogen chains (Figure 5.20). Moreover, some apolipoproteins (particularly Apo-C-III, and Apo-A-IV) showed distinct quantitative changes in the adsorbed amounts (Figure 5.21), indicating the influence of amphiphilic drugs, located in the emulsifier layer, on the resulting protein adsorption pattern. Identical results were obtained by investigating protein adsorption on Lipofundin N 20%, with and without amphotericin B [109].

5.4.4 Physiological Stability of o/w Emulsions

It is well known that lipid formulations such as o/w emulsions or SLNs are metastable systems, e.g. the chemical composition of the stabilizing lecithin layer may undergo changes during storage (e.g. lysolecithin and free fatty acid formation). This might lead to changes in the surface properties and consequently in the resulting adsorption patterns. Therefore, Lipofundin N and Lipofundin MCT of different ages were used as model emulsions to investigate whether there were changes in protein adsorption as a function of emulsion aging. Figure 5.22 shows close-ups of

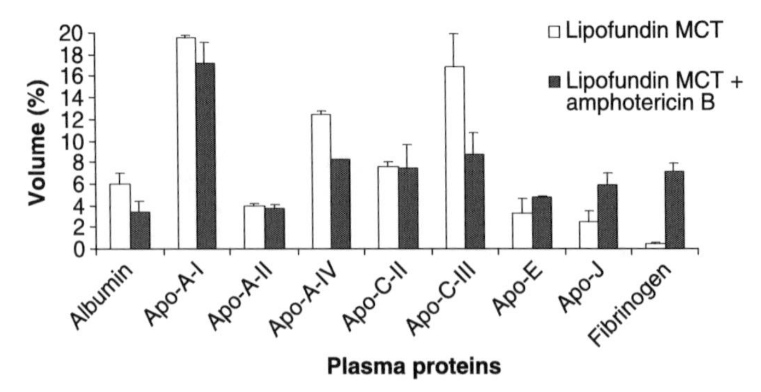

Figure 5.21 Major proteins adsorbed on Lipofundin MCT 20%, and Lipofundin MCT 20% + 0.1% amphotericin B, respectively. Error bars represent SD ($n = 2$). (Adapted from Schmidt [109].)

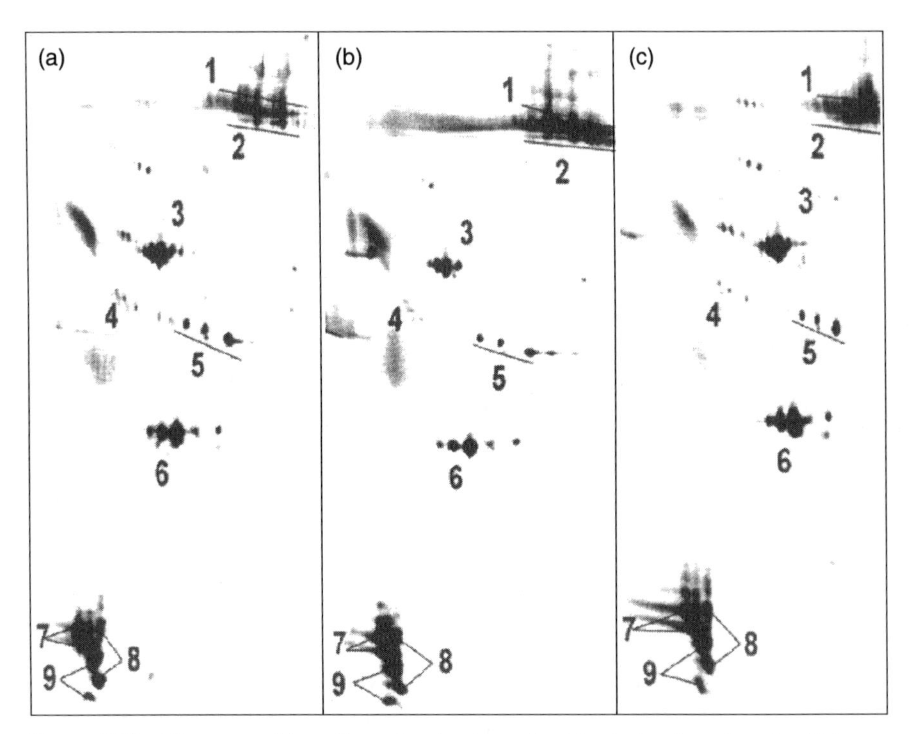

Figure 5.22 Close-ups of two-dimensional polyacrylamide electrophoresis gels of Lipofundin N 20%, showing the lower left part, p*I* 4.4–5.7 (from left to right, nonlinear) and M_r 200 to 6 kDa (top to bottom, nonlinear). (a) Freshly prepared (b), 16 months, (c) 28 months. (1) IgD, (2) albumin, (3) Apo-A-IV, (4) Apo-J, (5) Apo-E, (6) Apo-A-I, (7) Apo-C-III, (8) Apo-A-II, (9) Apo-C-II. (Adapted from Schmidt [109].)

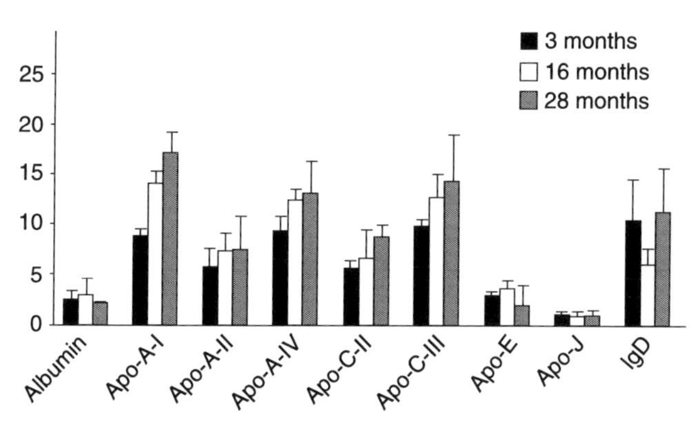

Figure 5.23 Major proteins adsorbed on Lipofundin N 20% emulsions of different age. Y axis: percentage of the overall detected protein amount. Error bars represent SD ($n = 3$). (Adapted from Schmidt [109].)

the 2-DE gels of Lipofundin N 20% emulsions. The adsorption patterns were found to be qualitatively almost identical. The overall detected protein amount on the gels was of the same order of magnitude, independent on the age of the emulsions. Moreover, comparing the percentages of each protein (Figure 5.23) showed that no or few changes were not significant considering the SD of the method. Of course, it is still not known, how pronounced a change in the adsorption pattern needs to be to cause a distinct change in the in vivo fate. However, from the data obtained to date, including some other emulsions [109], the patterns seem to be almost identical during storage. From this, it was concluded that, if o/w emulsions are physically stable long term, then they are also stable in surface properties and the resulting protein adsorption pattern that determines organ distribution. In line with the 'physicochemical stability' of emulsions, the results of this study show that there is also 'physiological stability'.

5.5 PLASMA PROTEIN ADSORPTION PATTERNS ON SOLID LIPID NANOPARTICLES

5.5.1 Influence of the Matrix Lipid of SLNs on Protein Adsorption

To determine the influence of the solid matrix lipid on the adsorption patterns, different SLN formulations were investigated, using 10.0% lipid and 1.2% Poloxamer 188 as stabilizers (Table 5.2) [116]. The resulting 2-DE gels were qualitatively (Figure 5.24) and quantitatively (Figure 5.25) very different, showing the importance of the used matrix lipid. The highest adsorption of immunoglobulin chains was found on stearic acid-SLNs (ST-SLNs – 31 vs 1–7% on the other types), whereas on Dynasan-SLNs (DY-SLNs) the highest adsorption of fibrinogen chains occurred (26 vs 6–9% on the other types).

TABLE 5.2 Size characterization data of solid lipid nanoparticles (SLNs) by photon correlation spectroscopy (PCS), ζ potential (ZP) and water contact angle (CA) on the bulk material

	1.2% poloxamer 188, 10.0% lipid, where lipid is			
	Stearic acid (ST-SLN)	Compritol 888 ATO (CO-SLN)	Cetyl palmitate (CP-SLN)	Dynasan 118 (DY-SLN)
PCS (nm)	296	239	275	306
ZP (mV)	–30.6	–26.2	–25.8	–26.6
CA (°)	77.8	86.5	87.8	100.4

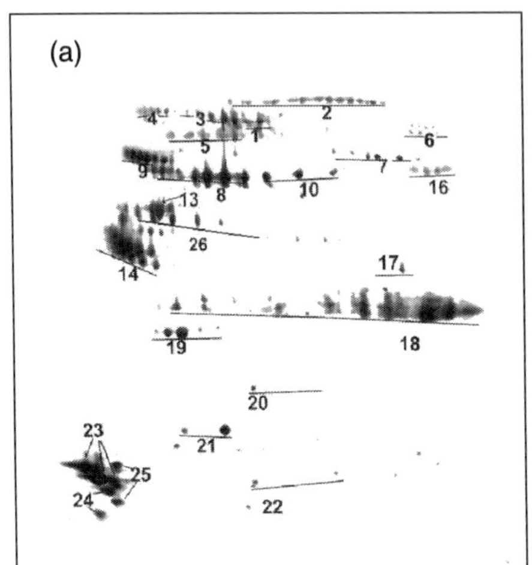

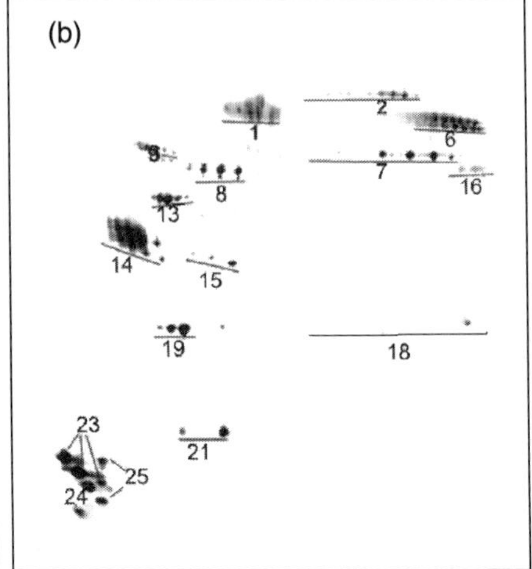

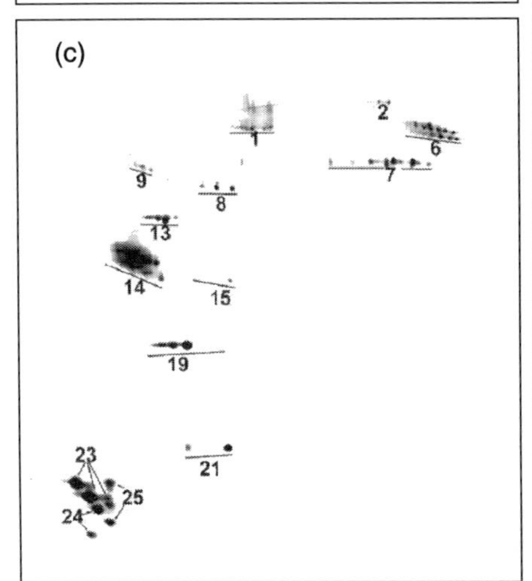

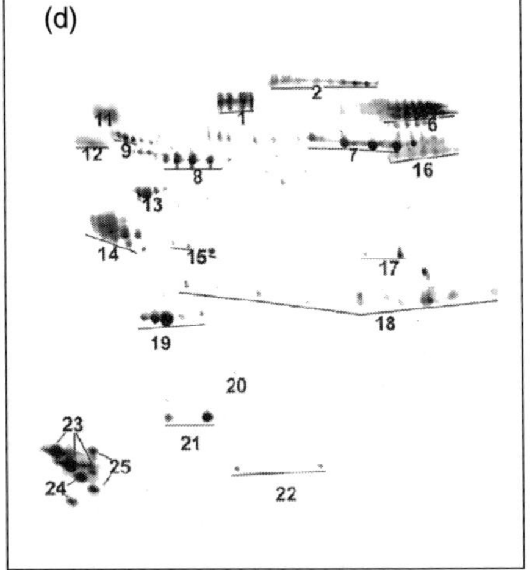

Figure 5.24 Plasma protein adsorption patterns on (a) ST-SLN, (b) CO-SLN, (c) CP-SLN, and (d) DY-SLN, respectively (the entire gels are shown; see text for abbreviations). (1) Albumin, (2) IgM μ chain, (3) IgD δ chain, (4) α1-B-glycoprotein, (5) IgA chain, (6) fibrinogen α chain, (7) fibrinogen β chain, (8) fibrinogen γ chain, (9) α_1-antitrypsin, (10) IgM μs intermediate chain, (11) α_1-antichymotrypsin, (12) α2-HS-glycoprotein, (13) Apo-A-IV, (14) Apo-J, (15) Apo-E, (16) IgG γ Kette, (17) C4γ, (18) Ig light chain, (19) Apo-A-I, (20) haptoglobin α2 chain, (21) transthyretin, (22) haptoglobin α1 chain, (23) Apo-C-III, (24) Apo-C-II, (25) Apo-A-II, (26) haptoglobin β chain.

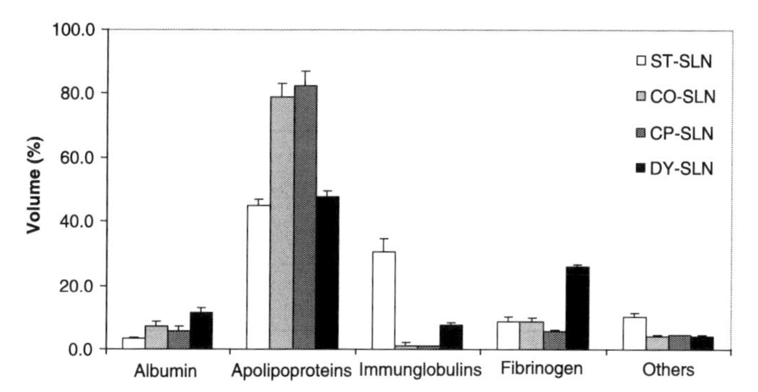

Figure 5.25 Major proteins adsorbed on (a) ST-SLN, (b) CO-SLN, (c) CP-SLN, and (d) DY-SLN, respectively (see text fro abbreviations). Y axis: percentage of the overall detected protein amount. Error bars represent SD ($n = 2$). SLN, solid lipid nanoparticles.

The adsorption patterns of Compritol-SLNs (CO-SLNs) and cetyl palmitate-SLNs (CP-SLNs) were, however, very similar (Figures 5.24 and 5.25). By measuring the contact angle of water on the several bulk media, which is one approach for quantifying hydrophobicity of surfaces [117], it was shown that these two types of SLNs were also very similar in this regard (about 87°), whereas stearic acid was more hydrophilic (contact angle or CA 77.8°) and Dynasan 118 more lipophilic (CA 100.4°) (Table 5.2). From this, it was concluded that, where there are similar surface hydrophobicities of the bulk media, the plasma adsorption patterns will also be similar (provided that size or charge effect of the particles on the adsorption patterns can be excluded) [82].

5.5.2 Influence of Surface Modification of SLNs on Protein Adsorption

To achieve the described goals (avoidance of recognition of the drug carriers by the MPS and site-specific drug delivery, especially brain targeting) SLNs were engineered by eight different block co-polymers, which were previously used for surface modification of o/w emulsions (compare Table 5.1 and section 5.4.2). Figure 5.26 is an exemplary case of the plasma protein adsorption patterns on four different types of such SLNs (SLN-184, SLN-235, SLN-407, and SLN-338, respectively). The priority objective was enrichment of Apo-E on the surface of SLNs, to investigate an in vivo well-tolerated carrier, which is able to transport drugs across the BBB. From this, it was interesting that the results were in good agreement with the Apo-E values obtained with o/w emulsions (compare Figure 5.18 and section 5.4.2) [73]: again, the highest Apo-E adsorption occurred when using polymers with short PEO chains in the molecule (i.e. P184 and P235). There was a good exponential relationship between adsorbed Apo-E and the PEO chain length (Figure 5.27) [114]. Moreover, typical opsonins were either not detected or in very low amounts, so it was concluded that these kinds of polymers (particularly P235 as an inhibitor of

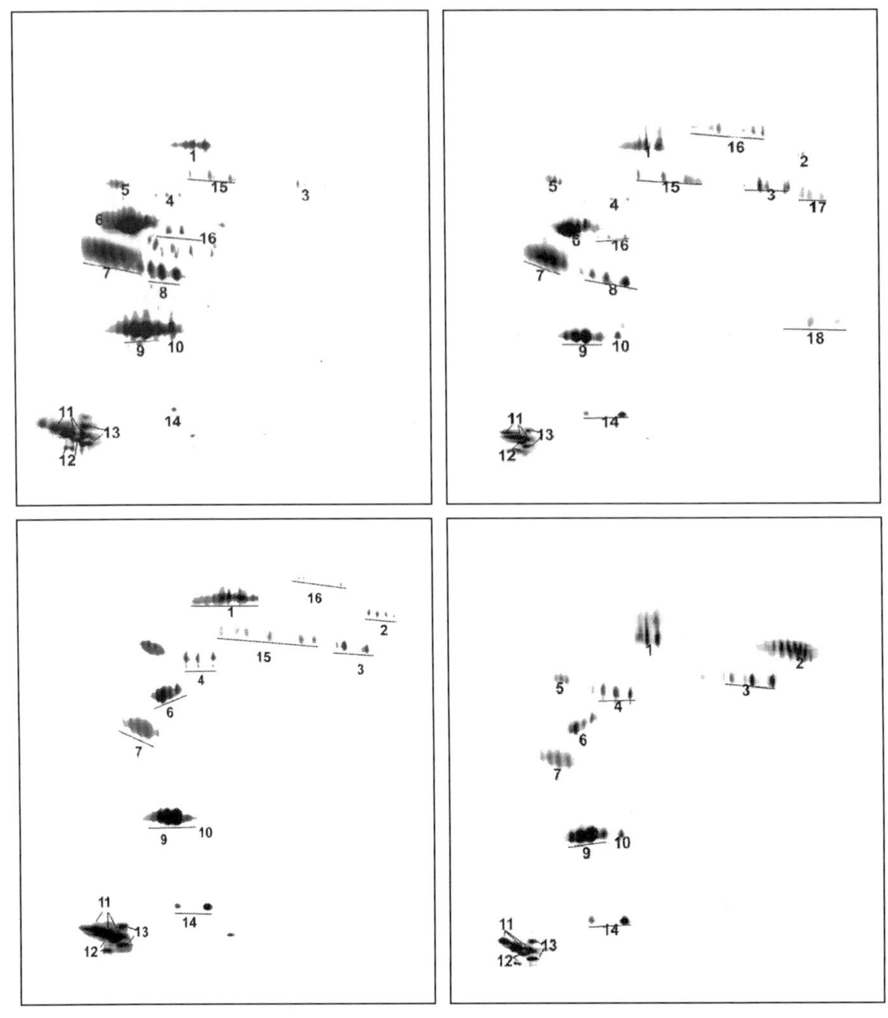

Figure 5.26 Two-dimensional polyacrylamide electrophoresis gels of solid lipid nanoparticles (SLNs) stabilized with block co-poloxamers. Top left, Poloxamer 184-stabilized SLN (P184-SLN); top right, P235-SLN; bottom left, P407-SLN; bottom right, P338-SLN. The entire gels are shown. (1) Albumin, (2) fibrinogen α, (3) fibrinogen β chain, (4) fibrinogen γ chain, (5) α_1-antitrypsin, (6) Apo-A-IV, (7) Apo-J, (8) Apo-E, (9) Apo-A-I, (10) pro-Apo-A-I, (11) Apo-C-III, (12) Apo-C-II, (13) Apo-A-II, (14) transthyretin, (15) β2-glycoprotein, (16) IgM μ chain, (17) IgG γ chain, (18) Ig light chain. (Modified after Göppert and Müller [114].)

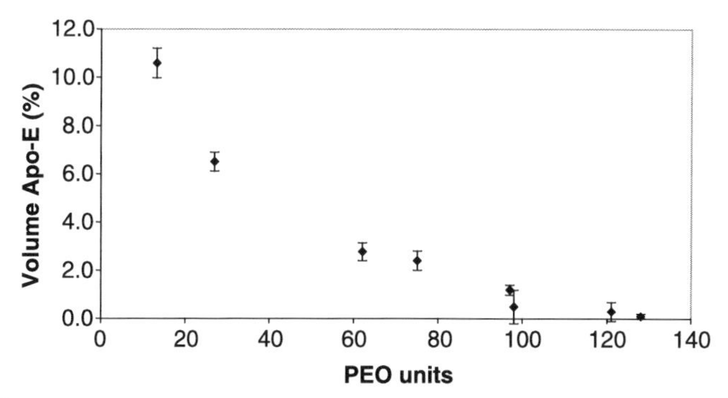

Figure 5.27 Percentages of adsorbed Apo-E on the solid lipid nanoparticles (SLNs) depending on polyethylene oxide (PEO) chain length of the used surfactants. Error bars represent SD ($n = 2$). (Modified after Göppert and Müller [114].)

P-glycoprotein [118, 119]) are promising surface modifiers to be used to deliver drugs to the brain by means of biofunctionalized SLNs [114].

In addition, adsorption patterns on polysorbate-modified SLNs were determined, and it was shown that the amount of adsorbed Apo-E on SLNs decreased with increasing hydrophilic lipophilic balance (HLB) values of the used polysorbates [120]. The most lipophilic Polysorbate 60 (HLB of 14.9) showed the highest adsorption of Apo-E (5.4%), whereas Polysorbate 20 with the highest HLB value (16.7) showed the lowest adsorption of Apo-E (2.3%) (Figure 5.28). The percentage of Apo-E on Polysorbate 80-modified SLNs was equal to Polysorbate 60; however, percentages of dysopsonins were higher and the percentage of Apo-C-II (inhibits binding of Apo-E to the LDL receptor [121]) was lower [120]. Thus, it was concluded that in this series Polysorbate 80 might have the highest potential to achieve brain targeting.

5.5.3 Physiological Stability of SLNs

Similar to fat emulsions, SLNs are metastable systems, as already described earlier. Therefore, the adsorption patterns were also determined on SLNs that had different physical long-term stability depending on their age [122]. Once more, no considerable changes were found (Figures 5.29 and 5.30), supporting the same organ distribution independent of storage time ('physiological stability'). This was a very important result for developing, in a controlled way, biofunctionalized SLNs for intravenous drug targeting.

5.6 CONCLUSION

In this chapter, we have reviewed state-of-the-art plasma protein adsorption patterns on parenteral lipid formulations, as well as giving a brief description of specifics of the analytic procedure. For the transfer of 2-DE analysis to SLNs, gel filtration was

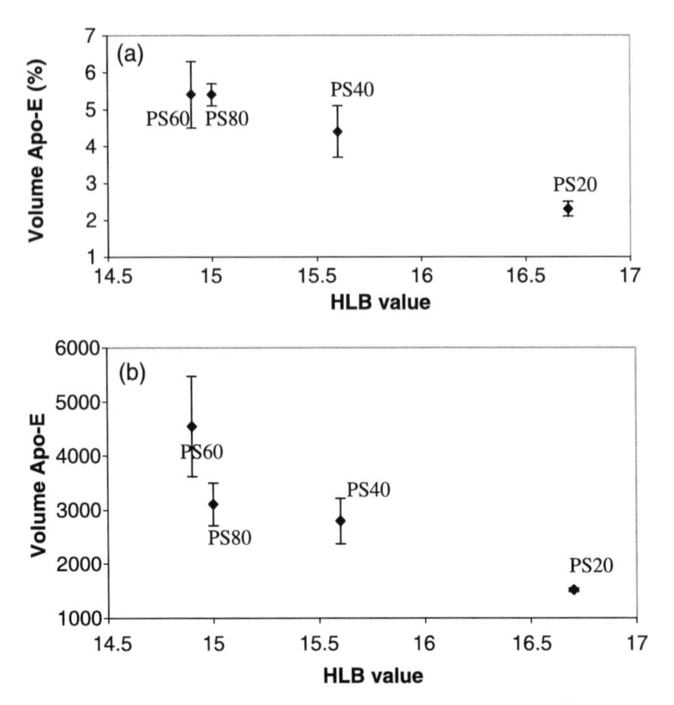

Figure 5.28 (a) Percentages of adsorbed Apo-E and (2b) total amount of adsorbed Apo-E on the different solid lipid nanoparticle (SLN) formulations depending on hydrophilic lipophilic balance (HLB) value of the used polysorbates (PS). Error bars represent SD ($n = 2$). (Modified after Göppert and Müller [120].)

established as a reliable method to separate all types of SLNs from excess plasma. It was shown that phosphate buffer pH 7.4 is the most suitable washing medium for lipid formulations and plasma is a reasonable compromise as an incubation medium. Besides this, it was shown that, in contrast to polymeric nanoparticles, there is no competitive displacement of apolipoproteins, on o/w emulsions or SLNs. From this, the stable patterns of lipid formulations may be better used for drug targeting than particles with patterns that depend on contact time with blood. Moreover, the adsorption patterns on lipid formulations that are physically stable long term, depending on their age, were investigated. No considerable changes were found, supporting the same organ distribution not dependent on storage time ('physiological stability'). These results provide an important basis for the development of lipid formulations for intravenous drug targeting via adsorbed blood proteins. Furthermore, it was shown that the matrix lipids play an important role for the resulting patterns of lipid formulations. Finally, Poloxamer 235 and Polysorbate 80 were investigated as promising surface modifiers for use in delivering drugs to the brain. To summarize, the 2-DE results obtained with parenteral lipid formulations are quite promising for further developments in this research field.

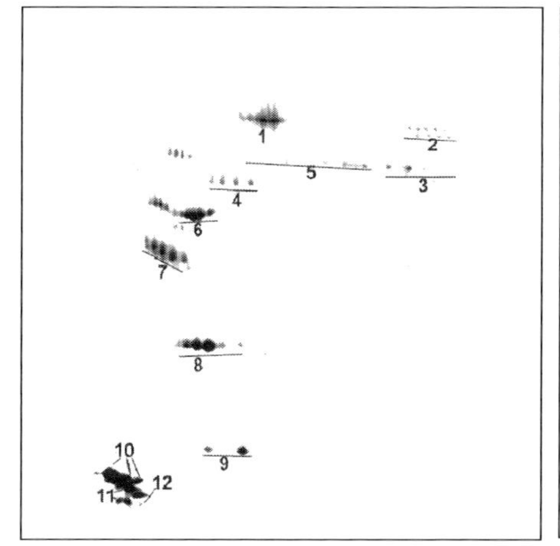

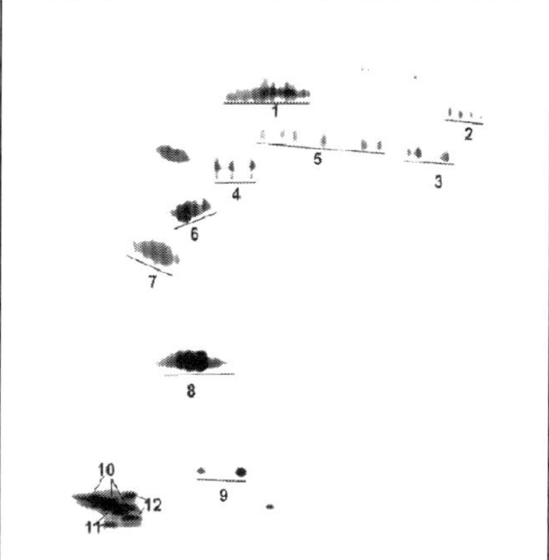

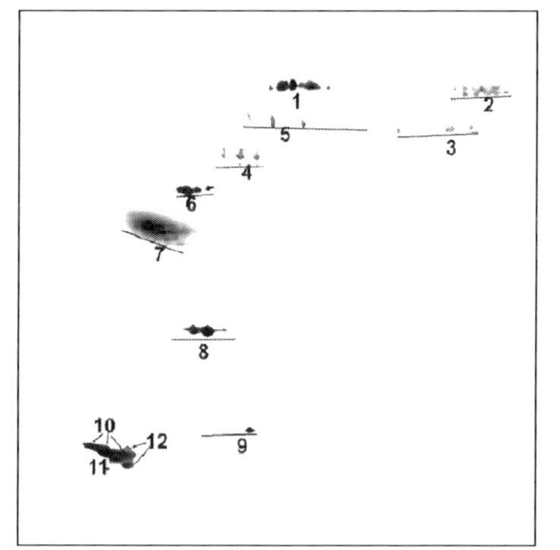

Figure 5.29 Protein adsorption patterns on solid lipid nanoparticles (SLNs) depending on their age: (a) freshly prepared, (b) at 12 months, (c) at 24 months. The entire gels are shown. (1) Albumin, (2) fibrinogen α chain, (3) fibrinogen β chain, (4) fibrinogen γ chain, (5) Apo-H, (6) Apo-A-IV, (7) Apo-J, (8) Apo-A-I, (9) transthyretin, (10) Apo-C-III, (11) Apo-C-II, (12) Apo-A-II. (Modified after Göppert et al. [122].)

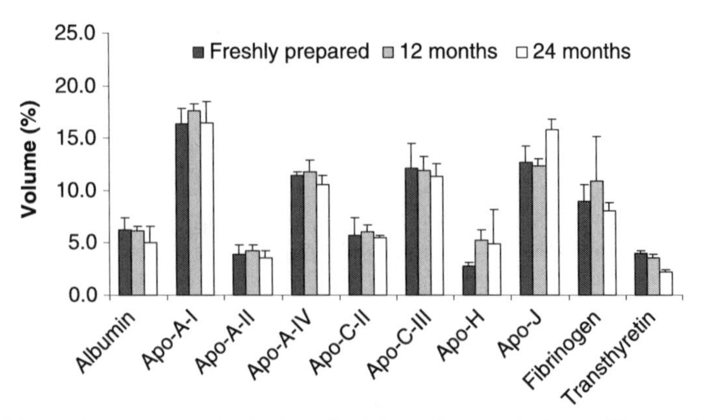

Figure 5.30 Major proteins adsorbed on SLN dependent on their age. The y axis: percentage of the overall detected protein amount ($n = 2$). (Modified after Göppert et al. [122].)

REFERENCES

1. Müller, R.H. *Colloidal Carriers for Controlled Drug Delivery and Targeting – Modification, characterization and in vivo distribution.* Boca Raton, FL: CRC Press, 1991.
2. Speiser, P.P. Nanoparticles and liposomes: a state of the art. *Methods Find Exp Clin Pharmacol* 1991;**13**:337–342.
3. Allen, T.M., Mehra, T., Hansen, C., Chin, Y.C. Stealth liposomes: an improved sustained release system for 1-beta-D-arabinofuranosylcytosine. *Cancer Res* 1992;**52**:2431–2439.
4. Marty, J.J., Oppenheim, R.C., Speiser, P. Nanoparticles–a new colloidal drug delivery system. *Pharm Acta Helv* 1978;**53**:17–23.
5. Kreuter, J. Nanoparticles and nanocapsules – new dosage forms in the nanometer size range. *Pharm Acta Helv* 1978;**53**:33–39.
6. Jeppsson, R., Rössner, S. The influence of emulsifying agents and of lipid soluble drugs on the fractional removal rate of lipid emulsion from the blood stream of the rabbit. *Acta Pharmacol Toxicol* 1975;**37**:134–144.
7. Müller, R.H., Mäder, K., Gohla, S. Solid lipid nanoparticles (SLN) for controlled drug delivery – a review of the state of the art. *Eur J Pharm Biopharm* 2000;**50**:161–177.
8. Juliano, R.L. Factors affecting the clearance kinetics and tissue distribution of liposomes, microspheres and emulsions. *Adv Drug Deliv Rev* 1988;**2**:31–54.
9. Bangham, A.D., Horne, R.W. Negative staining of phospholipids and their structural modification by surface-active agents as observed in the electron microscope. *J Mol Biol* 1964;**12**:660–668.
10. Bucke, W.E., Leitzke, S., Diederichs, J.E., Borner, K., Hahn, H., Ehlers, S., Müller, R.H. Surface-modified amikacin-liposomes: organ distribution and interaction with plasma proteins. *J Drug Target* 1997;**5**:99–108.
11. Cattel, L., Ceruti, M., Dosio, F. From conventional to stealth liposomes: a new frontier in cancer chemotherapy. *Tumori* 2003;**89**:237–249.
12. Lundberg, B., Hong, K., Papahadjopoulos, D. Conjugation of apolipoprotein B with liposomes and targeting to cells in culture. *Biochim Biophys Acta* 1993;**1149**:305–312.
13. Nobs, L., Buchegger, F., Gurny, R., Allemann, E. Current methods for attaching targeting ligands to liposomes and nanoparticles. *J Pharm Sci* 2004;**93**:1980–1992.
14. Senior, J.H. Fate and behavior of liposomes in vivo: a review of controlling factors. *Crit Rev Ther Drug Carrier Syst* 1987;**3**:123–193.
15. Lasic, D.D. Novel applications of liposomes. *Trends Biotechnol* 1998;**16**:307–321.

16. Clemons, K.V., Sobel, R.A., Williams, P.L., Pappagianis, D., Stevens, D.A. Efficacy of intravenous liposomal amphotericin B (AmBisome) against coccidioidal meningitis in rabbits. *Antimicrob Agents Chemother* 2002;**46**:2420–2426.
17. Gregoriadis, G. *Liposome Technology*. Boca Raton, FL: CRC Press, 1993.
18. Heath, T.D. Liposome dependent drugs. In: Gregoriadis G. (ed.), *Liposomes as Drug Carriers: Trends and progress*. Chichester: Wiley, 1988: pp. 709–718.
19. Tan, J.S., Butterfield, D.E., Voycheck, C.L., Caldwell, K.D., Li, J.T. Surface modification of nanoparticles by PEO/PPO block copolymers to minimize interactions with blood components and prolong blood circulation in rats. *Biomaterials* 1993;**14**:823–833.
20. Kreuter, J. Evaluation of nanoparticles as drug-delivery systems. III: materials, stability, toxicity, possibilities of targeting, and use. *Pharm Acta Helv* 1983;**58**:242–250.
21. Allemann, E., Gurny, R., Doelker, E. Drug loaded nanoparticles – preparation methods and drug targeting issues. *Eur J Pharm Biopharm* 1993;**39**:215–230.
22. Smith, A., Hunneyball, I.M. Evaluation of poly(lactic acid) as a biodegradable drug delivery system for parenteral administration. *Int J Pharm* 1986;**30**:215–220.
23. Hallberg, D., Holm, I., Obel, A.L., Schuberth, O., Wretlind, A. Fat emulsion for complete intravenous nutrition. *Postgrad Med* 1967;**42**:A149–152.
24. Wretlind, A. Development of fat emulsions. *JPEN J Parenter Enteral Nutr* 1981;**5**:230–235.
25. Wretlind, A. Recollections of pioneers in nutrition: landmarks in the development of parenteral nutrition. *J Am Coll Nutr* 1992;**11**:366–373.
26. Yamaguchi, T., Mizushima, Y. Lipid microspheres for drug delivery from the pharmaceutical viewpoint. *Crit Rev Ther Drug Carrier Syst* 1994;**11**:215–229.
27. Davis, S.S., Washington, C., West, P., Illum, L., Liversidge, G., Sternson, L., Kirsh, R. Lipid emulsions as drug delivery systems. *Ann NY Acad Sci* 1987;**507**:75–88.
28. Schmitt, J. Parenterale Fettemulsionen als Arzneistoffträger, In: Müller, R.H., Hildebrand, G.E. (eds), *Pharmazeutische Technologie: Moderne Arzneiformen*. Stuttgart: Wissenschaftliche Verlagsgesellschaft, 1998: pp. 189–194.
29. Caillot, D., Casasnovas, O., Solary, E., Chavanet, P., Bonnotte, B., Reny, G., et al. Efficacy and tolerance of an amphotericin B lipid (Intralipid) emulsion in the treatment of candidaemia in neutropenic patients. *J Antimicrob Chemother* 1993;**31**:161–169.
30. Müller, R.H., Schmidt, S., Buttle, I., Akkar, A., Schmitt, J., Bromer, S. SolEmuls – novel technology for the formulation of i.v. emulsions with poorly soluble drugs. *Int J Pharm* 2004;**269**:293–302.
31. Akkar, A., Namsolleck, P., Blaut, M., Muller, R.H. Solubilizing poorly soluble antimycotic agents by emulsification via a solvent-free process. *AAPS Pharm Sci Tech* 2004;**5**:E24.
32. Müller, R.H., Lucks, J.S. Arzneistoffträger aus festen Lipidteilchen, Feste Lipidnanosphären (SLN). European Patent EP 0605–497, 1996.
33. Wissing, S.A., Kayser, O., Müller, R.H. Solid lipid nanoparticles for parenteral drug delivery. *Adv Drug Deliv Rev* 2004;**56**:1257–1272.
34. Olbrich, C., Schöler, N., Tabatt, K., Kayser, O., Müller, R.H. Cytotoxicity studies of Dynasan 114 solid lipid nanoparticles (SLN) on RAW 264.7 macrophages-impact of phagocytosis on viability and cytokine production. *J Pharm Pharmacol* 2004;**56**:883–891.
35. Müller, R.H., Rühl, D., Runge, S., Schulze-Forster, K., Mehnert, W. Cytotoxicity of solid lipid nanoparticles as a function of the lipid matrix and the surfactant. *Pharm Res* 1997;**14**:458–462.
36. Gohla, S.H., Dingler, A. Scaling up feasibility of the production of solid lipid nanoparticles (SLN). *Pharmazie* 2001;**56**:61–63.
37. Dingler, A., Gohla, S. Production of solid lipid nanoparticles (SLN): scaling up feasibilities. *J Microencapsul* 2002;**19**:11–16.
38. Schwarz, C., Mehnert, W., Lucks, J.S., Müller, R.H. Solid lipid nanoparticles (SLN) for controlled drug delivery. I. Production, characterization and sterilization. *J Control Release* 1994;**30**:83–96.
39. Jenning, V., Gohla, S.H. Encapsulation of retinoids in solid lipid nanoparticles (SLN). *J Microencapsul* 2001;**18**:149–158.
40. Schwarz, C., Mehnert, W. Solid lipid nanoparticles (SLN) for controlled drug delivery. II. Drug incorporation and physicochemical characterization. *J Microencapsul* 1999;**16**:205–213.
41. Davis, S.S. Colloids as drug-delivery systems. *Pharm Technol* 1981;**5**:71–88.

42. Wilkens, D.J., Meyers, P.A. Studies on the relationship between the electrophoretic properties of colloids and their blood clearance and organ distribution in the rat. *Br J Exp Pathol* 1966;**47**: 568–576.
43. Müller, R.H., Heinemann, S. Surface modelling of microparticles as parenteral systems with high tissue affinity. In: Gurny, R., Junginger, H.E. (eds), *Bioadhesion – Possibilities and future trends*. Stuttgart: Wissenschaftliche Verlagsgesellschaft, 1989: pp. 202–214.
44. Davis, S.S., Illum, L., Moghimi, S.M., Davies, M.C., Porter, C.J.U., Muir, I.S., et al. Microspheres for targeting drugs to specific body sites. *J Control Release* 1993;**24**:157–163.
45. Price, M.E., Cornelius, R.M., Brash, J.L. Protein adsorption to polyethylene glycol modified liposomes from fibrinogen solution and from plasma. *Biochim Biophys Acta* 2001;**1512**:191–205.
46. Blunk, T., Hochstrasser, D.F., Sanchez, J-C., Müller, B.W., Müller, R.H. Colloidal carriers for intravenous drug targeting: plasma protein adsorption patterns on surface-modified latex particles evaluated by two-dimensional polyacrylamide gel electrophoresis. *Electrophoresis* 1993;**14**:1382–1387.
47. Hsu, M.J., Juliano, R.L. Interactions of liposomes with the reticuloendothelial system. II: Nonspecific and receptor-mediated uptake of liposomes by mouse peritoneal macrophages. *Biochim Biophys Acta* 1982;**720**:411–419.
48. Kazatchkine, M.D., Carreno, M.P. Activation of the complement system at the interface between blood and artificial surfaces. *Biomaterials* 1988;**9**:30–35.
49. Ogawara, K., Furumoto, K., Nagayama, S., Minato, K., Higaki, K., Kai, T., Kimura, T. Pre-coating with serum albumin reduces receptor-mediated hepatic disposition of polystyrene nanosphere: implications for rational design of nanoparticles. *J Control Release* 2004;**100**:451–455.
50. Patel, H.M. Serum opsonins and liposomes: their interaction and opsonophagocytosis. *Crit Rev Ther Drug Carrier Syst* 1992;**9**:39–90.
51. Vandorpe, J., Schacht, E., Dunn, S., Hawley, A., Stolnik, S., Davis, S.S., et al. Long circulating biodegradable poly(phosphazene) nanoparticles surface modified with poly(phosphazene)-poly(ethylene oxide) copolymer. *Biomaterials* 1997;**18**:1147–1152.
52. Gref, R., Minamitake, Y., Peracchia, M.T., Trubetskoy, V., Torchilin, V., Langer, R. Biodegradable long-circulating polymeric nanospheres. *Science* 1994;**263**:1600–1603.
53. Stolnik, S., Illum, L., Davis, S.S. Long circulating microparticle drug carriers. *Adv Drug Del Rev* 1995;**16**:195–214.
54. Moghimi, S.M., Muir, I.S., Illum, L., Davis, S.S., Kolb-Bachofen, V. Coating particles with a block co-polymer (poloxamine-908) suppresses opsonization but permits the activity of dysopsonins in the serum. *Biochim Biophys Acta* 1993;**1179**:157–165.
55. Illum, L., Davis, S.S., Müller, R.H., Mak, E., West, P. The organ distribution and circulation time of intravenously injected colloidal carriers sterically stabilized with a block copolymer – poloxamine 908. *Life Sci* 1987;**40**:367–374.
56. Moghimi, S.M. Re-establishing the long circulatory behaviour of poloxamine-coated particles after repeated intravenous administration: applications in cancer drug delivery and imaging. *Biochim Biophys Acta* 1999;**1472**:399–403.
57. Müller, R.H., Lück, M., Kreuter, J. Medicament excipient particles for tissue specific application of a medicament. PCT-application PCT/EP98/064299 (P53601), 2001.
58. Kreuter, J. Nanoparticulate systems for brain delivery of drugs. *Adv Drug Deliv Rev* 2001;**47**:65–81.
59. Kreuter, J., Shamenkov, D., Petrov, V., Ramge, P., Cychutek, K., Koch-Brandt, C., Alyautdin, R.N. Apolipoprotein-mediated transport of nanoparticle-bound drugs across the blood-brain barrier. *J Drug Target* 2002;**10**:317–325.
60. Dehouck, B., Dehouck, M.P., Fruchart, J.C., Cecchelli, R. Upregulation of the low density lipoprotein receptor at the blood-brain barrier: intercommunications between brain capillary endothelial cells and astrocytes. *J Cell Biol* 1994;**126**:465–473.
61. Dehouck, B., Fenart, L., Dehouck, M.P., Pierce, A., Torpier, G., Cecchelli, R. A new function for the LDL receptor: transcytosis of LDL across the blood-brain barrier. *J Cell Biol* 1997;**138**:877–889.
62. Kreuter, J., Alyautdin, R.N., Kharkevich, D.A., Ivanov, A.A. Passage of peptides through the blood-brain barrier with colloidal polymer particles (nanoparticles). *Brain Res* 1995;**674**:171–174.
63. Alyautdin, R.N., Petrov, V.E., Ivanov, A.A., Kreuter, J., Kharkevich, D.A. Transport of the hexapeptide dalargin across the hemato-encephalic barrier into the brain using polymer nanoparticles. *Eksp Klin Farmakol* 1996;**59**:57–60.

64. Alyautdin, R.N., Petrov, V.E., Langer, K., Berthold, A., Kharkevich, D.A., Kreuter, J. Delivery of loperamide across the blood-brain barrier with polysorbate 80-coated polybutylcyanoacrylate nanoparticles. *Pharm Res* 1997;**14**:325–328.
65. Alyautdin, R.N., Petrov, V.E., Langer, K., Berthold, A., Kreuter, J., Kharkevich, D.A. The delivery of loperamide to the brain by using polybutyl cyanoacrylate nanoparticles. *Eksp Klin Farmakol* 1998;**61**:17–20.
66. Alyautdin, R.N., Tezikov, E.B., Ramge, P., Kharkevich, D.A., Begley, D.J., Kreuter, J. Significant entry of tubocurarine into the brain of rats by adsorption to polysorbate 80-coated polybutylcyanoacrylate nanoparticles: an in situ brain perfusion study. *J Microencapsul* 1998;**15**:67–74.
67. Gulyaev, A.E., Gelperina, S.E., Skidan, I.N., Antropov, A.S., Kivman, G.Y., Kreuter, J. Significant transport of doxorubicin into the brain with polysorbate 80-coated nanoparticles. *Pharm Res* 1999;**16**:1564–1569.
68. Gelperina, S.E., Khalansky, A.S., Skidan, I.N., Smirnova, Z.S., Bobruskin, A.I., Severin, S.E., et al. Toxicological studies of doxorubicin bound to polysorbate 80-coated poly(butyl cyanoacrylate) nanoparticles in healthy rats and rats with intracranial glioblastoma. *Toxicol Lett* 2002;**126**:131–141.
69. Müller, R.H., Schmidt, S. PathFinder technology for the delivery of drugs to the brain. *New Drugs* 2002;**2**:38–42.
70. Jungblut, P., Wittmann-Liebold, B. Protein analysis on a genomic scale. *J Biotechnol* 1995;**41**:111–120.
71. Klose, J. Protein mapping by combined isoelectric focussing and electrophoresis of mouse tissues – A novel approach to testing for induced point mutation in mammals. *Hummangenetik* 1975;**26**:211–234.
72. Hochstrasser, D.F., Harrington, M.G., Hochstrasser, A.C., Miller, M.J., Merril, C.R. Methods for increasing the resolution of two-dimensional protein electrophoresis. *Anal Biochem* 1988;**173**:424–435.
73. Blunk, T. Plasmaproteinadsorption auf kolloidalen Arzneistoffträgern. Dissertation, Department of Pharmacy, Christian-Albrechts University of Kiel, 1994.
74. Cook, B.C., Retzinger, G.S. Elution of fibrinogen and other plasma protein from unmodified and from lecithin-coated polystyrene-divinylbenzene beads. *J Colloid Interf Sci* 1992;**153**:1–12.
75. Anderson, N.L., Anderson, N.G. A two-dimensional gel database of human plasma proteins. *Electrophoresis* 1991;**12**:883–906.
76. Appel, R.D., Hochstrasser, D.F., Funk, M., Vargas, J.R., Pellegrini, C., Müller, A.F., Scherrer, J.R. The MELANIE project: from a biopsy to automatic protein map interpretation by computer. *Electrophoresis* 1991;**14**:1223–1231.
77. Harnisch, S., Müller, R.H. Plasma protein adsorption patterns on emulsions for parenteral administration: establishment of a protocol for two-dimensional polyacrylamide electrophoresis. *Electrophoresis* 1998;**19**:349–354.
78. Thode, K., Lück, M., Semmler, W., Müller, R.H., Kresse, M. Determination of plasma protein adsorption on magnetic iron oxides: sample preparation. *Pharm Res* 1997;**14**:905–910.
79. Diederichs, J.E. Plasma protein adsorption patterns on liposomes: establishment of analytical procedure. *Electrophoresis* 1996;**17**:607–611.
80. Göppert, T.M., Müller, R.H. Alternative sample preparation prior to two-dimensional electrophoresis protein analysis on solid lipid nanoparticles. *Electrophoresis* 2004;**25**:134–140.
81. Thode, K., Lück, M., Schröder, W., Kresse, M., Semmler, W., Müller, R.H. The influence of the sample preparation on plasma protein adsorption onto macromolecule stabilized iron oxide particles and N-terminal microsequencing of unknown proteins. *J Drug Target* 1997;**5**:35–43.
82. Göppert, T.M. Plasmaproteinadsorptionsmuster auf parenteral applizierbaren kolloidalen Arzneistoffträgern zur Überwindung der Blut-Hirn-Schranke. Dissertation, Department of Pharmacy, Free University of Berlin, 2005.
83. Vroman, L., Adams, A.L., Fischer, G.C., Munoz, P.C. Interaction of high molecular weight kininogen, factor XII, and fibrinogen in plasma at interfaces. *Blood* 1980;**55**:156–159.
84. Vroman, L., Adams, A.L. Adsorption of proteins out of plasma and solutions in narrow spaces. *J Coll Interf Sci* 1986;**111**:391–402.
85. Brash, J.L. Protein adsorption at the solid-solution interface in relation to blood-material interactions. In: Brash, J.L., Horbett, T.A. (eds), *Proteins at Interfaces: Physicochemical and biochemical studies*. Washington DC: American Chemical Society, 1987:pp. 490–506.

86. Wojciechowski, P., Brash, J.L. The Vroman effect in tube geometry: the influence of flow on protein adsorption measurements. *J Biomater Sci Polym Ed* 1991;**2**:203–216.
87. Brash, J.L., Ten Hove, P. Protein adsorption studies on 'standard' polymeric materials. *J Biomater Sci Polym Ed* 1993;**4**:591–599.
88. Brash, J.L. Exploiting the current paradigm of blood-material interactions for the rational design of blood-compatible materials. *J Biomater Sci Polym Ed* 2000;**11**:1135–1146.
89. O'Mullane, J.E., Artursson, P., Tomlinson, E. Biopharmaceutics of microparticulate drug carriers. *Ann NY Acad Sci* 1987;**507**:120–140.
90. Boisson-Vidal, C., Jozefonvicz, J., Brash, J.L. Interactions of proteins in human plasma with modified polystyrene resins. *J Biomed Mater Res* 1991;**25**:67–84.
91. Blunk, T., Lück, M., Calvör, A., Hochstrasser, D.F., Sanchez, J-C., Müller, B.W., et al. Kinetics of plasma protein adsorption on model particles for controlled drug delivery and drug targeting. *Eur J Pharm Biopharm* 1996;**42**:262–268.
92. Harnisch, S., Müller, R.H. Adsorption kinetics of plasma proteins on oil-in-water emulsions for parenteral nutrition. *Eur J Pharm Biopharm* 2000;**49**:41–46.
93. Göppert, T.M., Müller, R.H. Adsorption kinetics of plasma proteins on solid lipid nanoparticles for drug targeting. *Int J Pharm* 2005;**302**:172–186.
94. Park, Y.J., Nah, S.H., Lee, J.Y., Jeong, J.M., Chung, J.K., Lee, M.C., et al. Surface-modified poly(lactide-co-glycolide) nanospheres for targeted bone imaging with enhanced labeling and delivery of radioisotope. *J Biomed Mater Res A* 2003;**67**:751–760.
95. Illum, L., Davis, S.S. Targeting of colloidal particles to the bone marrow. *Life Sci* 1987;**40**:1553–1560.
96. Porter, C.J., Moghimi, S.M., Illum, L., Davis, S.S. The polyoxyethylene/polyoxypropylene block co-polymer poloxamer-407 selectively redirects intravenously injected microspheres to sinusoidal endothelial cells of rabbit bone marrow. *FEBS Lett* 1992;**305**:62–66.
97. Cornelius, R.M., Archambault, J.G., Berry, L., Chan, A.K., Brash, J.L. Adsorption of proteins from infant and adult plasma to biomaterial surfaces. *J Biomed Mater Res* 2002;**60**:622–632.
98. Lück, M., Schröder, W., Harnisch, S., Thode, K., Blunk, T., Paulke, B.R., et al. Identification of plasma proteins facilitated by enrichment on particulate surfaces: analysis by two-dimensional electrophoresis and N-terminal microsequencing. *Electrophoresis* 1997;**18**:2961–2967.
99. Jahangir, R., McCloskey, C.B., Mc Clung, W.G., Labow, R.S., Brash, J.L., Santerre, J.P. The influence of protein adsorption and surface modifying macromolecules on the hydrolytic degradation of a poly(ether-urethane) by cholesterol esterase. *Biomaterials* 2003;**24**:121–130.
100. Archambault, J.G., Brash, J.L. Protein resistant polyurethane surfaces by chemical grafting of PEO: amino-terminated PEO as grafting reagent. *Colloids Surf B Biointerfaces* 2004;**39**:9–16.
101. Unsworth, L.D., Sheardown, H., Brash, J.L. Protein resistance of surfaces prepared by sorption of end-thiolated poly(ethylene glycol) to gold: effect of surface chain density. *Langmuir* 2005;**21**:1036–1041.
102. Leroux, J.C., Allemann, E., De Jaeghere, F., Doelker, E., Gurny, R. Biodegradable nanoparticles: from sustained release formulations to improved site specific drug delivery. *J Control Release* 1996;**39**:339–350.
103. Lück, M., Schröder, W., Paulke, B.R., Blunk, T., Müller, R.H. Complement activation by model drug carriers for intravenous application: determination by two-dimensional electrophoresis. *Biomaterials* 1999;**20**:2063–2068.
104. Lind, K., Kresse, M., Müller, R.H. Comparison of protein adsorption patterns onto differently charged hydrophilic superparamagnetic iron oxide particles obtained in vitro and ex vivo. *Electrophoresis* 2001;**22**:3514–3521.
105. Babensee, J.E., Cornelius, R.M., Brash, J.L., Sefton, M.V. Immunoblot analysis of proteins associated with HEMA-MMA microcapsules: human serum proteins in vitro and rat proteins following implantation. *Biomaterials* 1998;**19**:839–849.
106. Yamazaki, A., Winnik, F.M., Cornelius, R.M., Brash, J.L. Modification of liposomes with N-substituted polyacrylamides: identification of proteins adsorbed from plasma. *Biochim Biophys Acta* 1999;**1421**:103–115.
109. Leroux, J.C., De Jaeghere, F., Anner, B., Doelker, E., Gurny, R. An investigation on the role of plasma and serum opsonins on the internalization of biodegradable poly(D,L-lactic acid) nanoparticles by human monocytes. *Life Sci* 1995;**57**:695–703.

108. Gogos, C.A., Kalfarentzos, F. Total parenteral nutrition and immune system activity: a review. *Nutrition* 1995;**11**:339–344.
109. Schmidt, S. Parenterale O/W-Emulsionen: Plasmaproteininteraktion und Inkorporation von Arzneistoffen. Dissertation, Department of Pharmacy, Free University of Berlin, 2002.
110. Graham, D.J., Phillips, M.C. Proteins at liquid interfaces, I, Kinetics of adsorption and surface denaturation. *J Coll Interf Sci* 1979;**70**:403.
111. Rosseneu, M. *Structure and Function of Apolipoproteins*. Boca Raton, FL: CRC Press, 1992.
112. Brown, M.S., Goldstein, J.L. A receptor-mediated pathway for cholesterol homeostasis. *Science* 1986;**232**:34–47.
113. Goldstein, J.L., Basu, S.K., Brown, M.S. Receptor-mediated endocytosis of low-density lipoprotein in cultured cells. *Methods Enzymol* 1983;**98**:241–260.
114. Göppert, T.M., Müller, R.H. Protein adsorption patterns on poloxamer- and poloxamine-stabilized solid lipid nanoparticles (SLN). *Eur J Pharm Biopharm* 2005;**60**:361–372.
115. Harnisch, S. Vergleichende Untersuchungen zur Plasmaproteinadsorption auf O/W-Emulsionen und Polymer-Partikeln zur parenteralen Anwendung. Dissertation, Department of Pharmacy, Free University of Berlin, 1998.
116. Göppert, T.M., Müller, R.H. Influence of different lipids on plasma protein patterns of solid lipid nanoparticles. In AAPS Annual Meeting, Baltimore, MD, 2004.
117. Tröster, S.D., Kreuter, J. Contact angle of surfactants with a potential to alter the body distribution of colloidal drug carriers on poly (methyl methacrylate) surfaces. *Int J Pharm* 1988;**45**:91–100.
118. Batrakova, E.V., Li, S., Miller, D.W., Kabanov, A.V. Pluronic P85 increases permeability of a broad spectrum of drugs in polarized BBMEC and Caco-2 cell monolayers. *Pharm Res* 1999;**16**: 1366–1372.
119. Miller, D.W., Batrakova, E.V., Waltner, T.O., Alakhov, V., Kabanov, A.V. Interactions of pluronic block copolymers with brain microvessel endothelial cells: evidence of two potential pathways for drug absorption. *Bioconjug Chem* 1997;**8**:649–657.
120. Göppert, T.M., Müller, R.H. Polysorbate-stabilized solid lipid nanoparticles as colloidal carriers for intravenous targeting of drugs to the brain: Comparison of plasma protein adsorption patterns. *J Drug Target* 2005;**13**:179–187.
121. Weisgraber, K.H., Mahley, R.W., Kowal, R.C., Herz, J., Goldstein, J.L., Brown, M.S. Apolipoprotein C-I modulates the interaction of apolipoprotein E with beta-migrating very low density lipoproteins (beta-VLDL) and inhibits binding of beta-VLDL to low density lipoprotein receptor-related protein. *J Biol Chem* 1990;**265**:22453–22459.
122. Göppert, T.M., Souto, E., Müller, R.H. Comparison of plasma protein adsorption patterns on solid lipid nanoparticles (SLN) for intravenous drug targeting dependent on their age. In: CRS-Meeting, Miami, 2005.

CHAPTER 6

NANOPARTICLE TARGETING FOR DRUG DELIVERY ACROSS THE BLOOD–BRAIN BARRIER

James Egbert, Werner Geldenhuys, Fancy Thomas, Paul R. Lockman, Russell J. Mumper, and David D. Allen

6.1 INTRODUCTION TO DRUG DELIVERY

6.2 NANOPARTICLES AS A CURRENT DELIVERY SYSTEM

6.3 THE IMPORTANCE OF SPECIFICITY

6.4 DISCUSSION

6.1 INTRODUCTION TO DRUG DELIVERY

Drug discovery and development is a lengthy and costly business, generally requiring 15 years and $US880 million to put a new drug on the market [1]. The high failure rate (> 40%) seen in the industry results partly from inadequate adsorption, distribution, metabolism, excretion, and toxicology (ADME-Tox) properties [2, 3]. This failure rate exists despite techniques such as combinatorial chemistry and high-throughput screening, which are identifying vast numbers of compounds in a much shorter time [4]. Recently, there has been a paradigm change in the pharmaceutical industry, where key trends usually associated with the development side are now being incorporated earlier into the drug discovery process, with ADME screening in parallel with the traditional pharmacological assays [4]. The aim of companies is to discard compounds in the discovery phase as early as possible and to increase the success rate of compounds to the market (a 'fail fast, fail cheap' motto) [2, 5].

Property-based design is a new approach in drug discovery addressing good 'drug-like' properties for lead candidate compounds. A successful drug has a combination of biological activity and drug-like properties. Therefore, parallel screening of activity and property allows medicinal chemists to optimize biological

Role of Lipid Excipients in Modifying Oral and Parenteral Drug Delivery, Edited by Kishor M. Wasan

activities as well as 'drug likeness' [2]. Both pharmacodynamics (drug action) and pharmacokinetics (drug distribution in the body) are governed by physicochemical properties [6]. Unfortunately, for most existing drugs, changing the chemical composition is not feasible because of financial constraints or simply the resulting modification may lead to unexpected side effects, change distribution pharmacokinetics, and/or a loss of activity.

An alternative to chemical modification is the use of formulation sciences to compensate for, or overcome, the poor physicochemical properties of the compound. The past few years have seen an increase in the interest in the use of nanoparticles (NPs) to transport drug compounds to their sites of action. NPs are solid colloidal particles that range in size from 1 to 1000 nm [7–10]. In this discussion, NP formulation-containing drugs are distinguished from drug nanosuspensions, which are micrometer colloidal dispersions of the drug particles themselves [10].

6.2 NANOPARTICLES AS A CURRENT DELIVERY SYSTEM

Nanotechnology is an area of science that has recently gained momentum as an attractive means by which drugs can be delivered. The manipulation of atoms and molecules, leading to the construction of structures in the nanometer scale size range, is quickly showing returns, particularly in the areas concerned with drug solubility [11]. Research into the rational delivery of therapeutic agents via nano-sized particles is becoming the forefront of disease treatment and prevention.

Opponents of this technology raise concerns about its application to clinical use. Toxicity is a major concern, and varies from formulation to formulation. Disadvantages of some polymers involve their bioerosion in vivo compared with other, more favorable polymers, which are biodegraded. Bioerosion (e.g. polyalkylcyanoacrylates) involves the enzymatic attack of some side chains, inducing the solubilization of the polymer although the hydrocarbon backbone remains unchanged [12]. This leads to progressive nanoparticle disaggregation and complete excretion which cannot be completed unless the polymeric material is of low molecular mass [13]. This can create a substantial difference in toxicity compared with polymers that are biodegraded. The main chains of these types of polymers are hydrolyzed via enzymatic or chemical pathways, allowing the metabolized products to be filtered through the kidneys [13]. Ultimately however, the polymer's molecular weight, degradation mechanism, and the diffusibility of its metabolites will determine its risk in vivo [13].

Altered pharmacokinetic and distribution profiles may induce changes in tissue or receptor exposure, as well as new drug patterns of metabolization and drug action [13]. During intravascular administration, there are concerns about hemocompatibility, primarily in terms of clot formation caused by particle aggregation or hemolysis caused by erythrocyte damage by nanoparticles or their metabolites [13]. Similarly, inhalation of nanoparticles may result in sustained inflammatory effects [9] and detrimental impacts on the cardiovascular system, such as arrhythmias and coagulation [14]. Both advantages and disadvantages of nanotechnology need to be further investigated and weighed against the opportunities and consequences it presents.

6.2.1 Design

Nanoparticles are spherical colloidal particles comprising either natural or synthetic polymers or lipids, or combinations of both, and range in size from 1 to 1000 nm [11]. They are typically classified as either nanospheres or nanocapsules. Nanospheres are composed of a solid framework, whereas nanocapsules have a liquid central cavity surrounded by a wall [6]. Once constructed, the therapeutic agent of choice has a variety of means by which it can be associated with the polymer, e.g. the agent can be dissolved, entrapped, attached, or encapsulated throughout or within the polymeric matrix (Figure 6.1) [11]. The efficiency of the association of drug and polymer, as well as the release characteristics, varies depending on the method of preparation. Factors influencing drug release kinetics include, but are not limited to, forming materials, and their structure and stability.

Biodegradable polymers vary in purity and consist of either natural or synthetic materials [15]. Polysaccharides, polypeptides, and proteins such as albumin, fibrinogen, gelatin, and collagen are examples of natural polymers used in drug delivery. However, synthetic polymers are more commonly used as a result of the requirement that natural materials need to cross-link in the microencapsulation process, which can lead to denaturalization of the polymer and the embedded drug [15]. Synthetic polymers may have advantages over natural polymers because of their ability to impart an extended-release delivery property. This is accomplished by designing them to be biocompatible, lack immunogenicity, and have physical properties that allow them to be easily shaped [15]. They are, however, limited by the use of organic solvents and relatively harsher formulation conditions [16].

Although there are several synthetic polymers that are under investigation for clinical use, polylactides (PLAs) and poly(D,L-lactide-co-glycolide) (PLGA) are the most widely used polymers for drug delivery [17, 18]. These polymers work well because they are metabolized to lactic acid and glycolic acid and have metaboliz-

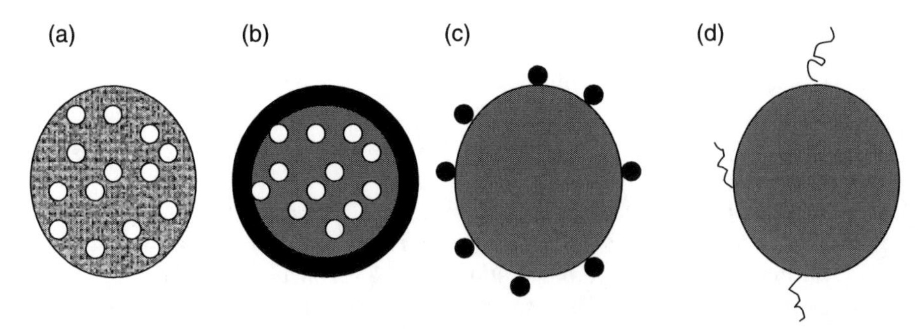

Figure 6.1 (a) Drug dissolved in polymeric matrix, (b) drug encapsulated in solid nanoparticle, (c) drug adsorbed to surface of nanoparticle, and (d) targeted nanoparticles with ligands attached. (Adapted from Lockman et al. [7].)

able moieties, which are eventually removed from the body via the citric acid cycle [16]. There are several methods that are employed when designing nanoparticles as drug carriers. These include: (1) emulsion polymerization, (2) interfacial polymerization, (3) desolvation evaporation, and (4) solvent deposition [7, 19]. We examine interfacial polymerization in more depth because of its demonstrated utility to date in applications for central nervous system (CNS) delivery.

Similar to emulsion polymerization, interfacial polymerization involves using monomers to create polymers. Interfacial polymerization is the merging of aqueous and organic phases by homogenization, emulsification, or micro-fluidization under high-torque mechanical stirring [7]. Peptides/proteins are prevented from being included in this step secondary to mechanical shearing, e.g. polyalkylcyanoacrylate nanocapsules are created because the monomer spontaneously forms 200–300 nm particles by anionic polymerization when dissolved in oil and added to an aqueous phase with constant stirring [7]. The drug is incorporated on addition to the monomer in the organic phase, allowing for drug encapsulation by the polymeric matrix [7, 20]. Encapsulation helps protect the drug until it reaches the target tissue by diminishing metabolism exposure.

There are multiple factors involved in drug release including: diffusion of the drug through the channels and pores of the polymeric matrix and across the polymer barrier, and degradation of the device [7, 15].

The size of the nanoparticle drug complex has a significant influence on organ distribution, reticuloendothelial system (RES) uptake, and liver metabolism. Specifically, nanoparticles with a size < 100 nm and those that present more of a hydrophilic surface exhibit less RES removal and metabolism, which increases the amount of drug available for brain exposure and uptake [7]. The type of polymer used, pH of the solution used during polymerization, and concentration of the monomer units all affect nanoparticle size in the 100–200 nm range [21, 22]. The most common method of sizing is photon correlation, which uses a laser beam directed at the particles in suspension and analyzing the time dependency of the scattered light changes. Other characteristics, such as molecular mass, density and crystallinity, can be distinguished by gel chromatography, helium compression pycnometry and x-ray diffraction, respectively [23].

6.3 THE IMPORTANCE OF SPECIFICITY

The physicochemical properties of a drug greatly influence the distribution of the drug in the body. The influence of barriers to drug distribution is especially evident for drugs where the drug target is located inside the CNS. For drugs that act in the CNS, a major rate-limiting step for uptake into the brain is passage through the blood–brain barrier (BBB). This is notable for hydrophilic compounds, charged molecules, and small proteins, for which the BBB is essentially impermeable [7, 24–27].

Anatomically, the BBB is made up of brain capillary endothelial cells connected by tight junctions (zonulae occludens) [28] that circumferentially surround the endothelial cell margin (Figure 6.2) and are responsible for preventing/retarding

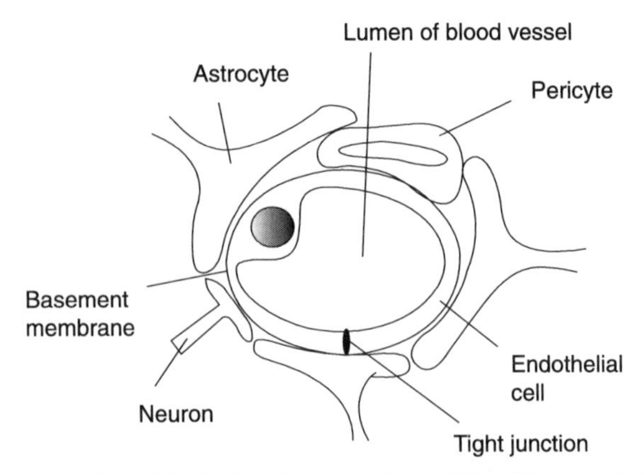

Figure 6.2 Cross-section of the brain microvasculature [30, 31]: the endothelial cells form the tight junctions excluding hydrophilic and charged compounds. The surface of the endothelial cells is covered with the foot processes of astrocytes, and also a few neurons are present.

paracellular transport across the BBB, thereby excluding hydrophilic compounds from the CNS [28]. Minimal pinocytosis is also observed at the BBB [28]. These tight junctions are considerably tighter (about 100) compared with those in other capillary endothelia [29]. Additional physiological characteristics of the BBB include a surface area of about $20\,m^2$, intracellular space of the brain capillary endothelium of about 0.8 μL/g (1 mL/1200 g human brain) and a distance between capillaries of 40 μm [30]. Four major cell types are found as part of the BBB microvasculature, as seen in Figure 6.2, including astrocytes, pericytes, foot processes, and nerve endings [30].

The tight junctions of the BBB endothelium eliminate movement of substances through the *paracellular* pathway, whereas the lack of pinocytosis across the endothelial cells of the BBB eliminates *transcellular* uptake [30, 31]. As a result of these physiological characteristics of the BBB, uptake of solutes or compounds into the brain is restricted to one of two processes, including lipid-mediated free diffusion or transport [30, 31]. Three types of transport systems are found in the BBB (Figure 6.3). These include carrier-mediated transport, active-efflux transport and receptor-mediated transport.

For a small drug molecule to gain entry to the brain (by means of passive transport), important physicochemical characteristics are needed, such as high lipid solubility, and a molecular mass of < 700 Da [31]. An inverse relationship exits for BBB permeability and hydrogen bonding potential of a compound [32]. This is expected, considering that hydrophilic compounds (i.e. highly polar/strong hydrogen-bonding moieties) are essentially excluded from the brain by the BBB [26].

6.3.1 Nanoparticle Drug Delivery to the Brain

The success of drug delivery to brain lies in improving the ability to penetrate the BBB, increase the concentration of biologically active molecules at desired target

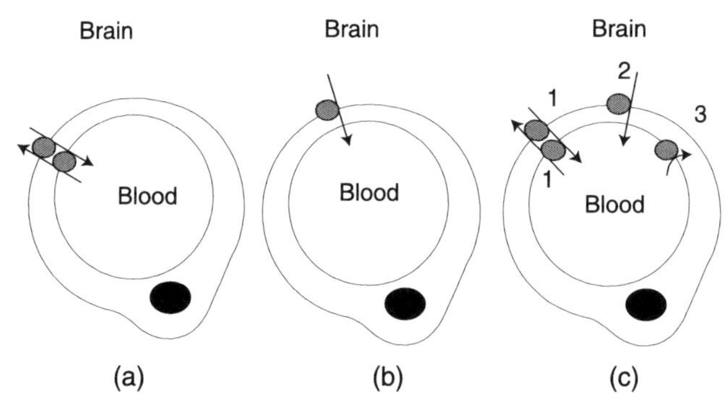

Figure 6.3 The different classes of blood–brain barrier (BBB) transport systems [30]. (a) Carrier-mediated transport is a bidirectional movement of solutes between the blood and brain, e.g. D-glucose, choline, amino acids. (b) Active efflux-mediated transport is unidirectional for metabolic byproducts as well as xenobiotics, e.g. P-glycoprotein and organic anion-transporting polypeptide type 2. (c) Receptor-mediated transport is involved with specific peptide uptake such as the insulin receptor, which takes up circulating insulin from the blood. Three separate systems are shown for receptor-mediated transport, e.g. the transferring receptor is shown as 1, where the transport is bidirectional. The BBB Fc transporter is an example of unidirectional transport, shown as 2 and the type I scavenger receptor is able to take up solutes only from the blood into the brain capillary, without transport to the brain.

sites, and avoid distribution into peripheral tissue [33]. As the BBB presents limitations in drug delivery to the brain, nanoparticle delivery systems may provide a significant advantage to current strategies, e.g. one such strategy involves artificially opening the barrier with various hyperosmotic agents or vasoactive molecules. However, barrier disruption is typically a homogeneous process with a generalized loss of barrier integrity. This potentially allows paracellular entry of endogenous molecules into the CNS, creating the potential for toxic and unwanted side effects [8, 34].

The encapsulation of drug in nanoparticles masks the inherent molecular characteristics that disallow drug entry into the brain [7]. It is the characteristics of the nanoparticle that govern penetration of nanoparticle drug complexes into the brain. It has been demonstrated that nanoparticles encapsulating drug may penetrate the brain via receptor-mediated endocytosis and/or passive diffusion. Specifically, nanoparticle systems using Polysorbate 80 as the monomer–polymer matrix cross the BBB via cellular endothelial endocytosis through surface apolipoprotein Apo-E adsorption and subsequent uptake via the low-density lipoprotein (LDL) pathway [7]. In addition, nanoparticles created from emulsifying wax have been shown to penetrate the BBB substantially within 60 seconds, suggesting penetration through a passive diffusion mechanism [8].

In addition to physical mechanical restrictions present at the BBB (i.e. lack of fenestrae and pinocytotic processes), a luminal electrostatic barrier at physiologic pH further impedes drug penetration into brain [34, 35]. Specifically, at the BBB there is an anionic boundary, created by the surface expression and adhesion of the glycocalyx residues: proteoglycans, sulfated mucopolysaccharides, and sulfated and sialic acid-containing glycoproteins and glycolipids [36, 37]. Although anionic molecules are retarded electrostatically, cationic molecules can occupy anionic areas of the BBB endothelium and are subject to increased permeability via presumed tight junction disruption [29]. However, cationic tight junction disruption, much like hyperosmotic agents or vasoactive molecules, creates the potential for toxic, unwanted side effects as a result of a loss of BBB integrity. Brain microvasculature that is exposed to prolonged or high-dose cationic particles may result in extravasation of luminal plasma contents in tight-junction compartments, and endothelial cellular swelling and death [29].

Nanoparticles of neutral or low anionic charge have no acute effect on BBB integrity whereas cationic nanoparticles have barrier disruption properties that occur in less than 60 s after intra-arterial perfusion [34]. However, apparently unique to anionic nanoparticles, low concentrations may penetrate the electrostatic BBB [38]. Overcoming the anionic barrier by anionic nanoparticle formulations should be studied further.

6.3.2 Cancer Therapy

Cancer is the uncontrolled proliferation of cells that have mutated as the result of genetics or environmental hazard. As cancerous cells proliferate, surrounding healthy tissue becomes unable to compete for available nutrients, is displaced by tumor cells, and has less physical ability to eliminate metabolic waste products.

Currently, cancer therapy is an intrusive, side effect-ridden administration of near-toxic doses of nonspecific agents. Chemotherapy and radiation are geared toward killing tumor cells that are more susceptible to eradication, based on their faster growth rates. Although research to improve chemotherapy has led to an increase in patient survival, the side effects are often so intense that the patient must discontinue therapy before all the cancer has been eliminated [39]. Therefore, the effectiveness of a treatment lies in the ability to target and kill cancer cells while affecting as few healthy cells as possible. As nanoparticles and similar carrier systems have the potential to target cancer or individual organs, separately from the drug that is encapsulated, they are an attractive medium to deliver both new and established agents against cancer [39].

Specific targeting through the use of nanoparticles is a complicated balance of design, to create a system not only to incorporate and retain a therapeutic agent and facilitate the transport of that agent to a specific area, but also to avoid the RES that so readily tries to destroy the carrier [39]. Degradation by the RES is influenced by size, surface charge, and morphology. Nanoparticles that are smaller and present a more hydrophilic surface are less likely to be taken up by the spleen and liver and reduce clearance by macrophages. This allows for longer circulating time in the bloodstream and greater exposure to the target site of interest [39].

Prolonged residence time of nanoparticle carrier systems is important in chemotherapy, given that necrotic and semi-necrotic regions of tumors have a slow and unpredictable blood flow, which may decrease exposure of the nanoparticle drug complex to cancerous tissue [39]. However, leaky tumor vasculature and poor surrounding lymphatic drainage may enhance nanoparticle permeation and retention

[39]. In spite of the natural retention of nanoparticles in tumor cores, ideally nanoparticles should be designed specifically to target individual cancer cells. Thus, the focus of current research is to define differences in cancerous tissues from those of normal tissue and exploit these differences, e.g. cancerous tissues often overly express particular antigens on their surface and nanoparticles can be designed to have corresponding antibodies presented on their surface [39].

Despite the potential increase in the specificity and efficacy of targeted nanoparticle systems, there are still issues about the toxicity of the treatment that does not attach to the cancerous tissue and remains circulating in the system [40]. Eventually, nanoparticles that do not attach will degrade throughout the host, releasing cytotoxic agents in a nonspecific manner. Depending on the drug of choice, this random administration may cause prolonged systemic toxicity.

Other, less intrusive, nanoparticle cancer therapies include: receptor/ligand-mediated target delivery. During angiogenesis, endothelial cells upregulate surface molecules [41, 42], e.g. integrin $\alpha v\beta 3$ (functions to maintain endothelial cell survival) can be utilized as an endothelial cell target for tumor regression gene delivery. This was accomplished by using cationic, polymerized, lipid-based nanoparticles covalently coupled to a small organic $\alpha v\beta 3$ ligand [43].

In an attempt to thwart unwanted toxicity, another method currently being researched to improve target therapeutics is the use of sequence-selective gene inhibition by siRNA (silencing) oligonucleotides [40]. This method may have advantages over drug-based systems in that it can target transcription factors, and has the ability precisely to select targets down to the level of single-nucleotide polymorphisms [40]. However, despite the high selectivity, systemic exposure can still induce nonspecific responses. Therefore, the problem lies not in the specificity of the treatment but with the specificity of the carrier. Schiffelers et al. report disulfide-stabilized arginine-glycine-aspartic acid (RGD) peptide ligands, which target integrin upregulated at sites of neovasculature, exposed on the surface of cationic polymer conjugated with polyethyleneglycol (PEG)–peptide conjugates with siRNA, showed evidence for sequence-specific inhibition of the target gene, reduction in angiogenesis, and inhibition of tumor growth [40].

Independent of the method applied, the effectiveness of treatment lies in the ability to specifically target and eliminate cancerous tissue, while affecting as few healthy cells as possible [39]. Advancements in the treatment of cancer are making headway in the creation of new agents as well as the delivery by which these new and old agents reach the cancer [39]. However, much more research is needed in isolating the differences between cancerous cells and normal tissue, and to utilize these differences to increase the selectivity of delivery.

6.4 DISCUSSION

Nanotechnology is an emerging and promising approach for the diagnosis, treatment, and prevention of diseases including CNS diseases. The potential of this system to expand the market for many drugs and form the basis of a highly profitable niche within the industry, has led to more than 30 years of research into the rational delivery and targeting of nanoparticles as pharmaceutical, therapeutic, and diagnostic agents. These systems will allow for prolonged or increased pharmacologic activity and decreased side effects and, in turn, provide increased patient convenience and improved quality of life.

REFERENCES

1. Yu, H., Adedoyin, A. ADME-Tox in drug discovery: integration of experimental and computational technologies. *Drug Discov Today* 2003;**8**:852–861.
2. Di, L., Kerns, E.H. Profiling drug-like properties in discovery research. *Curr Opin Chem Biol* 2003;**7**:402–408.
3. Prentis, R.A., Lis, Y., Walker, S.R. Pharmaceutical innovation by the seven UK-owned pharmaceutical companies (1964–1985). *Br J Clin Pharmacol* 1988;**25**:387–396.
4. Yamashita, F., Hashida, M. In situ approaches for predicting ADME properties of drugs. *Drug Metabol Pharmacokin* 2004;**19**:327–338.
5. Clark, D.E. In situ prediction of blood-brain barrier permeation. *Drug Discov Today* 2003;**8**:927–933.
6. Pinto-Alphandar, H., Andremont, A., Couvreur, P. Targeted delivery of antibiotics using liposomes and nanoparticles: research and applications. *Int J Antimicrob Agents* 2000;**13**:155–168.
7. Lockman, P.R., Mumper, R.J., Khan, M.A., Allen, D.D. Nanoparticle technology for drug delivery across the blood-brain barrier. *Drug Dev Ind Pharm* 2002;**28**:1–13.
8. Koziara, J.M., Lockman, P.R., Allen, D.D., Mumper, R.J. In situ blood-brain barrier transport of nanoparticles. *Pharm Res* 2003;**20**:1772–1778.
9. Hoet, P.H., Bruske-Hohlfeld, I., Salata, O.V. Nanoparticles – known and unknown health risks. J *Nanobiotechnol* 2004;**2**(1):12.
10. Rabinow, B.E. Nanosuspensions in drug delivery. *Nat Rev Drug Discov* 2004;**3**:785–796.
11. Moghimi, S.M., Hunter, A.C., and Murray, J.C. Nanomedicine: current status and future prospects. *FASEB J* 2005;**19**:311–330.
12. Lenaerts, V., Couvreur, P., Christiaens-Leyh, D., Joiris, E., Roland, M., Rollman, B., Speiser, P. Degradation of polyisobutylcyanoacrylate nanoparticles. *Biomaterials* 1984;**5**:65–68.
13. Couvreur, P., Dubernet, C., Puisieux, F. Controlled drug delivery with nanoparticles: current possibilities and future trends. *Eur J Pharm Biopharm* 1995;**41**:2–11.
14. Yeates, D.B., Muderly, J.L. Inhaled environmental/occupational irritants and allergens: mechanisms of cardiovascular and systemic responses: introduction. *Environ Health Perspect* 2001;**109**:479–481.
15. Reddy, K.R. Controlled-release, pegylation, liposomal formulations: new mechanisms in the delivery of injectable drugs. *Ann Pharmacother* 2000;**34**:915–923.
16. Langer, R. Tissue engineering: a new field and its challenges. *Pharm Res* 1997;**14**:840–841.
17. Kreuter, J. Nanoparticles and microparticles for drug and vaccine delivery. *J Anat* 1996;**189**:503–505.
18. Khouri, A.L., Fallouh, N., Roblot-Treupel, L., Fessi, H., Devissageuet, J.P.H., Puissieux, F. Development of a new process for the manufacture of polyisobutylcyanoacrylate nanoparticles. *Int J Pharm* 1986;**28**:125.
19. Douglas, S.J., Illum, L., Davis, S.S., Kreuter, J. Particle size and distribution of poly(butyl-2-cyanoacrylate) nanoparticles. II. Influence of stabilizers. *J Colloidal Interface Sci* 1985;**103**:154.
20. Ber, U.E., Kreuter, J., Speiser, P.P. Influence of the particle size on the adjuvant effects of polybutylcyanoacrylate nanoparticles. *Pharm Ind* 1986;**48**:75–79.
21. Kreuter, J. Nanoparticles and microparticles for drug and vaccine delivery. *J Anat* 1996;**189**:503–505.
22. Kreuter, J. Nanoparticle systems for brain delivery of drugs. *Adv Drug Del Rev* 2001;**47**:65–81.
23. Nagy, Z., Peters, H., Huttner, I. Charge-related alterations of the cerebral endothelium. *Lab Invest* 1983;**49**:662–671.
24. Tamai, I., Tsuji, A. Transporter-mediated permeation of drugs across the blood-brain barrier. *J Pharm Sci* 2000;**89**:1371–1388.
25. Allen, D.D., Lockman, P. R. (2003), "The blood-brain barrier choline transporter as a brain drug delivery vector." *Life Sci* 2003;**73**(13):1609–1615.
26. Clark, D.E., Pickett, S.D. Computational methods for the prediction of 'drug-likeness'. *Drug Discov Today* 2000;**5**:49–58.

27. Malan, S.F., Chetty, D.J., Du Plessis, J. Physicochemical properties of drugs and membrane permeability. *S Afr J Sci* 2002;**98**:385–391.
28. Brightman, M.W. Morphology of blood-brain interfaces. *Exp Eye Res* 1977;**25**(suppl):1–25.
29. Butte, A.M., Jones, H.C., Abbot, N.J. Electrical resistance across the blood-brain barrier in anaesthetized rats; a development study. *J Physiol* 1990;**429**:47–62.
30. Pardridge, W.M. Drug and gene targeting to the brain with molecular Trojan horses. *Nat Rev Drug Discov* 2002;**1**(2):131–9.
31. Pardridge, W.M. Brain drug delivery and blood-brain barrier transport. *Drug Deliv* 1996;**3**:99–115.
32. Van de Waterbeemd, H., Camenisch, G., Folkers, G., Chretien, J.R., Raevsky, O.A. Estimation of blood-brain barrier crossing of drugs using molecular size and shape, and H-bonding descriptors. *J Drug Target* 1998;**6**:151–165.
33. Soni, V., Kohli, D.V., Jain, S.K. Transferrin coupled liposomes as drug delivery carriers for brain targeting of 5-florouracil. *J Drug Target* 2005;**13**:245–250.
34. Lockman, P.R., Koziara, J.M., Mumper, R.J., Allen, D.D. Nanoparticle surface charges alter blood-brain barrier integrity and permeability. *J Drug Target* 2004;**12**:635–641.
35. Koziara, J.M., Lockman, P.R., Allen, D.D., Mumper, R.J. The blood-brain barrier and brain drug delivery. *J Nanosci Nanotechnol* 2005;in press.
36. Panyam, J., Labhasetwar, V. Biodegradable nanoparticles for drug and gene delivery to cells and tissue. *Adv Drug Deliv Rev* 2003;**55**:329–347.
37. Jain, R.A. The manufacturing techniques of various drug loaded biodegradable poly(lactide-co-glycolide) devices. *Biomaterials* 2000;**21**(23):2475–2490.
38. Vorbrodt, A.W. Ultracytochemical characterization of anionic sites in the wall of brain capillaries. *J Neurocytol* 1989;**18**:359–368.
39. Brannon-Peppas, L., Blanchette, J.O. Nanoparticle and targeted systems for cancer therapy. *Adv Drug Deliv Rev*. 2004;**56**:1649–1659.
40. Schiffelers, R.M., Ansari, A., Xu, J., Zhou, Q., Tang, Q., Storm, G., et al. Cancer siRNA therapy by tumor selective delivery with ligand-targeted sterically stabilized nanoparticle. *Nucl Acids Res* 2004;**32**(19):e149.
41. Yancopoulos, G.D., Klagsbrun, M., Folkman, J. Vasculogenesis, angiogenesis and growth factors: Ephrins enter the fray at the border. *Cell* 1998;**93**:661.
42. Eliceiri, B.P., Cheresh, D.A. Molecular basis of angiogenesis and vascular remodeling. *Curr Opin Cell Biol* 2001;**13**:563.
43. Brooks, P.C., Montgomery, A.M., Rosenfeld, M., Reisfeld, R.A., Hu, T., Klier, G., and Cheresh, D.A. Integrin $\alpha v\beta 3$ antagonists promote tumor regression by inducing apoptosis of angiogenic blood vessels. *Cell* 1994;**79**:1157.

CHAPTER 7

LIPID-COATED PERFLUOROCARBON STRUCTURES AS PARENTERAL THERAPEUTIC AGENTS

Evan C. Unger, Terry O. Matsunaga, and Reena Zutshi

7.1 INTRODUCTION

7.2 PHYSICAL CHARACTERISTICS OF LPSS

7.3 ULTRASOUND-MEDIATED BACKSCATTER, CAVITATION AND RADIATION FORCE/MICROSTREAMING, IMPLICATIONS FOR DIAGNOSTICS, SONOTHROMBOLYTIC THERAPY, AND DRUG DELIVERY

7.4 APPLICATIONS

7.5 OXYGEN DELIVERY

7.6 CONCLUSION

7.1 INTRODUCTION

Lipid-coated perfluorocarbon structures (LPSs) are a class of agents with a number of different therapeutic applications. In this chapter we describe the basis of the technology and some of these different therapeutic applications. LPSs are made up of lipids with perfluorocarbon (PFC) gas or liquid interiors. In some cases, hydrophobic drug reservoirs are also located in the interior. One of the features of LPSs is the PFC entrapped within these structures. As stated above, the PFC may be in a liquid or gaseous state and in some cases may convert from liquid to gaseous state on exposure to body temperature or on activation with energy such as ultrasound. By incorporating a hydrophobic reservoir into the structures, payloads of

Role of Lipid Excipients in Modifying Oral and Parenteral Drug Delivery, Edited by Kishor M. Wasan

hydrophobic drugs can be incorporated into LPSs for drug delivery [1]. Furthermore, by incorporating a bioconjugate ligand on the surface, the LPSs can be passively targeted to cell surface proteins/receptors for selective delivery [2]. The LPSs include responsiveness to ultrasound for active targeting, which can also be combined with passive targeting. The combination of passive with active targeting has shown increased drug delivery to selected cells or the intracellular space of selected cells [3]. Ultrasound can be used to activate bubble-based LPSs as therapeutic cavitation nuclei and this has applications for treating vascular thrombosis and drug delivery.

LPSs can be activated with external energy, particularly by ultrasound, through cavitation. Cavitation is the process of expansion and collapse of a bubble under ultrasound. LPSs with gaseous cores lower the threshold of energy of ultrasound for cavitation [4]. When cavitation occurs, energy can be concentrated and localized for targeted therapeutic effect. Potential applications include sonothrombolysis (use of ultrasound to hasten clot lysis in vascular thrombosis), and targeted drug and gene delivery for LPSs drugs/gene carriers. Cavitation can also be used to increase tissue permeability locally [5], e.g. for opening the blood brain barrier for drug delivery to the central nervous system [6]. LPSs can also be pushed by ultrasound using the radiation force of ultrasound [7] (discussed in greater detail in a subsequent section).

Aside from cavitation and radiation force, LPSs can also have applications in other therapeutic arenas as well, e.g. work by van Liew and Burkard [8] has determined that gaseous PFCs, on a per volume basis, have far more oxygen-carrying capacity than liquid PFCs. Hence, perfluoropentane LPSs are currently under development for oxygen delivery. Pre-clinical studies in the porcine model have demonstrated efficacy for oxygen delivery in a hemorrhagic shock model [9]. Alternatively, the delivery of other therapeutic gases such as anesthetics has shown promise using LPSs (CE Lundgren, personal communication). In addition, LPSs may aid in the elimination of dangerous gases from the body, e.g. excess nitrogen as occurs in nitrogen narcosis [10] and carbon monoxide in CO poisoning (CE Lundgren, personal communication) as well as for oxygen delivery.

In the subsequent sections of this chapter, we describe the physical properties of the PFC liquids and gases useful in making LPSs, the physics of ultrasonic interaction with LPSs (cavitation and radiation force), preclinical and clinical studies of sonothrombolysis, and applications in treating acute ischemic stroke and other ischemic conditions, targeted LPSs with bioconjugates, and drug and gene delivery applications.

7.2 PHYSICAL CHARACTERISTICS OF LPSS

7.2.1 Ultrasound Contrast Agents

It was known from early on that gas–water interfaces acted as highly efficient scatterers of ultrasound waves [11]. Initial interest in this phenomenon dates back to the 1940s and 1950s when the Department of Defense, and in particular the Navy, used sonar to detect the presence of submerged structures, such as submarines. Interest lay in the fact that submarines could avoid visibility from other ships by being submerged, but it was quickly observed that the propellers could not avoid the formation of cavitational nuclei (bubbles) when operating. Thus, it was noted that sound waves could easily detect these otherwise stealth submarines.

Lord Rayleigh, back in the early 1900s, wrote a treatise on the efficiency of backscatter of bubbles relative to the size/diameter of the body and noted that scattering efficiency varied as the sixth power of the relative diameter [12]. In other words, a bubble that was 2 μm in diameter, could exhibit 64-fold more ultrasound scattering than one that was 1 μm in diameter. For bubbles with larger diameters, the scattering efficiency followed Mie scattering, which varied as r^4 (twice the diameter yielded 16-fold more efficient scattering).

The use of gas-filled microstructures can be placed back four decades when it was realized by Gramiak and co-workers that the agitation and subsequent injection of radiocontrast dyes in the aortic root of the heart provided enhanced ultrasound imaging [13]. This concept was then followed shortly thereafter by the development of air-filled bubbles by Feinstein and co-workers [14]. Although the use of stabilized air bubbles provided promise for commercializing this as a contrast agent, bubbles were limited by their very short half-life. In circulation, air bubbles would be absorbed by the surrounding milieu [15], thus reducing their size, scattering efficiency, and circulating half-life.

The realization that bubbles could provide potential as ultrasound contrast agents, but that they were limited by short-circulating half-lives, prompted the search for a gas with just as efficient a scattering potential and yet a longer half-life. In the early 1990s it was realized that perfluorocarbon gases, primarily those of five carbons or less, would exist as a gas in the bloodstream, appeared to be biocompatible, and could be eliminated from the body via the respiratory process [16]. Perfluorocarbon gases had been used previously in ophthalmology for injection into the vitreous humor as a treatment for detached retinas [17]. It was noted that the perfluorocarbon gas, perfluoropropane, and a gas with properties similar to perfluorocarbons, sulfur hexafluoride, were used for this purpose because they could provide long-lived bubbles within the vitreous space. Eventually, the use of these agents as contrast agents expanded into the use of less volatile perfluorocarbons as well (i.e. perfluorohexane) [18].

Interestingly, it was first found that these perfluorocarbon gases would fortuitously form smaller bubbles when stabilized by lipids [19], proteins (albumin) [20], or polymers [21]. The formation of smaller bubbles was caused by the fact that perfluorocarbons possessed very low surface tensions. This was an important phenomenon because bubble stability was governed by a physical phenomenon known as LaPlace pressures [22], the formula being shown below:

$$\text{LaPlace pressure, } P = 2\gamma/R$$

where γ is the surface tension and R the bubble radius.

Thus, the critical pressure exerted on the bubble is reduced when the gas–water interface surface tension is low whereas the pressure increases when the radius is small. Thus, compared with air, the LaPlace pressure of perfluorocarbon bubbles is reduced for a bubble of similar size or, alternatively, the LaPlace pressure of perfluorocarbon bubbles compared with air bubbles is similar when the radius of the perfluorocarbon bubble is reduced.

It was later to be discovered that the perfluorocarbon gases were also ideal as contrast agents from a distribution and elimination perspective because they would not diffuse out of the bloodstream [23] (hence their utility as blood pool contrast agents secondary to intravascular distribution) and their facile elimination unchanged via respiratory exchange through the pulmonary vascular bed [24].

Coincident with this phenomenon is the fact that perfluorocarbon bubbles are virtually insoluble in aqueous milieu [25] so the volume of the bubble, relative to air, will not shrink secondary to diffusion of the gas out of the bubble, and subsequent absorption of the gas by gas-exchanging bodies as blood.

Despite the advantages of perfluorocarbon gases, perfluorocarbons alone are not sufficient to maintain a stable bubble. Work in the laboratory of Dr Unger at ImaRx Pharmaceutical Corporation focused on lipids as a stabilizing membrane for their bubble technology [26]. Lipids were an attractive choice as a membrane stabilizer because of their biocompatibility [27], minimal antigenicity [28], and known, relatively predictable, pharmacokinetic profiles [29] and routes of degradation and elimination [29]. In addition, lipids were excellent surfactants [30],®microspheres which is required to keep the gas–liquid interface low so that the bubbles would remain small. This effort led to the discovery that very small amounts of lipids were required to stabilize bubbles. As an indicator, in order to stabilize 10^9–10^{10} bubbles (average size <1 μm)/mL solution, <1 mg lipid is required [31].

The above technologies led to the development of ultrasound contrast agents. Currently there are two perfluorocarbon-filled contrast agents sold in the USA, Optison, a perfluoropropane gas-filled bubble with an albumin shell [32], and Definity®microspheres, a perfluoropropane gas-filled bubble with a lipid shell [33].

7.3 ULTRASOUND-MEDIATED BACKSCATTER, CAVITATION AND RADIATION FORCE/MICROSTREAMING, IMPLICATIONS FOR DIAGNOSTICS, SONOTHROMBOLYTIC THERAPY, AND DRUG DELIVERY

7.3.1 Backscatter

As described briefly above, a gas–water interface is an exquisitely efficient scatterer of sound. This is because of the impedance mismatch or physical property differences between a liquid and a gas. In diagnostic imaging, the phenomenon of highly efficient bubble-induced backscatter is the basis for its utility as an ultrasound contrast agent, especially for the vascular space (Figure 7.1). In addition, the fact that the bubbles range in size from 500 nm to approximately 2 μm [34], the bubbles can freely course through capillary beds. This ability to flow through capillary vessels has provided an opportunity to view blood supply through regions of tissue, which were previously poorly imaged. In particular, echocardiography has benefited to

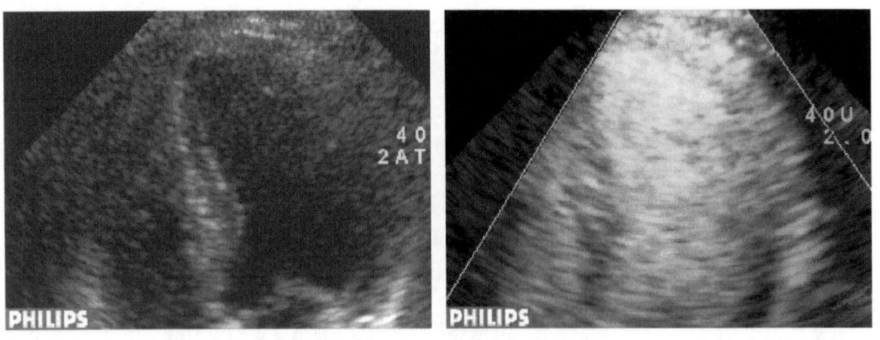

Figure 7.1 Myocardial contrast imaging with Definity: in the precontrast image it is not possible to see the entire myocardial border inpostcontrast it is possible to see the borders clearly and it is also possible to observe the degree of filling in the left ventricle.

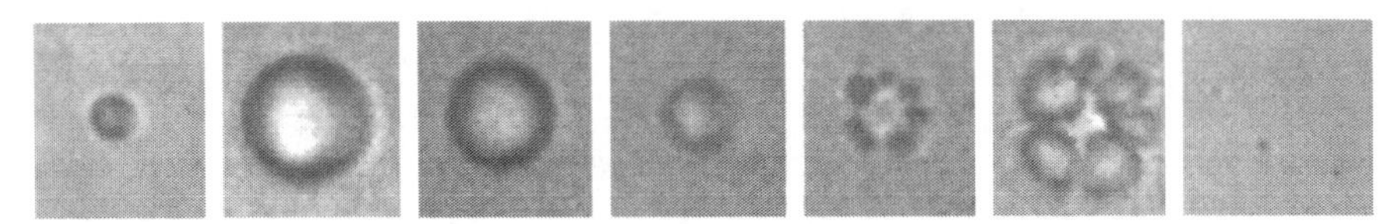

Figure 7.2 In the images above, a single 3 μm bubble is shown (far left) in the resting state. Insonation with a single pulse of ultrasound energy (2.4 MHz and 1.1 MPa) causes the bubble to expand, collapse, and fragment, yielding nanometer-sized fragments. As the bubbles expand and collapse, they generate a local shockwave on a nanoscale that can be used therapeutically. (Reproduced with permission from Chomas et al. [88].)

some extent from the ability to enhance the image of the myocardium and assess myocardial perfusion or blood supply [35].

7.3.2 Cavitation of Bubbles

Aside from the fact that gas–water interfaces act as reflectors of sound waves, another phenomenon that has been used to advantage for therapy is bubble cavitation. It is of interest that bubbles, in the size range described previously, fortuitously respond to frequencies in the range of clinical ultrasound. Figure 7.2 shows time-lapsed sequential photographs of a bubble that is exhibiting a single oscillation cycle at 2.4 MHz [36]. Bubbles undergoing ultrasound frequency-mediated oscillations will contract and expand, followed again by contraction. When the bubble reaches a critical minimum diameter, it will implode due to cavitation. A subsequent radiation force will be exerted on the surrounding environment. It has been found by Tachibana [37] and others that this radiation force is of sufficient magnitude to aid in the dissolution of clots. This phenomenon has been utilized in a number of in vitro and in vivo experiments to demonstrate dissolution of thrombotic occlusions in the femoral arteries [38], brain [39], canine femoral grafts [40], and coronary vessels [41].

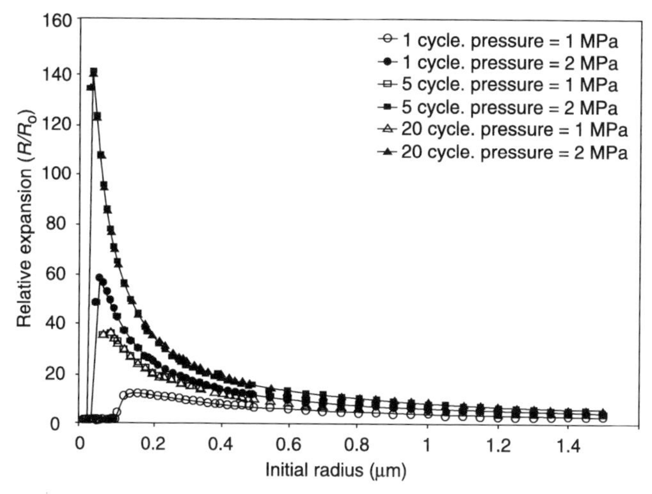

Figure 7.3 Smaller bubbles produce a greater relative expansion (R/R_0) in an ultrasound field. (Reproduced with permission from D. Patel et al. [43].)

An added advantage from a clinical perspective is the fact that bubbles also decrease the energy requirement for achieving the cavitational threshold [42]. This is an important observation because the decrease in ultrasound energy requirements also reduces the clinical possibility of inducing biorelated effects such as tissue heating or endothelial vessel compromise.

Submicron-sized bubbles (SMBs) (<1 μm in diameter) as opposed to microbubbles (> 1 μm in diameter) may have advantages as cavitation nuclei for therapeutic purposes such as sonothrombolysis. As shown in the Figure 7.3 from work by Ferrara et al. [43], the smaller the size of the bubble, the greater the relative expansion ratio of the bubble as it is insonated. The SMBs expand to a greater extent relative to their initial resting radius than the larger microbubbles. SMBs, as opposed to microbubbles, may also have advantages for targeting and drug delivery. It is well known that larger, micrometer-sized particles have accelerated clearance from the blood pool, predominantly by the reticuloendothelial system (RES) [44]. Smaller particles, particularly with diameters appreciably <1 μm in size, may have delayed clearance by the RES [44]. Surface chemistry of the particles is also quite important [45] but, as we discuss in subsequent sections of this chapter, it is possible to construct bubbles with surface chemistry appropriate to delay RES uptake and prolong circulation times.

7.3.3 Radiation Force/Microstreaming

Finally, radiation force or microstreaming is another effect that has developed clinical interest. Radiation force can be described as the force required to 'push' a body through a medium. Depending upon the density and composition of the medium, the ability to apply ultrasound-induced radiation force upon a body can be very efficient, as in the case of bubbles [46], or moderate, as in the case of lipid droplets

[47]. Nevertheless, it has been shown that radiation force can play a major role in the delivery of drug-loaded structures as described below. This phenomenon can potentially provide an advantage as a complementary mechanism to targeted drug delivery.

7.4 APPLICATIONS

Thrombosis is an important biomedical problem, the clinical implications of which are discussed in greater detail in the following section. The gaseous LPSs, as found in SMBs bubbles, have important therapeutic applications in treating thrombosis. In this section we discuss the physical basis for the use of bubbles in treating vascular thrombosis and some of the preclinical studies that have been performed using microbubbles or LPSs with ultrasound to dissolve thrombi.

7.4.1 Pathophysiology of Thrombus Formation

Thrombi are variably composed of red blood cells, fibrin, and platelets. Arterial thrombi tend to be platelet rich [48] and venous thrombi tend to have more red cells, more closely reflecting the components of whole blood, held together by a meshwork of fibrin and platelets [48]. Fibrin is formed when thrombin cleaves fibrinogen, which may then cross-link with other polymers of fibrin. Platelets may be activated by a number of biochemical or physical factors and bind to damaged endothelial cells or tissues. Fibrin, in turn, binds to the activated platelets, forming a plug at the site of a vascular injury, of obvious importance and utility in sealing a bleeding site in hemorrhage [49]. Unfortunately vascular thrombosis often occurs at inopportune times, resulting in pathological vascular occlusion. When a blood vessel is occluded, depending on the degree of occlusion and the presence or absence of collateral flow ischemia, the surrounding tissue oxygenated by that vessel will become ischemic. If ischemia is severe, this will result in cell death, necrosis of the tissue, and even death. Even if the cells do not die, cell damage can occur. which can lead to significant morbidity. Treatment of thrombosis is necessary to restore blood flow and prevent ischemic tissue damage. Clearly, the time to restoration of blood flow is critical in determining whether the tissue will be viable, severely compromised, or dead.

7.4.2 LPSs as Therapeutic Agents for Ultrasound-mediated Sonothrombolysis of Vascular Thrombi

As described above, bubbles of the size of SMBs oscillate in the presence of clinical and diagnostic ultrasound frequencies (1–10 MHz). The oscillation, cavitation, and subsequent radiation force or shock wave, has been utilized for dissolution of clot material in a variety of thrombotic occlusion studies.

It has been found by Apfel and Holland [50] that the energy requirements for cavitation are reduced in the presence of bubbles. Furthermore, in the presence of vascular thrombi (clots), one is able to dissolve thrombi using ultrasound and bubbles [51]. Note also that the cavitation threshold lowers with (1/√Frequency) [50]. Therapeutic ultrasound can be applied in frequencies as low as 20 kHz [52]. Cavitation will occur at ultrasound pressure levels about one-seventh that required at 1 MHz. The optimal frequency for therapeutic ultrasound has not been determined. Lower frequencies give better penetration into the tissues, but are more difficult to focus to a defined tissue region [53]. Probably more studies have been performed at 1 MHz than at any other frequency [54].

7.4.3 Research and Thrombosis

Thrombosis is often treated with lytic drugs. Lytic agents such as streptokinase [55], urokinase [56] or tissue plasminogen activator (t-PA) [57]; all work by activating plasminogen to form plasmin which, in turn, enzymatically cleaves fibrin. The activity of lytic agents or the rate of clot lysis is relatively slow because of rate-limiting accessibility of the substrate (fibrin in clot) to the plasmin. A number of studies have shown that ultrasound enhances the activity of lytic drugs in lysing clot [54] and this has been referred to as sonothrombolysis. Within the ranges of frequencies of ultrasound used for diagnostic imaging, the power levels are too low to cause cavitation without microbubbles. Diagnostic ultrasound can cause radiation force and acoustic streaming – relatively weak forces at these power levels – but strong enough to aid in dissolution of thrombi [58]. Radiation force and acoustic streaming can improve permeation of plasmin and/or lytic drug into the substrate of the clot to increase the rate of clot lysis. Studies performed by Francis have shown that ultrasound causes the fibrin fibers to disaggregate, a process that naturally reverses itself, which potentially exposes more plasmin-binding sites and changes fluid dynamics within the clot [59]. The above is but one mechanism proposed as to how ultrasound-mediated lysis of thrombi can occur.

The relatively weak force of ultrasound is increased by the presence of microbubbles [54], and perhaps even more so by submicron-sized bubbles, which may be more efficient cavitation nuclei than microbubbles and also may better permeate the substance of the clot as a result of their smaller size. Microbubbles within clots or adjacent to clots capture the ultrasound energy though cavitation. As the microbubbles expand and collapse in concert with ultrasound, this locally enhances acoustic streaming and effectively increases the force driving enzyme into the substrate of the clot. Figure 7.4 shows a diagram of an experimental clot flow-through system. Clots are placed into a polyethylene tube and physiological (0.9%) saline containing bovine plasminogen at a concentration of 180 μg/mL is circulated through the system. Saline, lytics (t-PA or urokinase), and/or ImaRx's proprietory LPSs are injected upstream of the clot and ultrasound is applied to the clot using a 1 MHz non-focused ultrasound probe. The clots are weighed pre- and post-treatment and the percentage clot lysis determined. As shown in Figure 7.5, on the bar graph chart the greatest clot lysis is obtained with LPSs + lytic drug + ultrasound (US). Somewhat greater clot lysis is obtained for UK + LPSs + US than for t-PA + LPSs + US. The combination of LPSs + ultrasound, referred to as bubble-assisted sonothrombolysis (BAST), in vitro shows promise for dissolving clots.

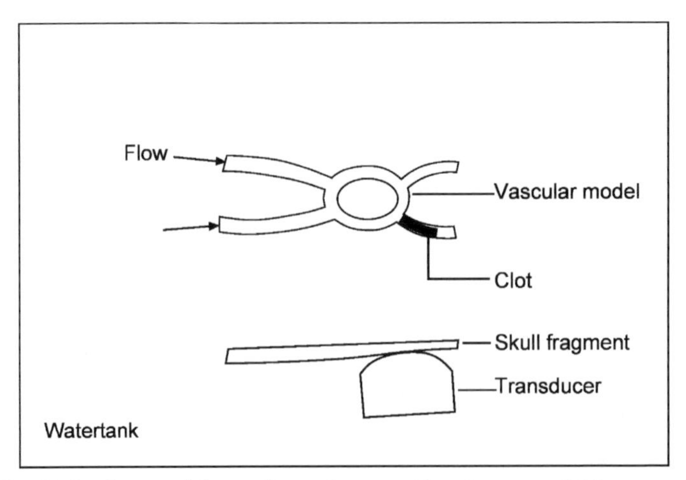

Figure 7.4 Stroke flow model experimental set-up. Vasculature model is anatomically accurate.

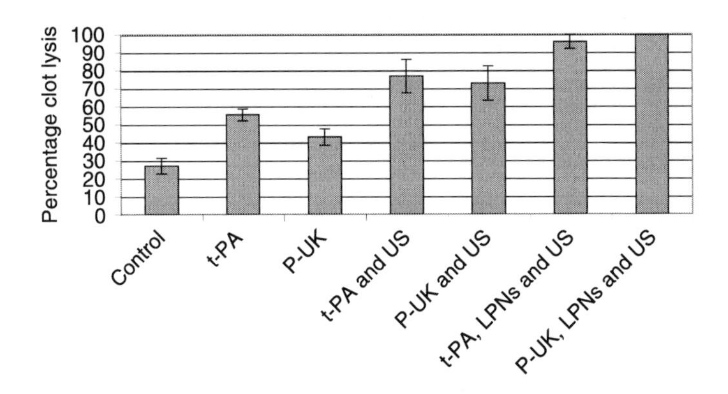

Figure 7.5 Results of clot model in vitro. t-PA, tissue plasminogen activator; US, ultrasound.

The in vitro experiments support the hypothesis that LPSs (submicron bubbles) function as cavitation nuclei to increase the rate of clot lysis. In vivo experiments have been performed with LPSs in a number of animal models. Figure 7.6 shows an arteriovenous graft in dogs. A 24-cm polytetrafluoroethylene (PTFE) graft is placed between the femoral artery and femoral vein in dogs. The venous outflow is ligated for at least 4 hours, causing thrombus to fill the graft from the venous outflow of the graft to the arterial limb. The ligature is removed and the thrombus exposed to US alone, US + LPSs and US + LPSs with image guidance, using low mechanical index (MI) diagnostic ultrasound. LPSs was administered via intravenous infusion. The bubbles can be seen on diagnostic ultrasound as bright reflections, which over time infiltrate the clot. Indeed an advantage of LPSs compared

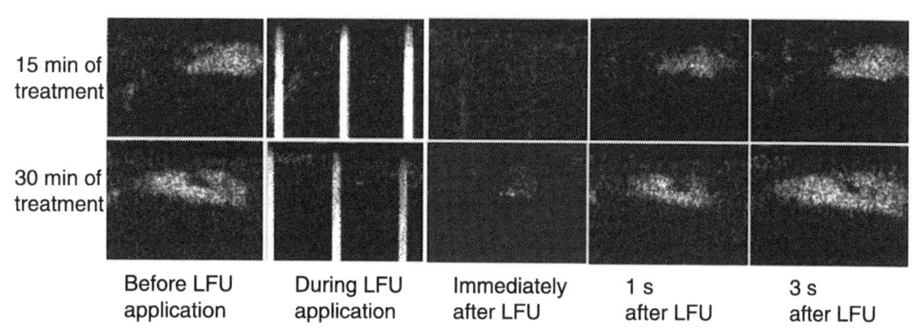

Figure 7.6 Arteriovenous graft in a canine model. Visualization of SMBs in the AV graft using low MI diagnostic ultrasound. Low frequency US (LFU) applied to completely cavitate bubbles. Reapplied only when nanobubbles refilled the graft. Resulted in rapid dissolution of the clot. Reprinted with permission Tsutsui et al., Radiology [60].

with microbubbles is that the smaller diameter, submicrometer LPSs appear to infiltrate the clot better from an intravenous administration. To maximize efficacy of BAST it is important to allow enough time for LPSs to permeate the treatment zone of the clot, and apply therapeutic ultrasound that cavitates and destroys the LPSs, and then to allow entry of fresh LPSs into the clot to coincide with timing of the next application of therapeutic US. Low MI US does not destroy the bubbles but readily detects their presence with high contrast to noise [60]. Using low MI imaging to guide the therapeutic ultrasound, it can be selectively applied to coincide with the timing of optimal concentrations of LPSs in the clot to increase the rate of BAST. Image-guided therapeutic US with intravenous LPSs affords rapid and effective declotting in this model without the need for a lytic drug [60].

If large particles are created in the process of clot dissolution, these might cause emboli that could block regions of arterial circulation distal to the clot or, in the case of venous clots, particles might move proximally to lodge in the pulmonary circulation. It is therefore desirable that a clot dissolution technique does not create fragments appreciably larger than red blood cells (up to about 7–10 μm in diameter). To study the effects of US and microbubbles on particle size, the effluent of clots treated in vitro with microbubbles + US were examined with particle sizing systems. Over 90% of the fragments were < 1 μm in size [61, 62]. In a follow-up study, dogs were also studied by pulmonary angiography pre- and post-BAST with LPSs. Post-BAST angiograms showed no evidence of vascular occlusions in the pulmonary circulation, i.e. no evidence of pulmonary emboli (T Porter, personal communication). An advantage of BAST appears to be its propensity to dissolve clots into very small particles, smaller than the size of the fragments created as a result many mechanical fragmentation/thrombectomy techniques.

7.4.4 Bubble-Assisted Sonothrombolysis – Clinical Studies

Vascular thrombosis is the most prevalent single cause of death in industrialized countries [63]. It is responsible for myocardial infarction, ischemic stroke, deep vein thrombosis (DVT), pulmonary emboli, peripheral arterial occlusions (PAOs), and occlusion of vascular grafts. Atheromatous disease (soft plaques) and, eventually, arteriosclerosis are manifested by plaque formation. Plaques are variably composed of inflammatory cells, lipids and fibrous tissue. The narrowing of arteries caused by plaque formation is responsible for hypoxic episodes in tissues, potentially resulting in episodic occurrences of angina, ischemic heart disease, claudication, transient ischemic attacks (TIAs), and peripheral artery disease. However, the most dangerous event is rupture of these vulnerable plaques. When plaques rupture this leads to platelet aggregation and thrombus formation. To restore blood flow it is necessary to lyse or remove the clot in order to recanalize the vessel. It may also be necessary to treat the underlying lesion, e.g. plaque in coronary artery disease, but effective rapid treatment of thrombosis is critical to restoring blood flow, particularly in stroke. BAST holds the potential to treat rapidly and less-invasively thrombotic occlusions. Targeted delivery of LPSs also holds potential to treat plaque itself (discussed below).

To treat vascular thrombosis, lytic drugs can be employed, and, as discussed above in the preclinical studies of BAST, lytic drugs often take too long to work, during which time the organ deprived of flow is ischemic, which may result in critical damage to the brain in the case of stroke, to the heart in myocardial infarction, or to other organs or tissues in other thrombotic events. Mechanical devices are sometimes employed to treat vascular thrombosis. These may consist of angioplasty balloons [64], stents [64], devices to fragment clot mechanically [65], aspirate thrombus [65], and others. Mechanical devices are used to remove clots in percutaneous, minimally invasive procedures in which a catheter is usually inserted into a blood vessel or in open surgical procedures.

Initial human studies using perfluoropropane-filled LPSs were performed in patients with thrombosed dialysis grafts. We chose this indication for initial study for proof of principle and to gain experience in a less complex model, than, for example, stroke. Thrombosed dialysis grafts are an important clinical problem, however, affecting 600 000 patients per year [63]. In renal dialysis, shunts and fistulae are placed to allow high blood flow between arteries and veins. This is necessary to allow sufficient blood flow to be removed for dialysis procedures to be performed. In current practice, when the dialysis grafts clot off, they are generally treated with a mechanical thrombectomy device that is placed into the graft through a catheter. In our BAST study, 11 patients were administered LPSs directly into the graft via a small-gauge (narrow) catheter accompanied by ultrasound and 12 patients also received a low dose of t-PA into the graft + LPSs and US. BAST was well tolerated and showed efficacy in all patients, both with and without t-PA [40].

Studies have also been conducted in patients with DVT and PAO, although the results are still being analyzed. In these studies LPSs were infused through a multi-side-hole catheter placed into the affected vein or arterial graft for 60 minutes and ultrasound was applied over the same time period across the intact skin. Some patients also received small doses of t-PA administered through the same catheter. If proven efficacious in this time period it will contrast to clinical experience with catheter-mediated infusion of lytic drugs such as t-PA, which often takes about 24 hours or more to recanalize a thrombosed arterial vessel [66].

Of all the organs in the body, the brain is probably most critically affected by ischemia. Although there may be time to schedule a patient for treatment in the angiography suite to place a catheter to treat PAO, for example, time is much more critical to salvage brain tissue after an ischemic stroke. In addition, it is technically more difficult to place a catheter, or mechanical thrombectomy device, into the brain than it is, for example, to place a catheter in the artery to the leg. Using intravenous LPSs and transcranial ultrasound therefore has the potential to offer a dramatic new method of treating stroke patients. BAST should be easier and quicker to perform than minimally invasive therapy for stroke treatment. In our clinical trials we have seen that BAST may be commenced in as little as 13 minutes after patients arrive in the emergency room (Andrei Alexandrov, personal communication). By comparison a catheter-based, minimally invasive treatment usually takes well over an hour to start after the patient arrives in the hospital. In stroke patients time is all important and the ability to institute effective therapy more quickly could be very important.

Stroke is the third most common cause of death in the USA and the single greatest cause of expenditure for long-term disability [67]. It costs the US health-care system over $US56 billion per year. The cost of taking care of a stroke patient with long-term disability is about $US77 000 per year [67]. There is therefore a very strong healthcare need to improve treatments and outcomes in the care of stroke patients.

Lytic therapy with t-PA is approved for treatment of acute ischemic stroke but only during the first 3 hours after stroke [68]. There is a roughly 20% rate of arterial recanalization at 2 hours after treatment of stroke with t-PA and as a complication of treatment with t-PA about a 6–8% rate of intracranial hemorrhage [68]. Less than 15% of patients with stroke present within 3 hours of stroke onset [64] and are therefore candidates to receive t-PA. In large part, because of fears about bleeding from t-PA, only about 20% of those patients who are candidates to receive t-PA are actually treated with the drug [64]. For patients presenting beyond 3 hours after stroke onset, they are generally treated only with aspirin [64].

Our clinical experience with BAST in stroke builds upon prior clinical experience with sonothrombolysis in stroke using ultrasound without bubbles and subsequently with bubbles. Alexandrov treated 126 patients in a randomized prospective study; 63 patients received 2 hours of continuous transcranial Doppler (TCD) US + standard dose of intravenous t-PA and 63 patients received intravenous t-PA alone. At 2 hours post-treatment the rate of arterial recanalization in the intravenous t-PA group was 30% and the rate of arterial recanalization in the US + intravenous V t-PA was 49% ($p = 0.03$) [69]. Molina treated 103 patients in three groups: t-PA alone ($n = 34$), t-PA + US ($n = 34$) and t-PA + US + microbubbles ($n = 35$). At 2 hours post-therapy the arterial recanalization rates were: t-PA alone = 38.9%, t-PA + US = 68%, and t-PA + US + microbubbles = 76% ($p = 0.004$). There were more responders in the t-PA + US + microbubbles group [70].

Molina's study showed that microbubbles significantly enhanced the rate of clot lysis from ultrasound [70]. The microbubbles used in the Molina study were lipid-coated air-filled galactose microspheres marketed as Levovist® by Schering AG in Europe as an earlier generation ultrasound contrast agent. Compared with LPSs,

Levovist has a shorter blood half-life [71]. Air is much more soluble in biological fluids than PFC gas, hence microbubbles of air have shorter half-lives than PFC-filled LPSs. Levovist bubbles have larger diameter than LPSs and, when suspended in saline for infusion, microbubbles tend to separate rapidly and float to the top. The small size of the LPSs enables them to be diluted and suspended in fluid for intravenous infusion. In the Molina study three boluses of Levovist were administered intravenously. The bubbles can be seen on the TCD ultrasound and last, at most, a minute or two after each injection. The LPSs last more than 5 minutes in the circulation after each injection, and create a much stronger signal on TCD than Levovist (University of Texas, personal communication). LPSs can also be maintained at steady-state concentrations by intravenous infusion during BAST because they are easier to suspend in intravenous solutions than microbubbles.

At the time of preparation of this chapter, a phase II trial was under way to test LPSs + US + lytic drug for treatment of acute ischemic stroke. Clinical trials are currently planned to see if it will be possible to develop LPSs for BAST treatment of acute stroke in combination with a lytic drug. However, plans are under way to develop BAST as a monotherapy for patients in whom lytic drugs are contraindicated. As a result of the potential complication of bleeding from lytic drugs, patients who are high risk of hemorrhage (e.g. post-surgical patients) are often not able to receive lytic therapy. Also the risk of bleeding increases particularly beyond 3 hours after a stroke [68]. Therefore many patients are not candidates for lytic therapy. As in vitro and preclinical studies show efficacy of BAST without lytic drugs, there is hope that BAST may be effective in stroke without a lytic drug.

7.4.5 Targeted LPSs

To make targeted LPSs, bioconjugates can be designed with targeting ligands to bind to selected receptors, cells, and tissues. Our group has produced a number of prototypical bioconjugages directed to integrins such as $\alpha_v\beta_{III}$, $\alpha_6\beta_I$ and GPIIb/IIIa. Most of the foregoing discussion has centered on BAST and the role of bubbles as cavitation nuclei for treatment of vascular thrombosis. It stands to reason that binding the cavitation nuclei directly to the treatment site should increase the therapeutic efficacy of the bubbles in treatment of vascular thrombosis (or efficacy of other targeted LPSs for other diseases). In the following we present an overview of our development of LPSs targeted to vascular thrombosis through the GPIIb/IIIa receptor on activated platelets.

Targeted bubbles can also be used to target blood clots site specifically. The targeting ligand, in this case, might be directed against fibrin or the GPIIb/IIIa receptor on the surface of activated platelets. As fibrin is present in old, usually stable clots, and acute thrombi and platelets are generally present in high concentrations only in acute thrombi, we chose to focus on ligands to the GPIIb/IIIa receptor of activated platelets. Ligands can be incorporated into the shell of LPSs by conjugating the ligand to a lipid through a PEG linker. Small cyclic RGD peptides, mim-

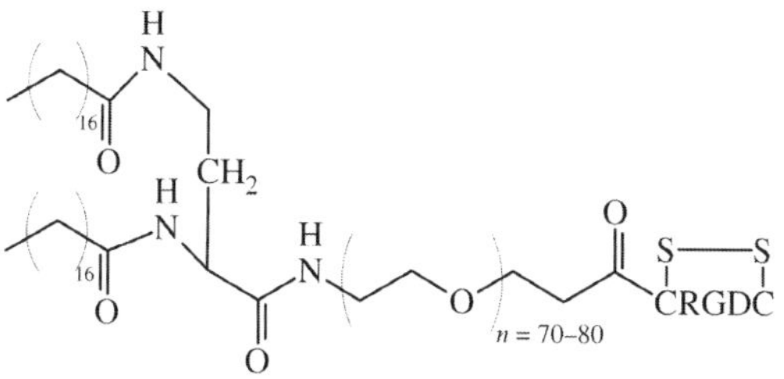

Figure 7.7 An RGD analog (RGD = arginine glycine aspartic acid). The structure of *N*, *N*-distearyl-diaminobutyryl, α-carboxy, ω-amino-polyethyleneglycol$_{3400}$-cyclo-CRGDC.

icking the recognition sequence on fibrinogen, have been used as targeting ligands for this purpose. The proposed structure is shown in Figure 7.7.

Perfluoropropane-filled RGD-containing bubbles have been shown to bind blood clots preferentially over non-targeted bubbles in an in vitro flow-through model shown previously (see Figure 7.4). These binding studies have been substantiated in vivo where a left atrial appendage clot could be clearly visualized post-contrast in a canine model [72].

Targeted microbubbles were formulated and then their binding avidity was tested. Bioconjugate ligands were used in either 1 wt%, (w/w), or 5 wt% relative to the lipids. The appropriate lipid, with chain length varying from C16 through C22, were formulated into microbubbles by first dissolving in propylene glycol followed by addition of glycerol and 0.9% saline. Samples were then purged with perfluoropropane followed by activation by shaking. Microbubbles were sampled on a Particle Sizing Systems Model 770 particle sizer (1 μm lower limit cutoff, Particle Sizing Systems, Santa Barbara, CA). All samples produced ensembles of microbubbles between 1.5 and 3.0 μm with a mean particle number between 5×10^8 and 1×10^9 particles/mL. The graphic depiction of the targeted microbubble is shown in Figure 7.8.

Note that the bioconjugate is made up of three components

1. A pseudoisosteric lipid mimic that inserts into the membrane
2. A polyethyleneglycol (PEG) tether
3. The targeting ligand (see also Figure 7.8).

The lipid mimic inserts into the membrane and the tether allows entropic freedom for the ligand to bind to the receptor. Multiple receptor binding allows the microbubble to be 'targeted' to the receptor.

The bioconjugates were characterized by MALDI mass spectrometry and high-performance liquid chromatography (HPLC) to confirm purity and then used for additional studies as well as for formulation of LPSs.

A platelet aggregation assay was performed; whole blood from healthy volunteers was collected and centrifuged at 150 ***g*** for 5 min to separate platelet-rich plasma (PRP). The platelet-poor plasma (PPP) was isolated by centrifugation of the remaining blood at 2500 ***g*** for 10 min. PRP was counted on an Accusizer Model

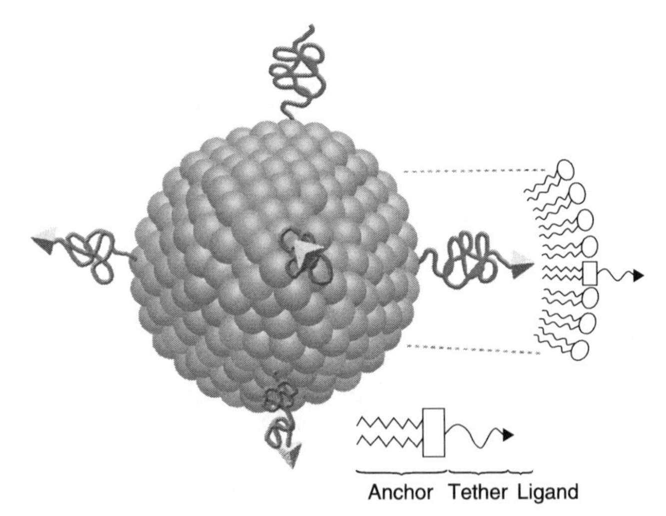

Figure 7.8 Graphic image of a targeted microbubble with bioconjugate ligand inserted on the lipid surface.

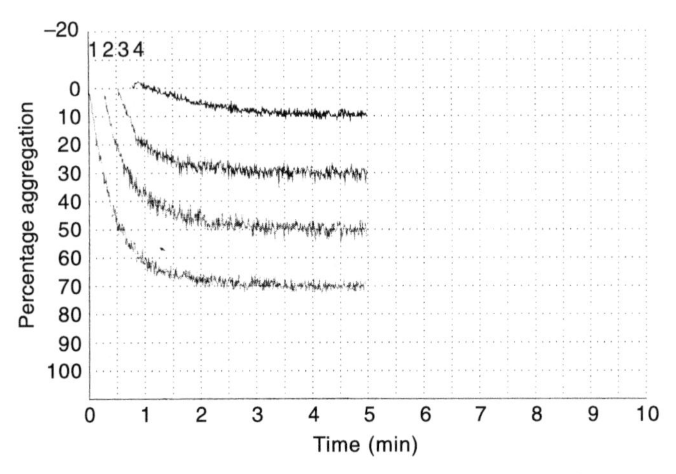

Figure 7.9 Graph of platelet aggregation inhibition for cyclo-CRGDC, where RGD = arginine-glycine aspartic acid.

770 (Particle Sizing Systems, Santa Barbara, CA). PRP was diluted with the appropriate amount of PPP to achieve a standard of 250 000 cells/μL. Peptide ligands were dissolved in 180 μL of platelet-rich plasma; 20 μL ADP and $CaCl_2$ were added to stimulate aggregation. Samples were then allowed to aggregate as a function of time. Percentage aggregation was then measured on a Bio/Data Corp. PAP-4 Platelet Aggregation Profiler, which measures aggregation as a function of light obscuration. Inhibition constants (IC_{50}) were determined as a function of the concentration able to inhibit 50% of the inherent aggregation relative to no peptide (control). Results are shown in Figure 7.9.

TABLE 7.1 IC_{50} values for all analogs

Compound	IC_{50} (μmol/L)
Cyclo-CRGDWPC	22
Cyclo-CRGDSPC	30
Cyclo-CRGDC	5.5
GGGRGDS	>200
KQAGDV	130
RGDS	500
RGDW	65
RGDF	80

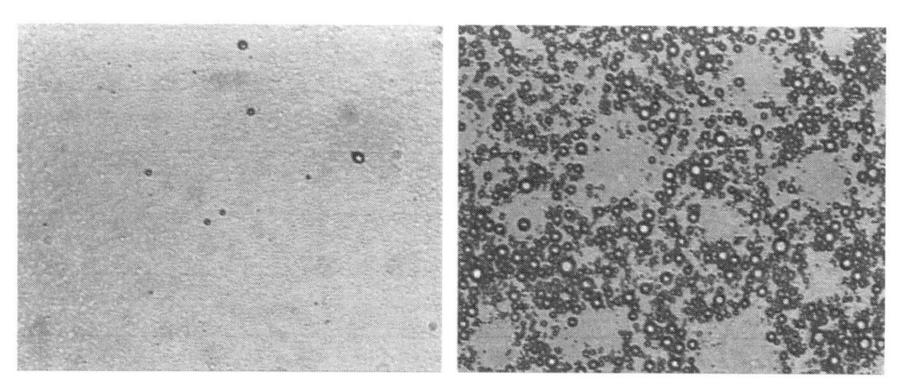

Figure 7.10 Targeted microbubbles bound to cell culture plate of activated platelets. The slide on the left was exposed to untargeted bubbles, on the right bubbles targeted to the GPIIb-IIa receptor. Both slides were washed and then imaged. (Reprinted with permission from Unger et al. [89].)

Based on the above data, IC_{50} values were reported for all analogs and a list is give in Table 7.1.

The appropriate ligand, cyclo-CRGDC was then selected based on its IC_{50} and synthesized as the bioconjugate. The bioconjugate was then incorporated into the microbubble and an in vitro microbubble binding assay was performed using fixed activated platelets: 50 μL of the standard platelet mixture (250 000 cells/μL) was added to 96-well plates followed by mild centrifugation of the coated plates at 150***g*** for 5 min. Plates were then incubated at 37°C for 5 min followed by three rinses with phosphate-buffered saline (PBS) on a plate washer. Platelets were activated by the addition of ADP and $CaCl_2$ (1 mmol/L). Following activation, 250 μL of PBS is added to each well followed by addition of aliquots of targeted or non-targeted microbubbles. Wells were then sealed with 250 μL PBS followed by sealing with pressure-sensitive film and inversion of the plate for 30 minutes. Plates were then washed with 60 μL aliquots PBS (×5) followed by addition of 250 μL PBS. Plates were then scanned on a visible light plate reader at 620 nm. Obscuration was also confirmed at 400x magnification (Figure 10). Microbubble density was then determined using NIH Image Software version 1.62.

Note that the untargeted microbubbles do not adhere significantly to the platelets while the targeted microbubbles bind with good avidity. Interestingly, similar results were obtained by Dayton and coworkers [73], who noted that the use of microbubbles with small molecule targeting ligands versus monoclonal antibody targeting ligands of different K_d values bound with similar avidity. This may be due to the polyvalency effect of multiple ligands on a single bubble affording a cooperative binding effect. This polyvalency has been described by Whitesides group at Harvard [74] and indicates that the effective binding potential of multiple ligands, as is the case with microbubbles, is synergistic and can provide an effectively greater binding avidity relative to individual ligands alone.

Thus, it can be concluded that using ligands with even low-micromolar binding inhibition constants (IC_{50}) may suffice as targeting ligands due to the polyvalency effect and more effective binding.

Arginine-glycine-aspartic acid (RGD)-targeted bubbles can potentially be used for ultrasound-mediated BAST as well. In this case, the binding of targeted bubbles to GPIIb/IIIa receptors increase the population of LPSs in the region of the clot, providing a higher concentration of cavitation nuclei for clot lysis.

In vitro experiments were performed using a clot model to quantify and determine the efficiency of sonothrombolysis with targeted versus control nontargeted microbubbles. In vitro ultrasound- and microbubble-mediated clot dissolution was determined by release of ^{125}I-labeled fibrinogen.

Blood was obtained from healthy volunteers 2 hours before the studies. Serum was obtained by centrifuging pooled blood at 3400 rev./min and decanting. The serum was mixed with a cocktail of factor IIa (thrombin), ^{125}I-labeled fibrinogen in a buffer of 0.1 mol/L Tris, 0.05 mol/L NaCl pH = 7.6, and 2 mol/L $CaCl_2$; 1 μg tissue plasminogen activator (t-PA), and 2 units heparin were also used.

Briefly, 160 μL aliquots of blood were added to 16 μL factor IIa/Ca^{2+} (thrombin) and 820 μL PPP. Microbubbles, untargeted or targeted, varying in concentration from 1 μL/ml (equivalent to 10^7 microbubbles) were added at varying intervals to the aliquot, followed by ultrasound exposure for a total of 60 min. After insonation, 50 μL aprotinin were added to the solution to prevent further reaction. Clots were then removed from the serum, and both fractions quantified by a γ counter.

The assay system was designed to determine the extent of clot dissolution by measuring the amount of solubilized, radiolabeled fibrinogen in the medium after ultrasound experiments. Ultrasound-mediated studies were performed at three different wavelengths: 40 kHz, 300 kHz, and 1 MHz. Figure 7.11 is a graph of experiments performed at 300 kHz using the above methodology.

Increase in clot lysis for targeted versus untargeted microbubbles of 10–25% was found in a series of six experiments ($n = 3$ per experiment). Similar results were obtained at a frequency of 40 kHz while, at 1 MHz, the difference between targeted and untargeted microbubbles was less significant. We concluded that, under the conditions employed, the intermediate frequency, 300 kHz, provided the best degree of lysis with targeted microbubbles.

To analyze for the presence of particulates, the effluent was centrifuged, and the ^{125}I counts (presumably from fragmented fibrin) were measured both in the supernatant and in the pellet (Figure 7.12). The supernatant showed the presence of radi-

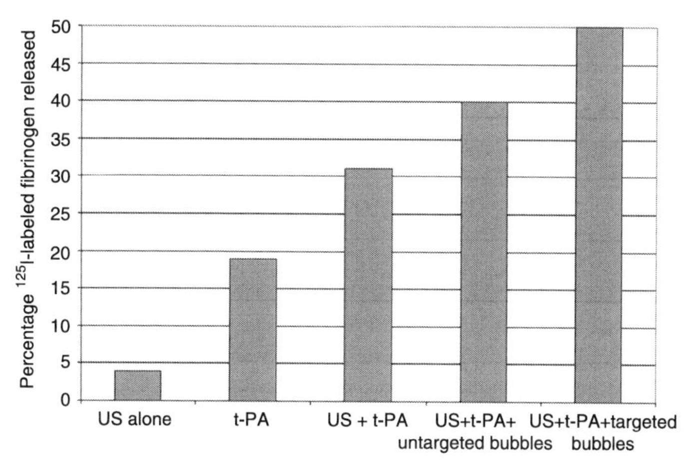

Figure 7.11 Results of the radiolabeled fibrinogen study. t-PA, tissue plasminogen activator; US, ultrasound.

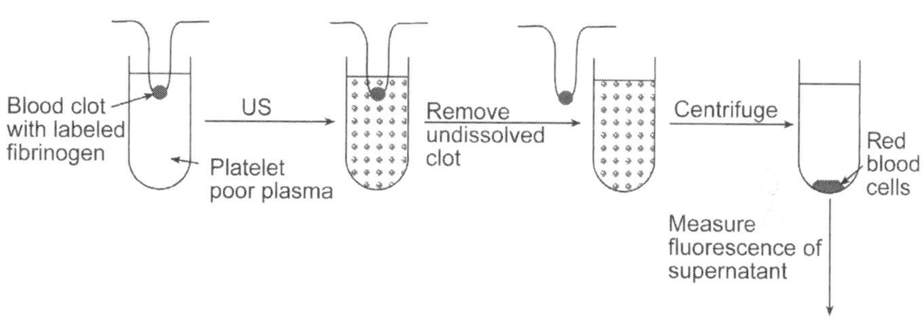

Figure 7.12 Schematic of experiment to analyze for the presence of particulates. The radiolabel was only found in the supernatant, indicating that any fragments were smaller than 7 μm. US, ultrasound.

olabeled iodine, whereas the pellet (primarily composed of red blood cells) contained no ^{125}I. This indicates that the size of fibrin (thrombus) fragments, if present at all, in the effluent must be <7 μm (size of a red blood cell = 7 μm). Fragments of this size will not occlude in capillary beds where the diameter is approximately 10 μm. This demonstrates that ultrasound treatment with microbubbles does not result in the release of large fragments that can potentially cause occlusions.

It should be noted that eptifibatide (Integrilin®), possesses an IC_{50} of the order of 140 nmol/L and still achieves effective binding to the GPIIb/IIIa receptor [75]. We rationalize that, although our single agents may possess higher IC_{50} values, the polyvalent effect of multiple ligands displayed on the microbubble surface will increase the effective avidity [74]. At the same time, we are currently exploring the use of cyclic ring modifications to optimize selectivity further for the GPIIb/IIIa receptor.

We plan on conducting in vivo studies in models of vascular thrombosis emphasizing models of myocardial infarction. If GPIIbIIIa-targeted LPSs show sufficient promise in preclinical models, then we will attempt to move the product into clinical studies for BAST.

7.4.6 Targeted LPSs as Ultrasound-mediated Drug Delivery Agents

A major challenge in chemotherapy is the drug delivery to tumors that are resistant to drug uptake, presumably as a result of the higher interstitial pressure in tumors [76]. The force that is imparted upon cavitation of bubbles can be potentially used for the vascular delivery of genes or drugs locally to a region of interest.

One such structure, called acoustically active lipospheres or AALs, contains an oily layer, which is capable of dissolving hydrophobic drug molecules. AALs encapsulating perfluorocarbon gas, when insonated, undergo cavitation into daughter nuclei, releasing drug at the site of insonation. Paclitaxel-carrying AALs that are less toxic than Taxol and respond to ultrasound have been developed in our lab [77]. In these experiments, paclitaxel was suspended in soybean oil and formulated with phospholipids with a headspace of perfluorobutane gas. The average particle size of the lipospheres was about 2.9 μm with > 95% of the particles being less than 10 μm in size. The AALs were acoustically active and, in a model in vitro experiment, released a hydrophobic dye on exposure to ultrasound. Acute toxicity studies in mice using the paclitaxel formulation showed a 10-fold decrease in toxicity when compared with Taxol.

Ultrasound-mediated delivery of genes from gene-carrying microbubbles has been demonstrated in an in vivo dog model by Vannan and co-workers [78]. Briefly, their work involved formulating SMBs using a mixture of cationic and neutral lipids with PFC gas to prepare LPSs that would bind plasmid DNA. The bubbles carrying a *CAT* gene were injected intravascularly, followed by ultrasound-mediated insonation of the apical region of the canine heart. Upon sacrifice, insonated and uninsonated regions of the heart were harvested and assessed for production of CAT protein. Results indicated that transfection of the CAT protein was increased in the regions where ultrasound insonation occurred regionally but no such increase occurred in the control or noninsonated regions of the heart. This demonstrated that bubbles and ultrasound could enhance the delivery of a gene selectively in the area of insonation.

In Morishita's group in vivo gene delivery was carried out using an echo contrast microbubble (Optison) [79]. Using a rat carotid artery model with balloon injury to test suppression of re-stenosis, they used ultrasound to treat the injury with anti-oncogene (*p53*) plasmid complexed to the Optison microbubble. After 2 weeks the intimal-to-medial area ratio showed a significant decrease over control (Figure 7.13).

7.4.7 Liquid PFC-filled Targeted LPSs as Ultrasound-mediated Drug Delivery Agents

As described previously for SMBs, this technology takes advantage of gaseous perfluorocarbon bubbles. However, one limitation to the use of bubbles is their size, e.g. 500–1000 nm, resulting in confinement to the vascular space. Indeed, ultra-

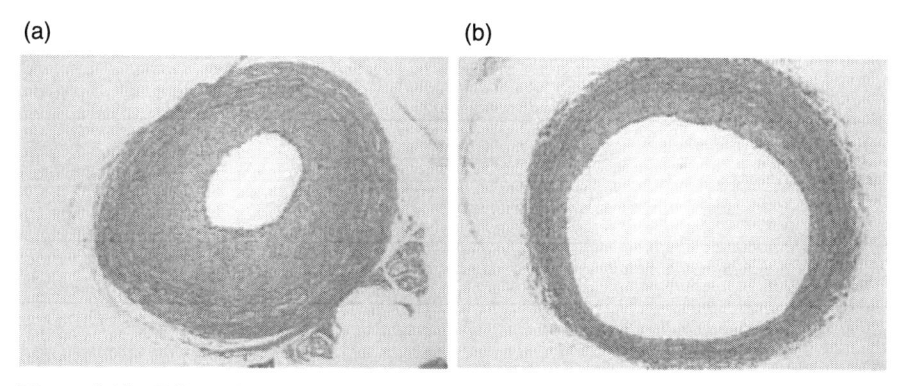

Figure 7.13 Effect of transfection of naked p53 plasmid DNA on neointimal formation in rat balloon injury model at 2 weeks after transfection. Representative cross-sections of balloon-injured vessels transfected with p53 vector at 2 weeks after injury (hematoxylin and eosin staining). (a) Naked p53 plasmid DNA by means of ultrasound (US) alone; (b) naked p53 plasmid DNA by means of ultrasound with Optison. (Reprinted with permission of Taniyama et al. [73].)

sound-mediated cavitation and delivery of drugs on bubbles can occur; however, penetration through the vascular endothelium could be an issue. Alternatively, the use of submicron-sized droplets, being smaller-sized structures of the order of 150–350 nm, may improve egress from the vascular space to provide an alternate method for delivery through the endothelium.

Work by Kuwatsuru [80] and Wu [81] has shown that the vascular endothelium in angiogenic vessels residing near tumors is somewhat 'leaky' and can allow the diffusion of structures <500 nm, through the endothelial bed. Consequently, structures with diameters less than this critical size could potentially act as delivery agents. Also, the addition of a targeting ligand directed toward cell surface proteins or receptors could increase the efficiency of delivery selectively to tumor cells or affected areas of pathology. Furthermore, once the LPSs are localized to the affected tumor cells, ultrasound can potentially be used to enhance the delivery of the drug-encapsulated droplets intracellularly into the affected cell, combining passive and active (energy-mediated) targeting.

We have conducted research using drug-loaded, perfluorocarbon and oil-filled targeted droplets for delivery to prostate tumors. Figure 7.14 displays the targeted droplets that we employ. Note that the oil- and perfluorocarbon-filled droplets are stabilized with a coating of lipids. In addition, the lipid membranes serve as an anchoring site for insertion of the lipid portion of the bioconjugate into the droplet. The bioconjugate also is comprised of a polyethyleneglycol (PEG) 'tether' to aid in 'extending' the ligand out into the milieu. At the far end of the PEG tether is the ligand with affinity to a specific cell-surface receptor – in the case of prostate cancer, the $\alpha_6\beta_1$ receptor on tumor cells. As can be seen in Figure 7.15, the droplets are small enough to diffuse through the endothelial gap junctions in the vasculature and 'leak' into the extravascular space or extracellular fluid (ECF). Once in the ECF, the targeting ligand has an opportunity to bind selectively to the $\alpha_6\beta_1$ receptor, which is

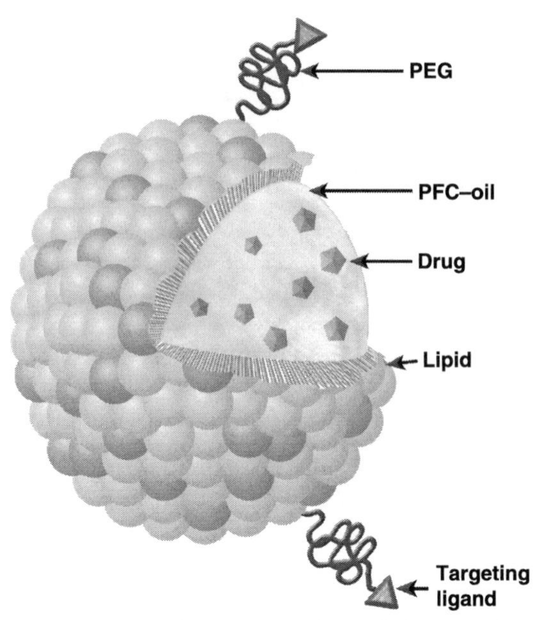

Figure 7.14 A targeted LPN capable of carrying hydrophobic drugs in the perfluorocation (PFC)–oil layer. PEO, polyethylene oxide.

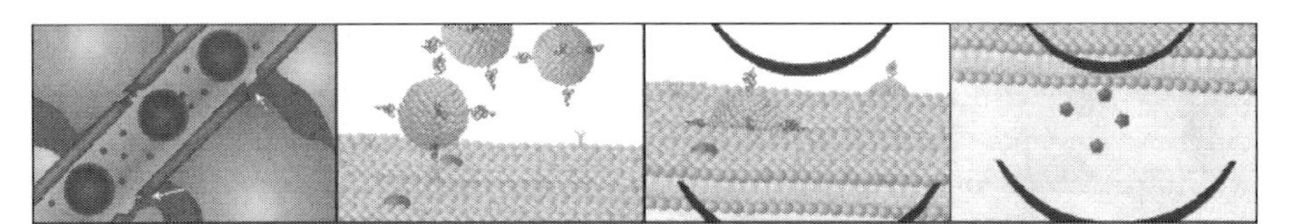

Figure 7.15 Submicron-sized droplet cartoon.

upregulated on prostate tumor cells. The droplets, now localized on the surface of the targeted tumor cell, can be subjected to ultrasound radiation force and driven intracellularly. This process, in principle, will increase the effective concentration of the drug to the target while, at the same time, decreasing the amount of drug exposure to normal cells. This should aid in providing a more effective delivery system while minimizing adverse effects.

In vitro experiments using fluorescently labeled targeted droplets with ultrasound have demonstrated that targeted droplets are able to bind to tumor cells and, after insonation with ultrasound at frequencies ranging from 1 to 10 MHz, the fluorescent label becomes internalized by the cells (PA Dayton, personal communication). Figure 7.16 is an image of five independent PC-3 prostate tumor cell populations grown on coverslips. All were incubated with droplets, however, only the cells in the center coverslip were insonated with ultrasound. Figure 7.17 is a graph quantifying the insonated vs uninsonated droplet control (n = 11). Note that the fluorescent intensity increases by almost an order of magnitude as a result of ultrasound-mediated delivery.

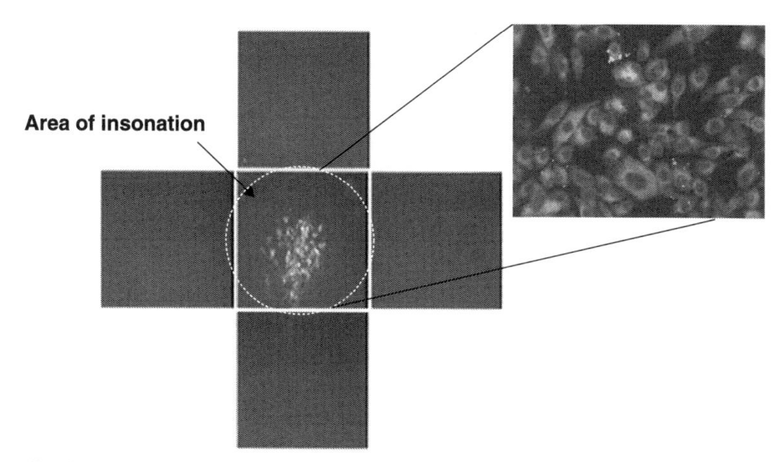

Figure 7.16 Fluorescent images of a PC-3 cell monolayer administered fluorescent paclitaxel-containing droplets. Cells exposed to ultrasound (center square) received substantial delivery of the fluorescent chemotherapeutic, in contrast to cells not exposed to ultrasound. The image on the left shows enlargement of the exposed area in fluorescent light. (Reprinted with permission of Dayton et al.)

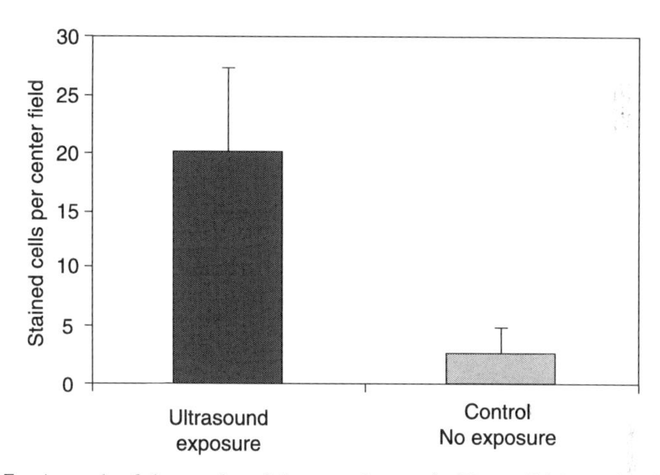

Figure 7.17 A graph of the results of the experiments in Figure 7.16 $n = 12$. (Reprinted with permission of Dayton et al. [46].)

Currently we are conducting in vivo studies in a DU-145H-inoculated mouse tumor model. Results to date appear promising.

7.5 OXYGEN DELIVERY

LPSs have shown promising results in preclinical studies as oxygen delivery agents and for transport and removal of other biologically active gases. Many of the pioneering studies have been performed by Lundgren and his colleagues at the State University in New York, Buffalo, using 2% dodecaperfluoropentane (DDFP) emulsion stabilized with PEG–Telomer B fluorosurfactant. DDFP boils at 28.5°C (converting from liquid to gaseous state). In this case the PEG–Telomer B functions as a lipid analog effectively to stabilize the droplets of DDFP. It is administered as a suspension of LPSs – droplets of perfluoropentane – and converts to bubbles and microbubbles of DDFP gas in vivo. DDFP emulsion was initially developed for use as an ultrasound contrast agent [82] and tested in clinical studies for that application, but Lundgren's group identified its potential for oxygen delivery and performed a number of animal studies to test this application. Pigs ventilated on room air were drained of half their blood volume and then administered 0.5 ml/kg of 2% perfluoropentane emulsion as a rapid intravenous infusion. Four of five control animals died and the one remaining survivor had renal insufficiency as well as other damage secondary to hypoxia. All five treatment animals receiving DDFP emulsion survived and had normal weight gain and function [9].

Compared with liquid PFCs or packed red blood cells, DDFP emulsion transports far more oxygen per unit volume. Studies in pigs suggest that a human dose necessary to provide half the oxygen-carrying capacity of the entire blood volume could be provided in only 30–40 mL [9]. DDFP is exhaled by the lungs and is eliminated entirely from the body after about 6 hours [83]. This is in contrast to liquid PFCs, which may have a longer plasma circulation time but which are retained long term by the liver and spleen [84]. The ability to transport large amounts of oxygen with small doses may enable clinical applications of DDFP emulsion for improving oxygenation in critical ischemia. In critical ischemia, in the absence of hemorrhage, administering additional red blood cells or other material to improve oxygen delivery usually results in volume overload, which can lead to pulmonary edema and congestive heart failure [85]. DDFP emulsion may afford a means of dramatically increasing oxygenation in the clinical setting while avoiding volume overload.

A preclinical study in a rat stroke model showed that DDFP emulsion decreased the volume of cerebral infarct (Lundgren, personal communication). Another preclinical study in a rat model of glioma showed that DDPF emulsion improved tumor response to radiotherapy [86]. Tumors are often hypoxic (i.e. low oxygen tension). Hypoxia is radioprotective [87]. A principal mechanism in response to radiation is creation of singlet oxygen and superoxide free radicals. Improved tumor oxygenation provided by DDFP emulsion could potentially be developed as a therapeutic adjunct to radiation therapy. Gliomas commonly recur or progress after radiation. Improved treatments are needed to help these patients.

Lundgren tested the potential application of DDFP emulsion in a model of nitrogen narcosis in pigs [10]. Pigs were subjected to hyperbaric atmospheres and rapidly decompressed to create a condition analogous to the 'bends'. Pigs were administered intravenous DDFP emulsion and the wash-out rate of nitrogen in the blood was determined. The treatment animals receiving DDFP had a significantly faster wash-out rate of DDFP than the control animals.

7.6 Conclusion

Lipid-coated perfluorocarbon structures are a class of agents with diagnostic and therapeutic applications. Phospholipid-coated, perfluoropropane microbubbles are approved in the U.S. as an ultrasound contrast agent for cardiac imaging and have been used in over a million patients. Clinical trials are presently under way to develop LPSs as therapeutic agents, initially using phospholipid-coated perfluoropropane submicron-sized bubbles. For this application the bubbles act as cavitation nuclei to increase the rate of clot lysis with ultrasound. LPSs, together with ultrasound, hold promise as a generalized platform for drug amplification and gene delivery. Without ultrasound, LPSs are promising oxygen delivery agents. LPSs can carry several hundredfold more oxygen per unit volume than packed red blood cells, and also hold the potential for delivery of other biologically active gases.

ACKNOWLEDGMENTS

The authors would like to acknowledge the National Institutes of Health for generous funding in supporting some of these projects. We acknowledge the following: NHLBI: HL071433-02, NCI: NO1 CO-37118.

REFERENCES

1. Unger, E.C., McCreery, T.P., Sweitzer, R.H., et al. Acoustically active lipospheres containing paclitaxel, a new therapeutic ultrasound contrast agent. *Invest Radiol* 1998;**33**:886–892.
2. Schumann, P.A., Christiansen, J.P., Quigley, R.M., et al. Targeted-microbubble binding selectively to GPIIbIIIa receptors of platelet thrombi. *Invest Radiol* 2002;37:587–593.
3. Unger, E.C., McCreery, T., Sweitzer, R., et al. MRX-501: a novel ultrasound contrast agent with therapeutic properties. *Acad Radiol* 1998;(suppl 1):S247–S249.
4. Apfel, R.E., Holland, C.K. Gauging the likelihood of cavitation from short-pulse, low-duty cycle diagnostic ultrasound. *Ultrasound Med Biol* 1991;**17**:179–185.
5. Price, R.J., Skyba, D.M., Kaul, S., Skalak, T.C. Delivery of colloidal particles and red blood cells to tissue through microvessel ruptures created by targeted microbubble destruction with ultrasound. *Circulation* 1998;**98**:1264–1267.
6. Hynynen, K., McDannold, N., Khotseva, N., Jolesz, F. Noninvasive MR imaging-guided focal opening of the blood-brain barrier in rabbits. *Radiology* 2001;**220**:640–646.
7. Shortencarrier, M.J., Dayton, P.A., Bloch, S.H., et al. A method for radiation force localized drug delivery using gas-filled lipospheres, IEEE *Trans Ultrasonics, Ferroelectrics Freq Control* 2004;**51**:822–830.
8. van Liew, H.D., Burkard, M.E. Relationship of oxygen content to PO_2 for stabilized bubbles in the circulation: Theory. *J Appl Physiol* 1996;**81**:500–508.
9. Lundgren, C.E., Bergoe, G.W., Tyssebotn, I. Microbubbles for treatment of hemorrhagic shock: A field usable concept? New Orleans: American Heart Association Scientific Sessions, 2004.
10. Lundgren, C., Bergoe, G., Olszowka, A., Tyssebotn, I. Tissue nitrogen elimination in oxygen-breathing pigs is enhanced by fluorocarbon-derived intravascular micro-bubbles, *Undersea Hyperb Med* 2005;**32**:215–226.
11. Wood, A.B. *Textbook of Sound*. London: Bell, 1964.
12. Rayleigh, L. On the pressure developed in a liquid during the collapse of a spherical cavity. *Phil Mag* 1917;**34**:94–98.

13. Gramiak, R., Shah, P.M. Echocardiography of the aortic root. *Invest Radiol* 1968:**3**:356–366.
14. Feinstein, S.B., Cherif, J., Ten Cae, F.J., et al. Safety and efficacy of a new transpulmonary ultrasound contrast agent: initial multicenter clinical results. *J Am Coll Cardiol* 1990;**16**:316–324.
15. Gould, J.L., Keeton, W.T. *Biological Science,* 6th edn. New York: WW Norton, 1996.
16. Tarter, V.M., Satterfield, R., Schumacher, D.J., et al. Comparison of Imagent BP (PFOB) and Fluosol-DA 20% as ultrasound contrasts. *J Ultrasound Med* 1991;**10**(suppl):S17–S18.
17. Haidt, S.J., et al. *IOUS Suppl* 1982;**22**:223.
18. Mattrey, R.F., Strich, G., Shelton, R.E., et al. Perfluorchemicals as US contrast agents for tumor imaging and hepatosplenography: preliminary clinical results. *Radiology* 1987;**163**:339–343.
19. Unger, E., Shen, D., Fritz, T., et al. Gas-filled lipid bilayers as ultrasound contrast agents. *Invest Radiol* 1994;**29**:S134–136.
20. Hilpert, P.L., Mattrey, R.F., Mitten, R.M., Peterson, T. IV Injection of air-filled human albumin microspheres to enhance arterial Doppler signal: A preliminary study in rabbits. *Am J Roentgenol* 1989;**153**:613–616.
21. Wheatley, M., Schrope, B., Shen, P. Contrast agents for diagnostic ultrasound: development and evaluation of polymer-coated microbubbles. *Biomaterials* 1990;**11**:713–717.
22. Webster, A.J., Cates, M.E. Osmotic stabilization of concentrated emulsions and foams. *Langmuir* 2001;**17**:595–608.
23. Miller, A.P., Nanda, N.C. Contrast echocardiography: New agents. *Ultrasound Med Biol* 2004; **30**:425–434.
24. Correas, J.M., Meuter, A.R., Singlas, E., et al. Human pharmacokinetics of a perfluorocarbon ultrasound contrast agent evaluated with gas chromatography. *Ultrasound Med Biol* 2001;**27**:565–570.
25. Budavari, S. (ed.). *The Merck Index*, 11th edn. Rahway, NJ: Merck & Co. Inc., 1989.
26. Unger, E., Shen, D., Fritz, T., et al. Gas-filled liposomes as echocardiographic contrast agents in rabbits with myocardial infarcts. *Invest Radiol* 1993;**28**:1155–1159.
27. Alving, C.R. Liposomes as safe carriers of drugs and vaccines. In: Baun-Falco O, Korting HC, Maibach HI (eds), *Liposome Dermatics*. Berlin: Springer-Verlag, 1992, pp 226–232.
28. Alving, C.R., Richards, R.L. Immunological aspects of liposomes. In: Ostro M (ed.), *The Liposomes* New York: Marcel Dekker, 1987.
29. Senior, J., Gregoriodis, G. Stability of small unilamellar liposomes in serum and clearance from the circulation: The effect of the phospholipid and cholesterol components. *Life Sci* 1982;**30**:2123.
30. Wu, P-S., Tin, G.W., Baldeschweiler, J.D., Shen, T.Y., Ponpipom, M.M. Effect of surface modification on aggregation of phospholipid vesicles. *Proc Natl Acad Sci USA* 1981;**78**:6211.
31. Unger, E., Shen, D., Fritz, T., et al. Gas-filled lipid bilayers as ultrasound contrast agents. *Invest Radiol* 1994;**29**:S134–136.
32. *Drug Facts and Comparisons*. Walters Kluwer Co., St Louis, MO, 2000, p. 2007.
33. *Drug Facts and Comparisons*. Walters Kluwer Co., St Louis, MO, 2001, Suppl. p. 2010.
34. Lindner, J.R., Song, J., Jayaweera, A.R., et al. Microvascular rheology of definity microbubbles after intra-arterial and intravenous administration. *J Am Soc Echocardiogr* 2002;**15**:396–403.
35. Gertz, E.W., Manspeaker, H.P., Fritz, T.A., et al. Unger, Aerosomes myocardial contrast echocardiography (MCE) accurately detects resting myocardial perfusion abnormalities post myocardial infarction, Ultrasound Med Biol 1997;**23**:23.
36. Chomas, J.E., Dayton, P., May, D., Ferrara, K.W. Threshold of fragmentation for ultrasonic contrast agents. *J Biomed Optics* 2001;**6**(2):141–150.
37. Tachibana, K. Enhancement of fibrinolysis with ultrasound energy. *J Vasc Interven Radiol* 1992;**3**:299–303.
38. Siegel, R.J., Atar, S., Fishbein, M.C., et al. Noninvasive transcutaneous low frequency ultrasound enhances thrombolysis in peripheral and coronary arteries, *Echocardiography* 2001;**18**:247–257.
39. Culp, W.C., Erdem, E., Roberson, P.K., Husain, M.M. Microbubble potentiated ultrasound as a method of stroke therapy in a pig model: preliminary findings. *J Vasc Interv Radiol* 2003;**14**:1433–6.
40. Culp, W.C., Porter, T.R., Xie, F., et al. Microbubble potentiated ultrasound as a method of declotting thrombosed dialysis grafts: experimental study in dogs. *Cardiovasc Interv Radiol* 2001;**24**:407–412.
41. Porter, T.R., Kricsfeld, D., Lof, J., Everbach, E.C., Xie, F. Effectiveness of transcranial and transthoracic ultrasound and microbubbles in dissolving intravascular thrombi. J *Ultrasound Med* 2001;**20**:1313–1325.

42. Apfel, R.E., Holland, C.K. Gauging the likelihood of cavitation from short-pulse, low-duty cycle diagnostic ultrasound. *Ultrasound Med Biol* 1991;**17**:179–85.
43. Patel, D., Dayton, P., Gut, J., et al. Optical and acoustical interrogation of submicron contrast agents. *IEEE Ultrasonics, Ferroelectrics, and Frequency Control* 2002;**49**(12):1641–1651.
44. Awasthi, V.D., Garcia, D., Goins, B.A., Phillips, W.T. Circulation and biodistribution profiles of long circulating PEG-liposomes of various sizes in rabbits. *Int J Pharm* 2003;253:121–32.
45. Woodle, M.C. Surface-modified liposomes: assessment and characterization for increased stability and prolonged blood circulation, *Chem Phys Lipids* 1993;**64**:249–262.
46. Dayton, P.A., Morgan, K.E., Klibanov, A.L., et al. A preliminary evaluation of primary and secondary radiation forces on acoustic contrast agents. *IEEE Transact Ultrasonics, Ferroelectrics and Frequency Control* 1997;**44**:1264–1277.
47. Shortencarier, M.J., Dayton, P.A., Bloch, S.H., et al. A method for radiation-force localized drug delivery using gas-filled lipospheres, *IEEE Transact Ultrasonics, Ferroelectrics Freq Control,* 2004;**51**:822–883.
48. Tan, K.T. Lip, G.Y.H. Red vs. white thrombi: Treating the right clot is crucial. *Arch Intern Med* 2003;**163**:2534–2535.
49. Rawn, J.D. *Biochemistry*. Neil Patterson Publishers, Burlington, NC, 1989, pp 201–207.
50. Apfel, R.E., Holland, C.K. Gauging the likelihood of cavitation from short-pulse, low-duty cycle diagnostic ultrasound. *Ultrasound Med Biol* 1991;**17**:179–85.
51. Birnbaum, Y., Luo, H., Nagai, T., et al. Noninvasive in vivo clot dissolution without a thrombolytic drug. *Circulation* 1998;**97**:130–134.
52. Nedelmann, M., Eicke, B.M., Lierke, E.G., et al. Low-frequency ultrasound induces nonenzymatic thrombolysis in vitro. *J Ultrasound Med* 2002;**21**:649–656.
53. Yin, X., Hynynen, K. A numerical study of transcranial focused ultrasound beam propagation at low frequency. *Phys Med Biol* 2005;**50**:1821–36.
54. Porter, T.R., Xie, F. Ultrasound, microbubbles, and thrombolysis, *Progr Cardiovasc Dis* 2001; **44**(2):101–110.
55. Mueller, R.L., Scheidt, S. History of drugs for thrombotic disease. Discovery, development, and directions for the future. *Circulation* 1994;**89**(1):432–449.
56. Pannell, R., Black, J., Gurewich, V. Complementary modes of action of tissue-type plasminogen activator and pro-urokinase by which their synergistic effect on clot lysis may be explained. *J Clin Invest* 1988;**81**:853–859.
57. Anglés-Cano, E. Overview on fibrinolysis: plasminogen activation pathways on fibrin and cell surfaces. *Chem Phys Lipids* 1994;**67/68**:353–362.
58. Porter, T.R., LeVeen, R.F., Fox, R., Kreisfeld, A., Xie, F. Thrombolytic enhancement with perfluorocarbon-exposed sonicated dextrose albumin microbubbles. *Am Heart J* 1996;**132**:964–8.
59. Veklich, Y., Francis, C.W., White, J., Weisel, J.W. Structural studies of fibrinolysis by electron microscopy. *Blood* 1998;**92**:4721–4729.
60. Tsutsui, J.M., Xie, F., Johanning, J., et al. Treatment of deeply located acute intravascular thrombi with low frequency guided ultrasound and intravenous microbubbles. *Radiology* 2006;in press.
61. Wu, Y., Unger, E.C., McCreery, T.P., et al. Binding and lysing of blood clots using MRX-408. *Invest Radiol* 1998;**33**:880–885.
62. Porter, T., Kricsfeld, D., Li, S., et al. Comparison of d-dimer activity and residual particle size produced following thrombus disruption by perfluorocarbon containing microbubbles versus urokinase in the presence of low frequency ultrasound. *J Am Coll Cardiol* 1998;**31**:219A (abstr).
63. World Health Organization. *World Health Report 2003 Shaping the Future*. Geneva: WHO.
64. American Heart Association Statistics Committee and Stroke Statistics Subcommittee. Heart Disease and Stroke Statistics – 2006 Update. A report from the American Heart Association Statistics Committee and Stroke Statistics Subcommittee.
65. Hirsh, J., Hoak, J. Management of deep vein thrombosis and pulmonary embolism. A statement for healthcare professionals from the council on thrombosis, American Heart Association. *Circulation* 1996;**93**:2212–2245.
66. Chan, J., Rees, C.R., Song, A.K., Pham, S. Usefulness of catheter-directed thrombolysis using alteplase in peripheral vascular occlusion. *BUMC Proceedings* 2001;**14**(1):3–7.

67. Sacco, R.L., Benjamin, E.J., Broderick. J.P., et al. Risk Factors Panel: American Heart Association Prevention Conference IV: prevention and rehabilitation of stroke: risk factors. *Stroke* 1997;**2**:1507–1517.
68. *Physicians Desk Reference*. Montvale, NJ: Thomson PDR, 2006: pp 1225–1229, 1232–1234, 1256–1259.
69. Alexandrov, A.V., Molina, C.A., Grotta, J.C., et al. Ultrasound-enhanced systemic thrombolysis for acute ischemic stroke, *N Engl J Med* 2004;**351**:2170–8.
70. Molina, C.A., Ribó, M., Arenillas, J.F., et al. Microbubbles administration accelerates clot lysis during continuous 2 MHz ultrasound monitoring in stroke patients treated with intravenous tPA. International Stroke Conference, 2005 (abstract).
71. Meuwly, J-Y., Schnyder, P., Gudinchet, F., Denys, A.L. Pulse-inversion harmonic imaging improves lesion conspicuity during US-guided biopsy. *J Vasc Intervent Radiol* 2003;**14**:335–341.
72. Takeuchi, M., Ogunyankin, K., Pandian, N.G., et al. Enhanced visualization of intravascular and left atrial appendage thrombus with the use of a thrombus-targeting ultrasonographic contrast agent (MRX-408A1): In vivo experimental echocardiographic studies, *J Am Soc Echocardiogr* 1999;**12**:1015–1021.
73. Zhao, S., Borden, M., Bloch, S.H., et al. Radiation-force assisted targeting facilitates ultrasonic molecular imaging. *Mol Imaging* 2004;**3**(3):135–48.
74. Mathai, M., Choi, S-K., Whitesides, G.M. Polyvalent interactions in biological systems: implications for design and use of multivalent ligands and inhibitors. *Angew Chem Int Ed* 1998;**37**:2754–2794.
75. Hantgen, R.R., Rocco, M., Nagaswami, C., Weisel, J.W. Binding of a fibrinogen mimetic stabilizes integrin αIIbβ3's open conformation. *Protein Sci* 2001;**10**:1614–1626.
76. Larina, I.V., Evers, B.M., Ashitkov, T.V., et al. Enhancement of drug delivery in tumors by using interaction of nanoparticles with ultrasound radiation. *Technol Cancer Res Treat* 2005;**4**:217–226.
77. Wu, Y., Unger, E.C., McCreery, T.P., et al. Binding and lysing of blood clots using MRX-408, *Invest Radiol* Dec 1998;**33**:880–885.
78. Vannan, M., McCreery, T., Li, P., et al. Ultrasound mediated transfection of canine myocardium by intravenous administration of cationic microbubble-linked plasmid DNA, *J Am Soc Echocardiogr* 2002;**18**:214–218.
79. Taniyama, Y., Tachibana, K., Hiraoka, K., et al. Local delivery of plasmid DNA into rat carotid artery using ultrasound. *Circulation* 2002;**105**:1233–1239.
80. Kuwatsuru, R., Liu, T., Cohen, F., et al. Acute liver transplantation rejection: Early detection of endothelial leak in a rat model using magnetic resonance imaging and a macromolecular contrast medium. *Invest Radiol* 1994;**29**(suppl 2):S297–S299.
81. Wu, N.Z., Da, D., Rudoll, T.L., et al. Increased microvascular permeability contributes to preferential accumulation of stealth liposomes in tumor tissue. *Cancer Res* 1993;**53**:3765–3770.
82. Robbin, M.L., Eisenffeld, A.J. Perflenapent emulsion: US contrast agent for diagnostic radiology – multicenter, double-blind comparison with a placebo. Radiology 1998;**207**:717–722.
83. Correas, J.M., Meuter, A.R., Singlas, E., et al. Human pharmacokinetics of a perfluorocarbon ultrasound contrast agent evaluated with gas chromatography. *Ultrasound Med Biol* 2001;**27**:565–570.
84. Miller, T.F., Milestone, B., Stern, R., et al. Effects of perfluorochemical distribution and elimination dynamics on cardiopulmonary function. *J Appl Physiol* 2001;**90**:839–849.
85. Koc, M., Toprak, A., Tezcan, H., Bihorac, A., Akoglu, E., Ozener, I.C. Uncontrolled hypertension due to volume overload contributes to higher left ventricular mass index in CAPD patients. *Nephrol Dial Transplant* 2002;**17**:1661–1666.
86. Koch, C.J., Oprysko, P.R., Shuman, A.L., et al. Radiosensitization of hypoxic tumor cells by dodecafluoropentane: a gas-phase perfluorochemical emulsion. *Cancer Res* 2002;**62**:3626–3629.
87. Fyles, A.W., Milosevic, M., Wong, R., et al. Oxygenation predicts radiation response and survival in patients with cervix cancer (Published erratum appears in *Radiother Oncol* 1999;**50**:371). *Radiother Oncol* 1998;**48**:149–156.
88. Chomas, J. E., Dayton, P. A., May. D., Allen, J. Optical observation of contrast agent destruction. *Applied Physics Letters* 2000;77(7):1056–1058.
89. Unger, E. C., Matsunaga, T. O. McCreery, T. P., et al. Therapeutic applications of microbubbles. *Eur J Radiol* 2002;42:160–168.

INDEX

Page references followed by t indicate material in tables.

A

ABC drug efflux transporters. *See* ATP-binding cassette (ABC) transporters
Absorptive transporters, 3–8
Accelerated stability testing, 116
Acetyl sulfisoxazole, bioavailability of, 33
Acoustically active lipospheres (AALs), 188
Active targeting, of selected cells, 104–108
Adenosine triphosphate (ATP), liposomal, 62. *See also* ATP entries
Adsorption, distribution, metabolism, excretion, and toxicology (ADME-Tox) screening, 160
Adsorption kinetics, 131–136
Affinity constants, 6
Allen, David D., ix, 160
Amino acid transporters, 3
Amphiphilic drugs, influence on protein adsorption pattern, 145
Amphiphilicity, polymer, 40
Amphotericin B, liposomal, 50
Angiogenesis inhibitors, 57
Angiogenic homing peptide, 54
Antibodies, 52, 70–71, 107
Antibody–peptide conjugation, onto long-circulating lipid emulsions, 106–108
Antibody-targeted liposomal drugs, 52
Anticancer antibodies, 70–71
Antioxidants, adding to emulsions, 112–113
Antisense oligonucleotide delivery, 55
Antitumor activity, emulsion size and, 98–99
Antitumor enzymes, 57
Apical membrane, transporter substrates expressed in, 9–10t
Apical Na^+-dependent bile acid transporter (ASBT), 4
Apo-E-adsorbing drug carriers, 127
Apolipoprotein E (Apo-E), lipid emulsions associated with, 106
Apolipoproteins, 94–95
  adsorption of, 134, 142–144
  affinity to lipid formulations, 140–141
Aqueous phase, in emulsion preparation, 112–113
Aramchol, 15
Asialofetuin liposomes, 61
L-Asparaginase, 57, 59
ATP-binding cassette (ABC) transporters, 2, 8–15. *See also* Adenosine triphosphate (ATP)
ATP-binding domain(s), 9
ATP depletion, intracellular, 16
Autoclaving, 113–114

B

Backscatter, ultrasound-mediated, 173–174
β-galactosidase, liposome-immobilized, 59
β-glucuronidase, liposome-immobilized, 58–59
Bioavailability, oral, 34–35
Biodegradable polymers, 162
Bioerosion, 161
Block AB-type copolymers, amphiphilic, 65
Blood-brain barrier (BBB)
  nanoparticle targeting for drug delivery across, 160–169
  therapeutic enzyme transfer through, 60
Brain, nanoparticle drug delivery to, 164–166
Breast cancer resistance protein (BCRP), 13–14
Bubble cavitation, 174–175

Role of Lipid Excipients in Modifying Oral and Parenteral Drug Delivery, Edited by Kishor M. Wasan

C

Caco-2 cells, 11, 16–18
Caelyx, liposomal, 50
Cancer, multidrug resistance in, 11. *See also* Tumor entries
Cancer cells, 16, 106–107
Cancer inhibition, 54
Cancer therapy, 59, 166–167
Carrier size, tumoral sites and, 98–99
Carrier systems, suitability of, 124–126
Cationic lipid-DNA complexes, developing, 101
Cationic lipid emulsions, as nonviral gene delivery vectors, 117
Cation transport, 8
Cavitation, of lipid-coated perfluorocarbon nanostructures, 171. *See also* Bubble cavitation
Cell-penetrating peptides (CPPs). *See* CPP-mediated molecule/particle delivery
Cells, active targeting of, 104–108
Centrifugation, in 2-DE analysis, 129
Cetyl palmitate-SLNs (CP-SLNs), 148, 149
Chemotherapy, 166
Cholesterol absorption, 8
Chylomicrons, lipid emulsions metabolized as, 94–95
Ciclosporin, 36, 62
Circulation, particulate drug carrier longevity in, 68–69
Circulation time, 101–103, 104
CNT transporters, 7–8
Coalescence, 91
Co-emulsification, 111, 117
Coenzyme Q10, liposomal, 62
Collagenase, 57
Colloidal dispersions, 64–65
Colloidal drug carriers, 2-DE analysis of, 128–139
Commercial emulsions, oils used in, 109t
Compritol-SLNs (CO-SLNs), 148, 149
Controlled drug delivery, 124
CPP-mediated molecule/particle delivery, 55–56
Creaming, floccule, 91
Cremophor EL, 35, 36
Critical micelle concentration (CMC), 36, 42, 65, 66
Cyclodextrins, 64
Cytokines, liposomal, 61
Cytoplasm, peptide and protein delivery to, 55
Cytotoxic drug transport, 13
Cytotoxic T lymphocyte (CTL) response, 63

D

Daunorubicin, liposomal, 53
Depsipeptides, 57
Diacyl-lipid-PEG conjugates, 66
Dialysis grafts, thrombosed, 180
Dialysis method, of preparing drug-loaded micelles, 42
Dietary fats, absorption and metabolism of, 95
Dietary sterol efflux, 14
Differential protein adsorption, 126–127
Digestible lipids, 33
DLVO (Derjaguin, Landau, Vervey, and Overbeek) theory, 92
DNA-complexed emulsions, 117
Dodecaperfluoropentane (DDFP) emulsion, oxygen transport by, 192–193
Doxil, targeting to tumor cells, 52
Doxorubicin, liposomal, 50, 53
Doxorubicin-loaded polymeric micelles, 69
Droplet size, control of, 114–115
Drug absorption, 16, 32–33
Drug delivery, 160–161. *See also* Lipid-based drug delivery entries
  across the blood-brain barrier, 160–169
  choosing lipids for, 33–34
  intracellular, 71–72
  liposomes in, 49–52
  via lipid-coated perfluorocarbon nanostructures, 171
Drug distribution, barriers to, 163
Drug efflux proteins, inhibition of ATPase activity of, 16
Drug-loaded micelles, 71
  anticancer effects of, 72
  preparation of, 42, 67–68
Drug release, factors in, 163
Drug release rate, liposomes and, 38–39
Drugs
  leakage from emulsions, 108
  localization of, 144–145

nucleic acid-based, 116–117
poorly soluble, 42
Drug transport, modes of, 2
Drug transporters, interaction with excipients and surfactants, 1–31
Dry emulsions, 36–37
Dynasan-SLNs (DY-SLNs), 147–149

E
Efflux transporters, intestinal, 8–15
Egbert, James, ix, 160
Elastase, 57
Electrostatic repulsion, 92, 93
Electrostatic stabilization, 92–93
Emulsifiers, 110–111
Emulsion-based drug delivery systems, 36
Emulsion formulations, stability measurement of, 115–116
Emulsion preparation
steps in, 113–114
triglycerides in, 108–110
Emulsions. *See also* Lipid emulsions
characterization of, 114–115
chemical stability of, 116
drug leakage from, 108
oil phase composition of, 99–100
phosphatidylcholine composition of, 100
preparation for intravenous administration, 108–116
stability of, 90–94
steric stabilization of, 94
sugar-coated, 106
Emulsion surface hydrophilicity, enhancing, 104
Encapsulation efficiency, 67
Endocytosis-mediated drug delivery, 71
Endogenous chylomicrons, lipid emulsions metabolized as, 94–95
Endosomal escape, 55
Enhanced permeability and retention (EPR) effect, 43, 49, 68, 69
Enkephalin, liposomal, 62
ENT transporters, 7–8
Enzymes, 56–57, 58–61
Epidermal growth factor receptor (EGFR)-targeted immunoliposomes, 54
Epirubicin transport, 18
Eptifibatide, 187
Ethanol, 35, 36
Excipients
interaction of drug transporters with, 1–31
for parenteral application, 35–36
traditional, vii

F
Familial hypertriglyceridemia (FHTG), 4
Fat emulsions, 37
Fenretinide, 52
Fexofenadine, 6
Flavonoids, stimulatory/inhibitory role of, 12
Flocculation, 92–93
emulsion destabilization and, 91–92
Fluorescently labeled micelles, 72
Folate homeostasis, 14
Folate-mediated targeting, 53
Folate-modified micelles, 70
Food effect, on absorption and bioavailability of drugs, 32–33
Fusogenic liposomes, 39

G
Galactosylated emulsions, 106
Galactosylated liposomes, 54
Gastric retention time, 34–35
GD2-targeted immunoliposomes, 52
Geldenhuys, Werner, ix, 160
Gel-liquid phase transition temperature (T_c), 110–111
Genes, ultrasound-mediated delivery of, 188
Gene therapy, 116–117
Genetic material, improved delivery of, 101
Glucocerebroside β-glucosidase, 59
Göppert, Torsten M., ix, 124
Granulocyte-macrophage colony-stimulating factor (GM-CSF), liposomal preparations of, 61
Griseofulvin, bioavailability of, 33
Growth factor receptors, in liposome targeting to tumors, 53

H
Hemodynamics, liposome-related changes in, 60–61
Hemosomes, long-circulating, 60
Hepatic sites, targeting delivery of micelles to, 70

Hepatitis A vaccine, virosomal, 54
Hepatitis B, hepatocyte targeting for, 106
Hepatocytes, 106
HER2-overexpressing tumors, immunotargeting, 52
Herbal products, effect on Pgp, 12
Hexose aminidase, 58
High-density lipoproteins (HDLs), 15, 94
Homogenization, high-pressure, 113
Hydrophile-lipophile balance (HLB) method, 111
Hydrophilic lipophilic balance (HLB) values, 151
Hydrophilic polymers, 40–41
Hyperbilirubinemia, 13

I
Immunoglobulins, 51, 134
Immunoliposomes, 51
  ATP-loaded, 62
  epidermal growth factor receptor (EGFR)-targeted, 54
  long-circulating, 52
Immunological adjuvants, liposomes as, 63
Immunomicelles, PEG-PE-based, 70
Immunotargeted liposomes, 52
Immunotargeted micelles, preparing, 70
Incorporated drugs, influence of liposomes on properties of, 56–64
Influenza vaccine, 54
Insulin, liposomal, 61
Interfacial polymerization, 163
Interferon, liposomal, 61–62
Intestinal absorptive transporters, 3–8, 18
Intestinal efflux transporters, 8–15
Intracellular drug delivery, lipid-core micelles for, 71–72
Intracellular targeting, 55–56
Intravenous administration, 64, 108–116
Intravenous lipid emulsions, 88–123
Ischemia, 179, 180, 181
Ischemia and reperfusion (I/R) injury, vitamin E and, 110
Ischemic conditions, liposomal ATP in, 62

J
Jeong, Seong Hoon, ix, 32

L
L-Lactate transport, 6
Lactoferrin, 106
LaPlace pressures, 172
LBA micelles, loading capacity of, 42–43
Lecithins, as emulsifiers, 110
Leroux, Jean-Christophe, ix, 88
Leupeptide, liposomal, 62
Levovist, 182
Ligands
  attaching to liposomes, 51
  incorporating on to the emulsion interface, 104–106
  for liposome targeting, 53–54
  single step attachment of, 52
Ligand-targeted micelles, 70–71
Lipid-based amphiphiles (LBAs), 42
Lipid-based drug delivery formulations, 35–43
  dry emulsions, 36–37
  fat emulsions, 37
  lipid-based micelles, 40–43
  liposomes, 38–39
  self-emulsifying and self-microemulsifying, 36
  solid dispersions, 37
  solid lipid nanoparticles, 38
Lipid-based drug delivery systems, viii
  advantages and disadvantages of, 43
  oral and parenteral, 32–47
Lipid-based micelles, 40–43
Lipid-coated perfluorocarbon nanostructures (LPNs). *See also* Targeted LPNs
  applications of, 171, 176–191
  development of, 193
  oxygen delivery by, 191–193
  perfluoropropane-filled, 180
  physical characteristics of, 171–173
  as therapeutic agents, 170–196
  in ultrasound-mediated sonothrombolysis, 176–177
Lipid-containing co-polymers, 68t
Lipid-core micelles, 66–67
  for intracellular drug delivery, 71–72
  as pharmaceutical carriers, 64–72
Lipid emulsion composition, effect of, 99–100
Lipid emulsion–plasmid DNA complex, transfection efficiency of, 117

Lipid emulsions. *See also* Emulsions
associated with apolipoprotein E, 106
benefits of, 89
biodistribution of, 96–108
for delivering nucleic acid-based drugs, 116–117
effect of surface charge on, 100–101
elimination mechanisms for, 94–96
intravenous, 88–123
long-circulating, 101–104, 118
metabolized as endogenous chylomicrons, 94–95
storing as a lyophilized powder, 116
uses for, 89
Lipid emulsion size, effect on biodistribution, 96–99
Lipid excipients/surfactants
development of, vii
drug transporter modulation by, 15–16
effect on MRPs, 18
pharmaceutical-grade, viii
Lipid formulations, protein adsorption patterns on, 124–159
Lipid nanocapsules, 96–97
Lipid nanoparticles, solid, 38
Lipid–polymer conjugates, micelle preparation of, 67
Lipids
choosing for drug delivery applications, 33–34
effect on drug absorption and bioavailability, 32–33
as membrane stabilizers, 173
Lipofundin MCT, 139–140, 142, 145–146
Lipofundin N, 145–147
Lipolysis, influence of lipid emulsion size on, 97
Lipophilic drugs, 1–2
Lipoprotein lipase (LPL), 94
Liposomal antigens, 63–64
Liposomal DNAase I, 60
Liposomal drugs, 50t
Liposomal peptides, 61–62
Liposome delivery, Tf-mediated, 53
Liposome formulations, development of, 125
Liposome-immobilized enzymes, 59, 60–61
Liposomes, 38–39
as adjuvants, 63–64
defined, 48–49
in drug delivery, 49–52
folate, 53
influence on incorporated drug properties, 56–64
long-circulating, 51–52
PEGylated, 39
pH-sensitive, 54–55
PTD-modified, 56
rapid removal from circulation, 39
Liposome targeting, 52–56
Liu, Lichuan, 1
Liver X receptors (LXRs), 14–15
L-lactate transport, 6
Lockman, Paul R., ix, 160
Long-chain triglycerides (LCTs), 109
Long-circulating immunoliposomes, 52
Long-circulating lipid emulsions, 101–104, 118
antibody-peptide conjugation onto, 106–108
Long-circulating liposomes, 51–52
Lurtotecan, liposomal, 50
L-valyl ester prodrugs, 5
Lysosmal storage diseases, 56, 57
Lytic drugs, 177, 180, 181

M
Macromolecules, entry into tumor tissues, 49
Mannosylated liposomes, 61
Matsunagi, Terry O., ix
Medium-chain triglycerides (MCTs), 99–100, 109
Micellar carriers
advantages of, 65, 69
biological characteristics of, 66
Micelle charge, controlling, 71
Micelles, 64–65. *See also* Lipid-core micelles; Mixed micelles
drug-loaded, 67–68
ligand-targeted, 70–71
lipid-core, 66–67
as pharmaceutical carriers, 49
polymeric, 65–66
Micelle stability, 41, 42
Micelle targeting, 'passive,' 68–69
Micellization, 64–65
Microbubbles, targeted, 183, 184, 185, 186

Microemulsions, emulsion destabilization, 91
Microfluidization process, 113
Microparticulate drug carriers, coupling of PTDs to, 56
Microreservoir-type systems, 48
Microstreaming, 175–176
Mixed micelles, 43, 67
Monocarboxylate transporter isoform type 1 (MCT1), 6–7
Mononuclear phagocyte system (MPS), 89, 96
  avoiding recognition by, 126
  lipid emulsion elimination by, 95–96
  polyethyleneglycol lipids and, 103–104
Mucosal immune response, enhancement of, 63
Müller, Rainer H., ix, 124
Multidrug resistance-associated protein (MRP) transporter family, 4, 13
Multidrug resistance protein 1 (MDR1) transporter, 11–12
Mumper, Russell J., ix, 160

N
Nanobubbles
  arginine-glycine-aspartic acid (RGD)-targeted, 186–187
  as cavitation nuclei, 175
  function of, 178–179
  perfluoropropane-filled, 183
  targeted, 182
Nanocapsules, 96–97, 162
Nanodroplets, targeted, 189–190
'NanoInvasive' therapy, 180, 181
Nanoparticle carrier systems, residence time of, 166–167
Nanoparticle drug delivery, to the brain, 164–166
Nanoparticles
  as a current delivery system, 161–163
  design of, 162–163
  entry into tumor tissues, 49
  solid lipid, 147–151
Nanoparticle targeting, for drug delivery across the blood-brain barrier, 160–169
Nanospheres, 162
Nanostructures, diffusion of, 189
Nanotechnology, concerns related to, 161
Niemann–Pick C1-like 1 protein (NPC1L1), 8
Nondigestible lipids, 33
Nucleic acid-based drugs, lipid emulsions for delivering, 116–117
Nucleoside transport processes, 7
Nutrient transporters, 4

O
OATP transporters, 6
OCT transporters, 8
Oil-in-water emulsion drug loading, influence on protein adsorption, 144–145
Oil-in-water emulsion lipid composition, influence on protein adsorption, 139–141
Oil-in-water (o/w) emulsions, 88
  advantages of, 125
  physiological stability of, 145–147
  plasma protein adsorption patterns on, 139–147
Oil-in-water emulsion surface modification, influence on protein adsorption, 141–144
Oil phase, in emulsion preparation, 108–110
Oils, for oral administration, 35
Oligopeptide transporter 1 (PEPT1), 5–6
170-kDa P-glycoprotein (Pgp), 11–12, 16
Opsonins, 95, 126–127, 149
Oral bioavailability, improved, 34–35
Oral drug absorption, 1
  low bioavailability in, 11–12
Oral lipid-based drug delivery, viii
Orphan nuclear receptors, 2–3
Osmolarity, of emulsions, 112
Ostwald ripening, 91
Oxygen delivery, by lipid-coated perfluorocarbon nanostructures, 191–193
Oxygen-transporting systems, artificial, 60

P
Paclitaxel
  delivery of, 43
  encapsulation efficiency of, 67
  liposomal, 50
  solubility of, 109
Palmitoyl-L-Asparaginase, 59

Pang, K. Sandy, ix, 1
Paracellular drug absorption, 1
Parenteral application, excipients for, 35–36
Parenteral lipid formulations, 33
  protein adsorption patterns on, 124–159
Park, Jae Hyung, x, 32
Park, Kinam, x, 32
Particulate drug carriers, longevity of, 68–69
'Passive targeting,' 49, 68–69
PEG chains, antibodies linked to, 107–108. *See also* Polyethyleneglycol (PEG)
PEG-coated liposomes, 52, 61
PEG-DSPE, 104, 107–108
PEG hemosomes, 60
PEG-lipid derivatives, as co-emulsifiers, 103–104
PEG–PEA conjugate, 41–42
PEG–PEA micelles, 43
PEG–PE micelles, 66–67
  circulation times and accumulation pattern for, 68–69
  negative charge on, 71–72
  tumor accumulation of, 70–71
PEGylated liposomes, 39
Peptide drugs
  liposome influence on, 56–58
  low stability of, 57–58
Peptide hormones, 57
Peptides, 55–56, 61–62
Peptidomimetic drugs, 6
Perfluorocarbon gases, as contrast agents, 172–173
Perfluorochemicals, emulsions containing, 110
170-kDa P-glycoprotein (Pgp), 11–12, 16
pH, of final emulsions, 112
Phagocyte system, lipid emulsion elimination by, 95–96
Pharmaceutical carriers, lipid-core micelles as, 64–72
Phosphate transporters, 7
Phosphatidylethanolamine (PEA), 41–42. *See also* PEG–PEA entries
Phospholipid liposomes, 'plain,' 50
Phospholipid residues, 66
pH-responsive polymeric micelles, 69
pH-sensitive liposomes, 54–55
Plasma, as an incubation medium, 136–138
Plasma protein adsorption patterns, 130, 132, 139–151
Platelet aggregation assay, 183–184
Pluronic block copolymers, inhibition of transporters by, 16
Poloxamer 235, 138
Poloxamer 338, 101
Poloxamer 407, 132, 137, 138, 139
Polyethyleneglycol (PEG). *See also* PEG entries
  coupling of Tf to, 53
  properties of, 40–41
Polyethyleneglycol lipids, 103–104
Polyethyleneglycol liposomes, 50
Polyethylene oxide (PEO) chains
  lengths of, 149
  steric stabilization of, 117
Polylactic acid (PLA) nanoparticles, 136
Polymeric amphiphiles, lipids as hydrophobic parts of, 41–42
Polymeric micelles, 40, 43, 65–66
Polymeric nanoparticles, advantages of, 125
Polymer–lipid hybrid nanoparticle (PLN) system, 18
Polymers, 51, 162
Polysorbate 80 emulsions, 104
Polyvalency effect, 186
Poorly soluble drugs, 64–72
'Poorly water soluble' chemical entities, viii
Pronase, 57
Property-based drug design, 160–161
Propofol, 37
Proteases, 57
Protein adsorption. *See also* Protein adsorption patterns
  influence of o/w emulsion drug loading on, 144–145
  influence of o/w emulsion surface modification on, 141–144
  influence of SLN matrix lipid on, 147–149
  influence of SLN surface modification on, 149–151
Protein adsorption patterns, 124–159
  analytical procedure for, 127–139
  differential protein adsorption, 126–127
  on o/w emulsions, 139–147
Protein antigen delivery, 54
Protein drugs, 56–58

Proteins
intracellular delivery of, 55–56
liposomal, 58–61
preferential adsorption of, 126
PTD (protein transduction domain) peptides, protein delivery by, 56
Purine nucleoside absorption, 7
Pyrene incorporation method, 42

R
Radiation force, 175–176
Reticuloendothelial system (RES), 39
RGD (arginine-glycine-aspartic acid) peptides, 54
Rhodamine transport, 18
Ritonavir, 36
Rossi, Joanna, x, 88
RS-1541 emulsion formulations, antitumor activity of, 99t

S
Saquinavir, 36
Sedimentation, floccule, 91
Selected cells, active targeting of, 104–108
Self-emulsifying drug delivery systems (SEDDSs), 36
Self-microemulsifying drug delivery systems (SMEDDSs), 36
Sequence-selective gene inhibition, 167
Serum, as an incubation medium, 136–138
Short-chain fatty acid transport, 6
Short-chain triglycerides (SCTs), 109
Side effects, reduction by liposomes, 38–39
siRNA oligonucleotides, sequence-selective gene inhibition by, 167
Sitosterolemia, 14
SLC transporters, 4–8
SLN matrix lipid, influence on protein adsorption, 147–149. *See also* Solid lipid nanoparticles (SLNs)
SLN surface modification, influence on protein adsorption, 149–151
Small intestine, drug-metabolizing enzymes of, 1–2
Sodium citrate, 136
SolEmuls technology, 125
Solid dispersions, 37
Solid lipid nanoparticles (SLNs), 38, 125–126. *See also* SLN entries
adsorption patterns on, 132–133, 135, 136
plasma protein adsorption patterns on, 147–151
protein adsorption patterns on, 153
separation of, 129
stability of, 151
two-dimensional polyacrylamide electrophoresis gels of, 150
Solute carrier transporters (SLCs), 2
Somatostatin analogs, 57
SonoLysis, 179–182
Specificity, importance of, 163–167
Sphingomyelin, effect on circulation time, 101–103
Stealth liposomes, blood concentration-time profile of, 105
Steam sterilization, 116
Stearic acid-SLNs (ST-SLNs), 147–149
Steric hindrance, 103
Steric stabilization, 94
Sterol efflux, 14
Stroke, 181
Submicrometer emulsions, 92
Sugar-coated emulsions, to target hepatocytes, 106
Sun, Huadong, 1
Superoxide dismutase, liposome-encapsulated, 59
Surface charge, 100–101, 115
Surface modifiers, 141–142
Surfactants, 110
drug transporter modulation by, 15–18
emulsion stability and, 90
lipid, 173
non-phospholipid, 112t
Systemic circulation, lipid emulsion escape from, 97–98

T
Targeted liposomes, 51
Targeted LPNs, 182–191. *See also* Lipid-coated perfluorocarbon nanostructures (LPNs)
liquid PFC-filled, 188–191
as ultrasound-mediated drug delivery agents, 188
Targeting. *See also* Active targeting
folate-mediated, 53

intracellular, 55–56
transferrin-mediated, 53
Targeting ability, liposomes with, 39
TAT-mediated intracellular delivery, 55–56
TAT PTD delivery, 56
Taxol, 109
Tay–Sachs disease, 58
Therapeutic agents
lipid-coated perfluorocarbon nanostructures as, 170–196
polymeric micelle improvement of, 40
for vascular thrombi, 176–177
Therapeutic enzymes, 56–57, 58
Thermoresponsive polymeric micelles, 69
Thomas, Fancy, x, 160
Thrombolytic therapy, 57, 60
Thrombosis, therapy for, 177–179
Thrombus formation, pathophysiology of, 176
TO-CO-cholesterol emulsions, 103
D-α-Tocopheryl poly(ethylene glycol) 1000 succinate (TPGS), 16–18
TOCOSOL-paclitaxel, 109–110
Topotecan, 14
Torchilin, Vladimir P., x, 48
Toxicity, of nanoparticle systems, 167
Transcellular drug absorption, 1
Transduction phenomenon, 55
Transferrin (Tf)-mediated targeting, 53
Transferrin-modified micelles, 70
Transporter-inhibiting excipients/surfactants, effects of, 17t
Transport systems, in the blood-brain barrier, 164, 165
Triglyceride oils, in emulsion preparation, 108–110
Tumoral sites, carrier size and, 98–99
Tumor immunotherapy, 53
Tumor necrosis factor α (TNF-α), liposomal preparations of, 61
Tumors, 188, 192. *See also* Cancer entries
Tumor site-selective liposomes, 39
Tumor targeting, 52, 53
Tumor vasculature, 49
2C5 immunomicelles, radiolabeled, 71
Two-dimensional polyacrylamide gel electrophoresis (2-DE) analysis, 127–139
sample preparation in, 128–139
steps in, 127–128

U
Ultrasound, clot lysis from, 181–182
Ultrasound contrast agents, 171–173
Ultrasound-mediated backscatter, 173–174
Ultrasound-mediated drug delivery agents, 188–191
Ultrasound-mediated sonothrombolysis, LPNs as therapeutic agents for, 176–177
Unger, Evan C., x

V
Vaccines, virosome-based, 54
L-Valyl ester prodrugs, 5
Van der Waals' forces, 92
Vascular thrombosis, 179–180
Vasoactive intestinal peptide (VIP), 53–54
Vector molecules, 55
Vegetable oils, in emulsion preparation, 108–110
Virosomes, 54
Vitamin E, 109, 111
Vitamin receptors, in liposome targeting to tumors, 53
Vitamin transporters, 4
Vroman effect, 132, 136

W
Washing media, effect on adsorption pattern, 131
Water, drug solubility in, 35
Water-soluble drug, in liposomes, 38

Z
Zutshi, Reena, x

Printed and bound by CPI Group (UK) Ltd, Croydon, CR0 4YY